AF594933

Manufacturing and Application of Stainless Steels

Manufacturing and Application of Stainless Steels

Special Issue Editor

Andrea Di Schino

MDPI • Basel • Beijing • Wuhan • Barcelona • Belgrade

Special Issue Editor
Andrea Di Schino
Dipartimento di Ingegneria,
Università di Perugia
Italy

Editorial Office
MDPI
St. Alban-Anlage 66
4052 Basel, Switzerland

This is a reprint of articles from the Special Issue published online in the open access journal *Metals* (ISSN 2075-4701) from 2018 to 2020 (available at: https://www.mdpi.com/journal/metals/special_issues/application_stainless_steels).

For citation purposes, cite each article independently as indicated on the article page online and as indicated below:

LastName, A.A.; LastName, B.B.; LastName, C.C. Article Title. *Journal Name* **Year**, *Article Number*, Page Range.

ISBN 978-3-03928-650-8 (Pbk)
ISBN 978-3-03928-651-5 (PDF)

© 2020 by the authors. Articles in this book are Open Access and distributed under the Creative Commons Attribution (CC BY) license, which allows users to download, copy and build upon published articles, as long as the author and publisher are properly credited, which ensures maximum dissemination and a wider impact of our publications.
The book as a whole is distributed by MDPI under the terms and conditions of the Creative Commons license CC BY-NC-ND.

Contents

About the Special Issue Editor

Andrea Di Schino After completing his degree in Physics at the University of Pisa and PhD in Materials Engineering at the University of Napoli, Andrea Di Schino joined Centro Sviluppo Materiali SpA (CSM), a European leading centre for Materials Research. After spending more than 10 years in CSM, he joined the University of Perugia as Professor of Metallurgy. His main interest is to combine fundamental theoretical research and experiments to explain mechanisms and discover new insight in the field of steel metallurgy. He is author of more than 200 papers relating to physical metallurgy and product development. Prof. Di Schino is member of the Editorial Board of several Journals.

Editorial

Manufacturing and Applications of Stainless Steels

Andrea Di Schino

Dipartimento di Ingegneria, Università di Perugia, Via G. Duranti, 06125 Perugia, Italy; andrea.dischino@unipg.it

Received: 24 February 2020; Accepted: 26 February 2020; Published: 1 March 2020

1. Introduction and Scope

Stainless steels represent quite an interesting material family, both from a scientific and commercial point of view, owing to their excellent qualities in terms of strength and ductility, combined with corrosion resistance. Thanks to such properties, stainless steels have been indispensable for technological progress during the last century and their annual consumption has increased faster than other materials. They find application in all fields requiring materials with good corrosion resistance, together with the ability to be worked into complex geometries. Despite their diffusion as a consolidated material, many research fields are active regarding the possibility of increasing stainless steel's mechanical properties and corrosion resistance by grain refinement or alloying by interstitial elements. At the same time, innovations are coming from the manufacturing process of the stainless steel family of materials, including the possibility to manufacture them from metal powder for 3D printing. The scope of this Special Issue embraces interdisciplinary work covering physical metallurgy and processes, reporting about experimental and theoretical progress concerning microstructural evolution during processing, microstructure–properties relationships and various applications, including automotive, energy and structural.

2. Contributions

The book collects manuscripts from academic and industrial researchers with stimulating new ideas and original results. It consists of four review and ten research papers. The review papers focus on the state of the art and perspectives in repairing and reinforcing historic timber structures by stainless steels [1], on duplex and super-duplex stainless steel microstructures and the evolution of their properties by surface modification processes [2], on the process and properties of reversion-heated austenitic stainless steels [3] and on the 3D printing of stainless steels by the laser powder bed fusion technique [4]. A group of research papers deal with physical metallurgy and advancedcharacterization techniques [5], others with process aspects [6–8] or property application items related to stainless steels [9–14]. The paper by Fava et al. [5] presents and discusses the results of mechanical spectroscopy tests carried out on Cr martensitic steel. The study regards the following topics: (i) embrittlement induced by Cr segregation; (ii) the interaction of hydrogen with C–Cr associates; (iii) the nucleation of Cr carbides. This technique permitted the authors to characterize the specific role played by point defects in the investigated phenomena. Due et al. [6] investigate non-metallic inclusions in 316L stainless steel bars with and without Ca treatments. The inclusions are extracted using electrolytic extraction. After that, the characteristics of the inclusions, such as morphology, size, number, and composition, are investigated by using a scanning electron microscope in combination with energy-dispersive X-ray spectroscopy. Chen et al. [7] report the formation and characteristics of non-metallic inclusions in 304L stainless steel during the vacuum oxygen decarburization refining process, using industrial experiments and thermodynamic calculations. The compositional characteristics indicated that two types of inclusions with different sizes (from 1 to 30 μm) existed in 304L stainless steel during the refining process, i.e., CaO–SiO2–Al2O3–MgO external inclusions, and CaO–SiO2–Al2O3–MgO–MnO endogenous inclusions. Luu et al. [8] analyze the post-annealing mechanical behavior of 316L austenitic stainless steel (SUS316L) after electrically assisted annealing with a single pulse of electric current and evaluated the feasibility of a two-stage forming process of the selected SUS316L with rapid electrical assisted annealing. Peng et al. [9] report an experimental data assessment and fatigue design recommendation for stainless steel welded joints. Juuti et al. [10] present a new family of ferritic stainless steels for service temperatures up to 1050 °C, utilizing intermetallic phase transformation. Cianetti et al. [11] report a novel method for the evaluation of the surface fatigue strength of a stainless steel component. The proposed approach is a hybrid method, numerical–theoretical, which allows us to estimate the surface fatigue strength in a very short time without having to resort to finite element models that often are so complex that they starkly contrast industrial purposes. Prosviryakov et al. [12] present a novel corrosion-resistant steel with a high boron content. The positive influence of Zr addition on the microstructure and mechanical properties after hot deformation is shown. The Zr-alloyed steel demonstrates hot deformation without fracturing in the temperature range of 1273–1423 K, and in the strain rate range of 0.1–10 s^{-1}, despite the high volume of brittle borides. Gennari et al. [13] present a paper about the microstructural and corrosion properties of cold-rolled laser-welded UNS S32750 duplex stainless steel. Zhang et al. [14] report the temperature dependence of phase transformation kinetics during tempering in 13Cr super-martensitic stainless steel.

Acknowledgments: As Guest Editor, I would like to especially thank Kinsee Guo, Assistant Editor, for his support and active role in the publication. I am also grateful to the entire staff of the *Metal* Editorial Office for the precious collaboration. Last but not least, I wish to express my gratitude to all the contributing authors and reviewers: without your excellent work, it would not have been possible to accomplish this Special Issue that I hope will be a piece of interesting reading and reference literature.

Conflicts of Interest: The author declares no conflict of interest.

References

1. Corradi, M.; Osoforo, A.I.; Borri, A. Repair and reinforcement of historic timber structures with stainless steel—A review. *Metals* **2019**, *9*, 106. [CrossRef]
2. Tahchieva, A.B.; Llorca-Isern, N.; Cabrera, J.M. Duplex and super duplex stainless steels: Microstructure and property evolution by surface modification processes. *Metals* **2019**, *9*, 347. [CrossRef]
3. Järvenpää, A.; Jaskari, M.; Kisko, A.; Karjalainen, P. Process and properties of reversion-treated austenitic stainless steels. *Metals* **2020**, *10*, 281. [CrossRef]
4. Zitelli, C.; Folgarait, P.; Di Schino, A. Laser powder bed fusion of stainless steel grades: A review. *Metals* **2019**, *9*, 731. [CrossRef]
5. Fava, A.; Montanari, R.; Varone, A. Mechanical spectroscopy investigation of point defect-driven phenomena in a Chromium martensitic steel. *Metals* **2018**, *8*, 870. [CrossRef]
6. Du, H.; Karasev, A.; Sundqvist, O.; Jonsson, P. Modification of non-metallic inclusions in stainless steels by addition of CaSi. *Metals* **2019**, *9*, 74. [CrossRef]
7. Chen, X.; Cheng, G.; Li, J.; Hou, Y.; Pan, J.; Ruan, Q. Characteristic and formation mechanism of inclusions in 304L stainless steel during the VOD refining process. *Metals* **2018**, *8*, 1024. [CrossRef]
8. Luu, V.T.; Nguyen, T.A.N.; Hong, S.T.; Jeong, H.J.; Han, H.N. Feasibility of a two stage forming process of a 316L austenitic stainless steel with rapid electrically assisted annealing. *Metals* **2018**, *8*, 815. [CrossRef]
9. Peng, Y.; Chen, J.; Dong, J. Experimental data assessment and fatigue design recommendation for stainless steel welded joints. *Metals* **2019**, *9*, 723. [CrossRef]
10. Juuti, T.; Manninen, T.; Uusikallio, S.; Komi, J.; Porter, D. New ferritic stainless steels for service temperatures up to 1050 °C utilizing intermetallic phase transformation. *Metals* **2019**, *9*, 664. [CrossRef]
11. Cianetti, F.; Ciotti, M.; Palmieri, M.; Zucca, G. On the evaluation of surface fatigue strength of stainless steel aeronautical component. *Metals* **2019**, *9*, 455. [CrossRef]
12. Prosviryakov, A.; Mondoloni, B.; Churyumov, A.; Pozdniakov, A. Microstrcture and hot deformation behavior of a novel Zr-alloyed high-Boron steel. *Metals* **2019**, *9*, 218. [CrossRef]

13. Gennari, C.; Lago, M.; Bogre, B.; Meszaros, I.; Calliari, I.; Pezzato, L. Microstructural and corrosion properties of cold rolled laser welded UNS S32750 duplex stainless steel. *Metals* **2018**, *8*, 1074. [CrossRef]
14. Zhang, Y.; Yin, Y.; Li, D.; Ma, P.; Liu, Q.; Yuan, X.; Li, S. Temperature dependent phase transformation kinetics of reverted austenite during tempering in 13Cr supermartensitic stainless steel. *Metals* **2019**, *9*, 1203. [CrossRef]

© 2020 by the author. Licensee MDPI, Basel, Switzerland. This article is an open access article distributed under the terms and conditions of the Creative Commons Attribution (CC BY) license (http://creativecommons.org/licenses/by/4.0/).

Review

Laser Powder Bed Fusion of Stainless Steel Grades: A Review

Chiara Zitelli [1], Paolo Folgarait [1] and Andrea Di Schino [2,*]

[1] Seamthesis Srl, Via IV Novembre 156, 29122 Piacenza, Italy
[2] Dipartimento di Ingegneria, Università degli Studi di Perugia, Via G. Duranti 93, 06125 Perugia, Italy
* Correspondence: andrea.dischino@unipg.it

Received: 11 May 2019; Accepted: 25 June 2019; Published: 28 June 2019

Abstract: In this paper, the capability of laser powder bed fusion (L-PBF) systems to process stainless steel alloys is reviewed. Several classes of stainless steels are analyzed (i.e., austenitic, martensitic, precipitation hardening and duplex), showing the possibility of satisfactorily processing this class of materials and suggesting an enlargement of the list of alloys that can be manufactured, targeting different applications. In particular, it is reported that stainless steel alloys can be satisfactorily processed, and their mechanical performances allow them to be put into service. Porosities inside manufactured components are extremely low, and are comparable to conventionally processed materials. Mechanical performances are even higher than standard requirements. Micro surface roughness typical of the as-built material can act as a crack initiator, reducing the strength in both quasi-static and dynamic conditions.

Keywords: stainless steel; laser powder bed fusion; additive manufacturing; innovation

---

1. Introduction

Additive manufacturing (AM), also known as 3D-printing, is an emerging technology [1], which is in the spotlight for its unique capability to produce near-net-shape components, even geometrically complex ones, without part-specific tooling being needed. AM is particularly suited for small batch production [2], weight reduction [3,4], part-customization [5,6], and functional integration [4,7–9], which is why it first emerged as a rapid prototyping technology. The adoption of AM technologies resulted in a new production paradigm [10,11]: the designer can project a component, or optimize the geometry of an already-existing one, according to its service conditions, and free from production-related constraints (e.g., undercuts, straight cuts, internal ducts with sharp edges). At the same time, AM made it possible to simplify component assembly, merging different parts into one single monolith. An emblematic example is the fuel nozzle shown in [12,13], which went from being an assembly of 20 parts to being a single unit. This allowed a 25% weight reduction.

A wider adoption of AM took place thanks to the possibility of processing metal alloys with mechanical properties that are comparable to the equivalent wrought alloys. Since 2000, AM technology has been assisting in a fast acceleration [14,15] due to the degree of development being gained in several sectors, i.e., lasers, computers, computer-aided design (CAD) technologies, programmable logic controllers (PLCs), and data storage systems [16,17]. The first commercially available AM system, back in 1987, was SLA-1 by 3D Systems, and it was based on the stereolithography technique (SLA stands for stereolithography apparatus): the desired piece is obtained through the superimposition of thin layers of ultraviolet light-sensitive liquid polymer solidified by an ultra-violet (UV)-laser source. Growing interest in the field led several companies and researchers to work simultaneously on developing systems for metal alloy handling. In this field, EOS Gmbh presented its first prototype (EOS M160) for metal processing in 1994, and the following year, the EOS M250 system was launched on the market [18]. In the meantime, Deckard filed a patent [19] concerning an apparatus capable of sintering powders thanks to a laser source. The cited manufacturing devices are the precursors of modern powder bed fusion (PBF) technology. PBF is a layer-wise production technology accomplishing material consolidation through a heat source, which can be a laser (in this case we refer to L-PBF) or an electron beam (EB-PBF): the spot of the heat source, impinging on the previously spread powder bed, releases the quantity of energy necessary to melt metal particles. Other AM technologies for metal alloys include: direct energy deposition (DED), binder jetting (BJ), sheet lamination (SL) and bound metal deposition (BMD). Besides the general AM advantages, the success of PBF (independently on the actual heat source) against other metal AM technologies, lies in its:

- capability of obtaining the best geometrical and dimensional tolerances;
- low waviness (the low-frequency roughness component) of surfaces, thus minimizing the need for machining allowance;
- capability of achieving the highest relative densities, up to 100%, with respect to wrought or forged metals;
- capability of producing both thin structures, e.g., lattice and trabecular, and heavy cross sections; and
- capability of minimizing oxide impurities, as it works under controlled atmospheres (usually nitrogen, or argon for reactive alloys).

The latter is extremely relevant for stainless steel alloys, as oxides could adversely affect their corrosion resistance and act as crack initiators [20,21]. Within the PBF subclass, L-PBF has gained much more interest due to its lower technological complexity and required production times [22], resulting in lower capital investment and production costs. Hence, the range of laser-based manufacturing systems is much wider than that of electron beam-based systems, and building envelopes have reached 1 m^3, allowing for the production of larger components than EB-PBF. EB-PBF is to be preferred when processing high-cost crack-susceptible alloys (such as TiAl alloys), since residual stresses are suppressed and higher production costs are balanced by the higher added-value output. However, L-PBF processed materials can be affected by some process-related defects [23,24], including:

- high-levels of residual stresses, which can cause distortions, cracks and delamination;
- porosities and incomplete fusion-related defects;
- cracks (in susceptible alloys) and metastable microstructures, as a consequence of high cooling rates;
- balling phenomenon, at the origin of discontinuous scan tracks;
- micro roughness due to partially sintered metal particles, especially experienced on inclined surfaces.

In the following section, the causes of defects will be addressed, together with countermeasures. Focusing on metal alloys, it is, in principle, possible to manufacture every weldable metal alloy once the proper setting of working parameters is defined (e.g., laser power, layer thickness, gas fluxing). L-PBF techniques have been successful in producing functional components from a wide range of metal alloys [25], with working parameter identification and validation being the real know-how of L-PBF technology developers. Today, the most established and L-PBF verified alloys include: Aluminum alloys (AlSi10Mg, AlSi7Mg0.6), Cobalt alloys (CoCrMo), Nickel alloys (Haynes HX, Inconel 625, Inconel 718), Iron alloys (Maraging steels, AISI 304, AISI 3016L, Tool steels) and Titanium alloys (Ti6Al4V, Ti6Al4V ELI, CP-Titanium Grade 2).

Stainless steels are nowadays used in almost every application field. In fact, thanks to their peculiar combination of properties—namely, strength and corrosion resistance—since its discovery in the early 19th century [26], they have been adopted in automotive [27,28], construction and building [29–31], energy [32–34], aeronautical [35,36], medical [37] and food [38–41] applications. The implementation of stainless steel grades in L-PBF systems, together with a deeper understanding of the technology, could definitely result in a wide adoption of the technology itself. Currently, L-PBF is already being used for stainless steel component production in engineering applications, but a fundamental lack in standardization and proper metrology definition [42–47] is limiting the spread of technology: for this reason, international committees are joining together to accelerate the process [48,49]. The challenge is to consistently define proper processing routes and requirements, standard mechanical requirements, proper heat treatments, dynamic performances, post-processing and qualification needs. This review aims at reporting the state of the art of the application of stainless steel alloys in L-PBF systems; starting from the working conditions of the cited process (and their effect on process performance), a list of stainless steel grades already tested on L-PBF systems is shown, together with their relative density, microstructure, quasi-static tensile properties and fatigue performances. In conclusion, the not yet exploited potential of making stainless steel via L-PBF systems will be discussed, along with future challenges.

2. Laser Powder Bed Fusion Working Principles and Process-Related Defects

The melting and fusion of material in L-PBF is obtained in a discrete way: layer-by-layer, the laser melts new powder, which solidifies as soon as the laser moves on to the neighbouring powder portion, leading to a continuous solid. The final part is an ensemble of micron-size welding lines overlapping in the horizontal plane and superimposed in the vertical plane. A representation of L-PBF working principle is reported in Figure 1, and the overall process can be summarized as follows:

1. CAD manipulation and model "slicing", the latter being performed by specific software properly subdividing the geometry into *n* slices with a height equal to the selected layer;
2. loading data to the L-PBF hardware;
3. production stage:
    a. spreading the powder layer, thanks to a rake or roller (Levelling System in Figure 1);
    b. switching on the laser, for melting and subsequent solidification;
    c. lowering of the building platform, which is retractable;
    d. previous steps 3a–3c repeated until all layers are fused;
4. removal of unused powder ("metal powder" in Figure 1) and extraction of the final part.

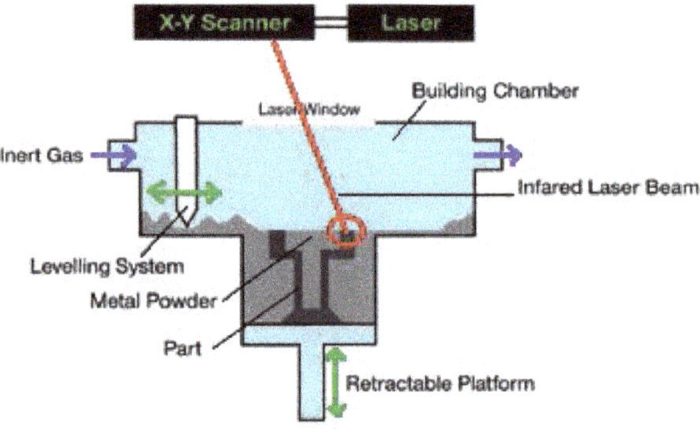

Figure 1. Schematics of a generic laser powder bed fusion system, adapted from [50], with permission from Elsevier, 2019.

The peculiar nature of L-PBF involves the interaction at the micron-sized level between photons (generated by the laser source) and a discrete metal substrate (the powder bed), consisting of metal particles dispersed in an inert gas atmosphere. Complicated physics are involved, such as absorption, transmission and reflection of laser energy, adhesion of micron-scale particles, rapid melting and solidification, molten metal flow, metal evaporation and microstructural evolution [10,23,51–57]. L-PBF relevant parameters can be classified into two groups: some of them directly set by users (e.g., laser power and speed, scan pattern, layer thickness and powder properties), while others are process-related (e.g., powder bed density, powder bed temperature). The latter can be indirectly controlled, using the first set of parameters, but it is harder to address. Several research groups are working to develop physical models to describe, and consequently control, the process, with some examples being described in [58–69]. Developed models also aim at defect forecasting and mitigation [70–75]. It is of major relevance to remember that, unlike conventional processes, all AM technologies are producing the alloy and the specific geometry to be put in service simultaneously. This means that process parameters usually related to the material quality (e.g., laser parameters determining melting and solidification behavior) now directly affect the performance [76] of the final components, as shown in Figure 2. The result is that, even if some post processing is projected (e.g., heat treatments, machining), material-related defects arising during 3D printing will cause the artefacts to be discarded at the end of the whole cycle, which can take up to weeks of work for large components.

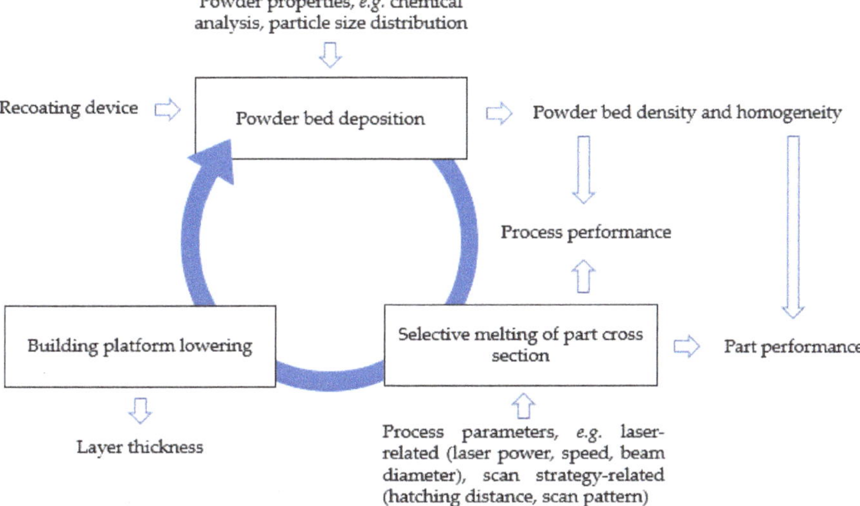

Figure 2. Main factors influencing process and part performances, and their inter-correlation.

To obtain a fully dense material, it is necessary to understand the effect of various process parameters. L-PBF machinery typically installs a system of lenses and a scanning mirror—or galvanometer—to move the position of the laser beam spot on the powder bed. The laser spot is used like a pen tip that, while rastering the surface of the powder bed, solidifies the powdered metal, releasing the energy necessary to achieve complete melting. Some laser characteristics are listed in Table 1: properties are reported in ranges, as they can vary between different system providers or as an effect of specific metal alloys. Scanning strategy parameters are particularly relevant for the quality of the manufactured component as they affect the heat cycle experienced by the alloy, determining, for example, final microstructure, residual stresses [77], and other defects.

The most relevant parameters to scanning strategy are:

- laser power;
- laser speed;
- hatch distance—the distance from the middle line of two consecutive lines, determining the effective hatch overlap, according to Figure 3;
- layer thickness;
- scan pattern.

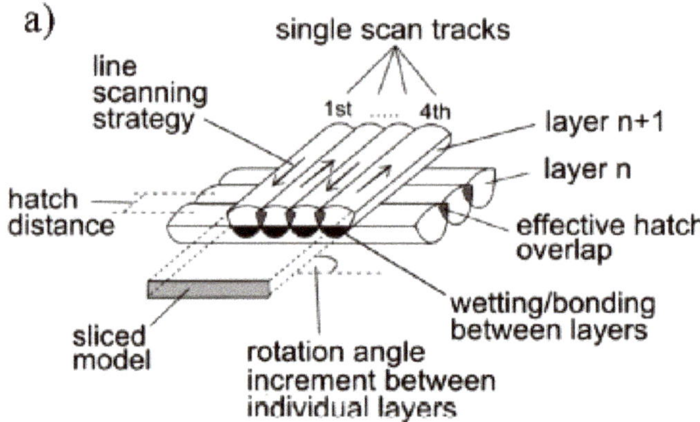

Figure 3. Schematic drawing of a generic pattern of single scan lines [78].

A short list of L-PBF machinery properties and related manufactured components features are shown in Table 1.

Table 1. L-PBF products and main properties.

L-PBF Machinery Characteristics	
Heat source	One fiber laser, or more
Laser power [W]	50–1000
Laser speed [mm/s]	10–15,000
Laser beam diameter [µm]	30–500. Most common, 80–100
Building chamber atmosphere	Inert gas—Typically, Nitrogen or Argon
Building rate [cm^3/h]	2–120
Building volume [mm]	Up to 800 × 400 × 500 (width × depth × height)—most common, 250 × 250 × 300
Ref.	[2,10,79–85]
L-PBF Produced Components Features	
Relative density [1] [%]	Up to 100
Upper surfaces roughness ($R_{a,X\text{-}Y}$) [µm]	4–10
Lateral surfaces roughness ($R_{a,Z}$) [µm]	>20
Minimum feature size [µm]	75–250
Geometric tolerance [mm]	±0.05–0.1
Impurities	Risk of contamination by process gas (nitrogen) or moisture
Effect on chemical composition	Minimum loss of low vapor pressure alloying elements
Powder size requirements [µm]	10–60
Ref.	[10,51,79,86–88]

[1] Relative density is evaluated as a ratio between the density of L-PBF produced material and the density of the same metal alloy processed with conventional technologies (e.g., rolling, forging).

The combined effect of the listed parameters can be calculated through the heat input released by the laser source: it can be quantified by the energy density parameter E, according to Equation (1):

$$E = P/v \cdot \varphi \ [\text{J/mm}^2] \tag{1}$$

Where P is laser power (W), v is laser speed (mm/s) and φ is the laser beam diameter at the powder bed surface. In Table 1, typical values for the mentioned parameters are reported. An alternative formulation is given by Equation (2):

$$E = P/v \cdot h \cdot t \text{ [J/mm}^3\text{]} \quad (2)$$

where h refers to the hatch distance, as pointed out in Figure 3, and t is the layer height.

At a micro scale, it is important to set the proper parameters to guarantee the highest achievable density. In Figure 4, different kinds of porosities (detrimental for density) are shown:

- irregularly shaped lack-of-fusion defects, as in Figure 4a,b, could be related to low E values or balling phenomena [89];
- spherical pores, like the one in Figure 4c, could be related to high E values, causing alloying element vaporization, a low packing density powder bed, which is in turn full of gas around metal particles, or small gas pores entrapped inside the metal particles themselves [23,90];
- layered pores, as shown in Figure 4d, could be caused by cracks at the melt pool boundaries [89].

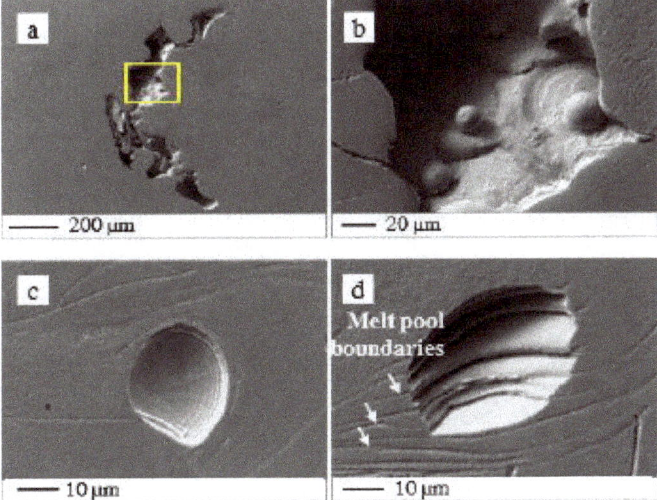

Figure 4. SEM images of porosity defects observed in L-PBF 316L samples: (**a**) low and (**b**) high magnification of incomplete fusion defect containing unmelted particles partially sintered; (**c**) gas pore; (**d**) cavity associated with residual stresses. From [89], with permission from Elsevier, 2019.

The temperature inside the laser spot usually reaches thousands of degrees Kelvin (or Celsius), and it is usually higher than the melting point of metal. The molten metal is strictly in contact with the substrate or the previously consolidated layers, determining a steep temperature gradient [10] that is at the basis of residual stresses phenomenon [91,92]. Scan pattern refers to the path followed by the laser beam in order to accomplish the complete melting of every slice or layer. At a macro level, the laser can raster the whole length of the slice in unidirectional or bidirectional mode, called the meander strategy in Figure 5 (consequent lines are scanned in the same direction or opposite), or it can subdivide the area in islands—also known as the chess strategy—or stripes (according to Figure 5). Furthermore, it handles a rotation angle between consequent layers, as can be seen in Figure 3, where the drawing represents a 90° rotation between the *i-th* and the *i+1-th* layer. The chess or stripe strategies and rotation angles are selected to reduce thermal gradients along the single layer (and, in turn, residual stresses [93]), while raising the level of anisotropy inside the material.

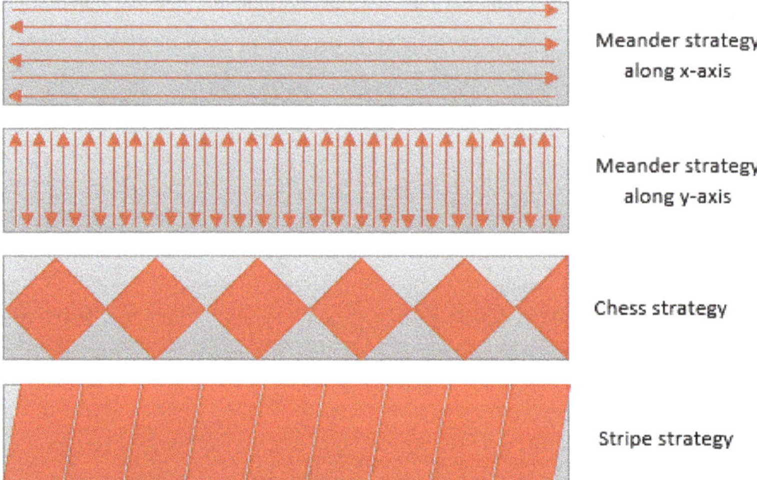

Figure 5. Examples of different scanning strategies on a rectangular cross section—the direction of scan lines inside the red areas can vary from layer to layer, selecting a non-zero rotation angle (according to Figure 3) between subsequent layers.

Residual stresses, together with porosities, are of major concern in L-PBF as they can determine the distortion of artefacts after cut or the failures of a production job, even at its final stage. To resolve the first problem, it is common practice to heat treat components still attached to the building platform; this helps in relaxing residual stresses before the removal. In the second case, it is necessary to reduce the temperature gradients experienced by the materials.

The laser interacts with a discrete substrate formed by metal powder particles that are used as feedstock; these are micron-sized particles of the selected alloy, obtained using specific processes. It is necessary for L-PBF to flow easily during the recoating phase to guarantee a proper powder bed density; various studies [94–97] demonstrated the impact of different particles characteristics on the quality (in particular, in terms of density and surface roughness) of the final component. The majority of metal powders used in L-PBF systems are, nowadays, produced through an atomizing process that involves the interaction of a stream of molten metal with a high energy jet, usually gaseous (e.g., nitrogen of argon) [98]; this process is called gas atomization (GA). Alternatively, it is possible to use plasma atomized (PA) powders, produced using plasma torches instead of the gas stream. In Figure 6, the main differences are appreciable: GA particles are characterized by the presence of satellites (small particles attached to the surface of bigger ones) and sometimes they show gas pores (Figure 6c), PA powder is mostly spherical, with smooth surfaces and a narrower Particle Size Distribution (PSD).

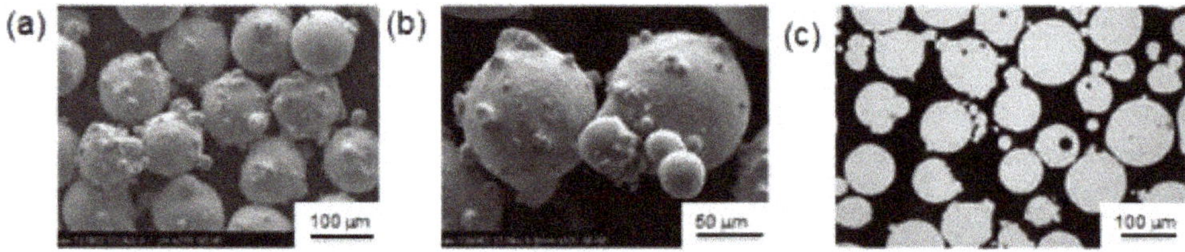

Figure 6. *Cont.*

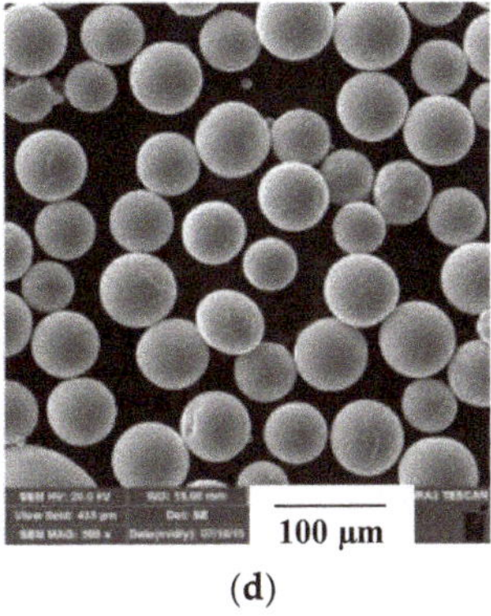

(d)

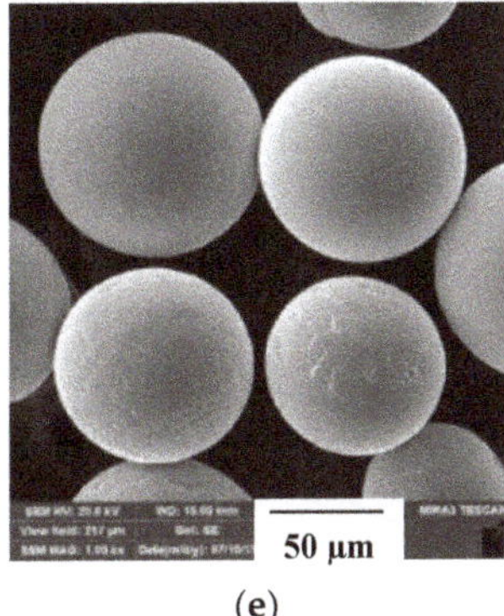

(e)

Figure 6. Comparison of metal powder obtained from different production processes. Gas atomized particles [99]: (**a**) SEM image 100×; (**b**) SEM image at 500×; (**c**) LOM of GA particles in the metallographic section. Plasma atomized particles [100]: (**d**) SEM image at 500×; (**e**) SEM image 1000×; with permission from John Wiley and Sons, 2019.

To summarize the effect of the cited process aspects:

- laser energy density E directly impacts melting behavior: at low scan speed and high laser power, E is high, causing evaporation, porosities and potential denudation of near surfaces, while high scan speed and low power determine low E density, which is insufficient to fully melt a proper powder volume and interlayer bonding [23,101];
- increasing scan speed, at constant laser power and layer thickness, lowers the maximum temperature reached in the melt pool, reducing residual stresses [102], until an unacceptable level of porosities is produced due to powder insufficient melting [103];
- increasing laser power, at constant laser speed and layer thickness, increases the maximum temperature and residual stresses [23,102];
- increasing the layer thickness, keeping the other parameters unmodified, can result in residual stresses mitigation [102], and a more economical process (i.e., with reduced production times), but it is necessary to evaluate the potential lack-of-fusion defects;
- the effect of laser power on melt pool defects and residual stresses is greater than that of laser speed [104];
- Guan et al., in [105], demonstrated that low overlapping between lines is not relevant for final density, as subsequent layer remelting allows metal filling, as can be seen in Figure 7;
- metal powder size and morphology determine the formation of porosities: spherical and small metal particles are to be preferred over non-spherical particles, as they form a denser powder bed [23,106];
- moreover, surface roughness is affected by the width of the melting track, which in turn is controlled by laser power and scan speed values. Inclined surfaces are the most disadvantaged, because heat conduction in the powder bed below is less efficient than the areas over the consolidated material [107];
- porosities must be carefully controlled, as they are detrimental for fatigue resistance of alloys; in particular, pore size has been demonstrated to be the most relevant parameter [108].

Considering the previously described aspects, it seems evident that parameter selection is the result of modelling and experimental validation aiming at reducing production time and costs while guaranteeing the melting of a proper quantity of metal (both powdered and consolidated). One challenge is that the best set of parameters (even if validated) is no longer valid as soon as you change the laser you are working with (i.e., when using a different manufacturing machine), the powder batch (e.g., in terms of chemical analysis, particle size distribution, powder morphology) [109], or the geometry (e.g., sharp features, heavy sections or portion of inclined surfaces facing loose powder). A lot of work has been performed in order to investigate the optimal working conditions of L-PBF and guidelines for the selections of parameters [110–116].

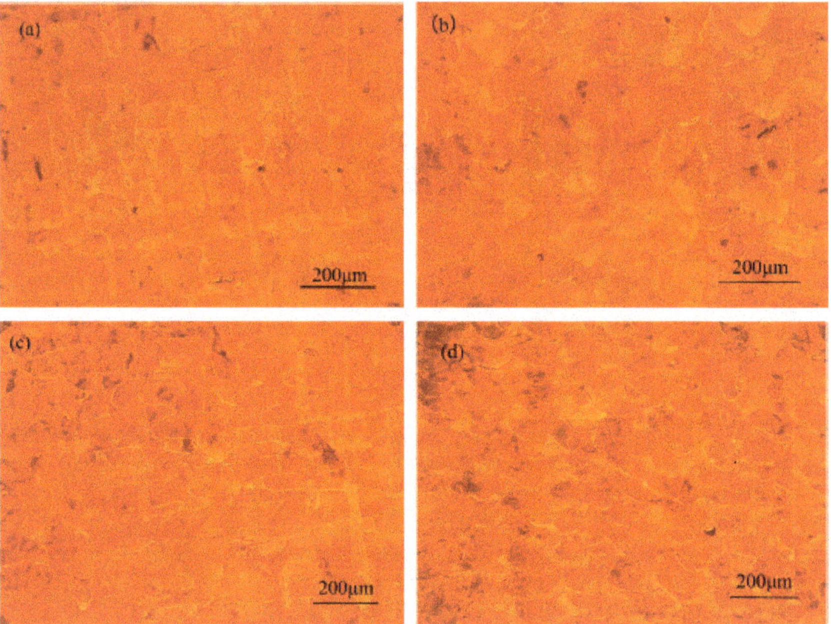

Figure 7. OM of L-PBF stainless steel 304 samples on cross-section and vertical section: (**a**,**b**) at an overlap rate of 0% showing full density, like (**c**,**d**) obtained with 50% overlap between parallel lines. From [105], with permission from Elsevier, 2019.

3. Stainless Steel Grades Processed in L-PBF Systems

Stainless steels are widely employed for their unique performances at room and high temperatures, owing to their chemical and microstructural features. In L-PBF, the right chemical composition is guaranteed by wise manipulation and correct storage of metal powders, mainly to avoid oxygen, moisture or oil pick up [117]; meanwhile, the microstructural properties (e.g., grain size, phases) are determined by the processing parameters. Specifically, the final microstructure depends on the local heat flow direction, competitive growth of grains, and laser scanning strategy. Typical cooling rates are in the range 10^5–10^6 K/s [10,118,119] due to the heat exchange with the gas atmosphere, the unfused powder, and the material already consolidated underneath; the resulting solidification microstructure is fine and far from that provided by thermodynamic equilibrium [120]. The laser scanning strategy has an impact on texture; for example, when the scanning strategy is set with no rotation between subsequent layers, the as-built material exhibits a fibrous aspect with a <001> direction of grain growth (building direction, z, normal to the building platform). This situation leads to a strongly orthotropic behavior of the material [81]. On the other hand, selectively scanning the powder bed in small islands, sometimes non-consecutively, and rotating the direction of the laser between the different layers, makes it possible to obtain an almost untextured microstructure. An elongated grain structure along the z-direction (as shown in Figure 8), is due to both prevalent heat extraction from the bottom side of the melt pool (i.e., the building substrate) and epitaxial grain growth, like fusion welding. In L-PBF, the existing base-metal grains (i.e., the grains existing in the last melted layer) act as a substrate for nucleation [53]. Moreover, the melting of subsequent layers causes a reheating of the already consolidated material, determining the solid-state phase transformations.

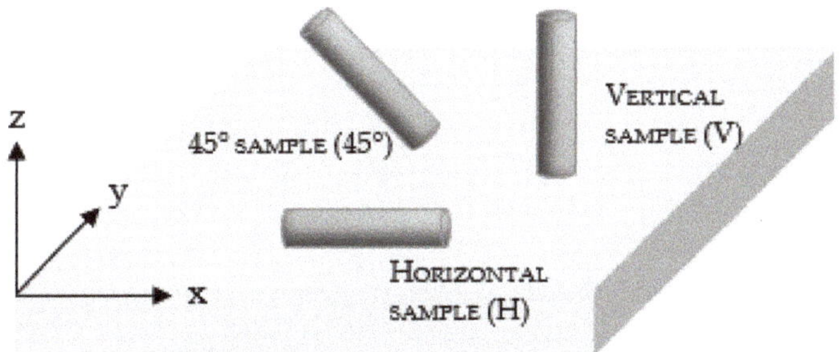

Figure 8. Schematics of commonly projected building directions for tensile specimens.

The peculiar microstructure arising from the described physical phenomena determines the different mechanical performances depending on the tested direction, or, in the case of the final components, on the service loading condition. This is the why the best course of action, as a rule, is to produce tensile specimens with their main axis oriented along different directions, as schematized in Figure 8; vertical specimens (V) are representative of the commonly identified longitudinal direction, while horizontal specimens (H) are representative of the transversal direction. The described unique metallurgical behavior of materials produced via AM led Murr et al. [121] to affirm that such a methodology could extend traditional materials science and engineering as far as making it possible to plan application-specific microstructural architectures in as-built components.

In the following sections, a review of mechanical performances obtained from L-PBF processed alloys is reported. For ease of reading, the correlated process parameters are not shown, but can be consulted in referenced papers. The authors would like to underline that tensile results reported under the as-built conditions in Tables 2 and 4–6 involved testing samples with no stress-relieving heat treatment; the presence of residual stresses inside the specimen can result in deformation (if not properly machined) and yield strength values affected by the presence of a tension status of compression or tension inside the material itself, even before load is applied. This observation should aid readers in understanding the primary importance of stress-relief in L-PBF.

3.1. Austenitic Stainless Steel Grades

Austenitic stainless steels commonly processed in L-PBF systems are essentially alloy 304, alloy 304L and AISI 316L, the latter being the only one of these to have been commercialized by systems manufacturers [122–126].

- In Table 2, tensile results are reported and compared to the standard minimum requirements:
- tension tests performed at room temperature showed good performance, apart from fracture elongation, with results being higher than the minimum requirements usually applied for the selected stainless steel grades processed with conventional technologies;
- fracture elongation is the most negatively affected parameter, for samples tested under the as-built conditions;
- analyzing the listed tensile properties for 316L stainless steel powders, we can state that the experimental research performed in the cited papers achieved comparable results. It can be assumed that they all used proper parameter sets.

Table 2. Tensile strength of austenitic stainless steel grades obtained from L-PBF, compared to standard reference values.

Grade	Equipment	Relative Density [%]	Cond.	BD	Test Cond.	*YS* [MPa]	*UTS* [MPa]	*El.* [%]	Ref.
304	SD	NR	As built	H 45° V	RT	530 370 450	700 540 550	38 29 58	[105]
304L	3D Systems ProX-300	99.99	As built	-	RT	485	712	61	[127]
316L	SLM Solutions 125HL	95.99–99.30	HT–1040 °C/4h	V	RT	376	637	32.4	[128]
316L	SLM Solutions 280HL	>99	As built	H 45° V	RT	528 590 439	639 699 512	38.0 34.1 11.8	[78]
316L	Sisma MYSINT100	99.3–100	As built	45° V	RT	505–515 430–495	650 550-575	41 66–72	[89]
316L	Renishaw AM250	NR	As built	H	RT	554	685	36	[129]
316L	NR	NR	As built	- - - -	RT 250 °C 1100 °C 1200 °C	456 376 - -	703 461 300 150	45 31 15–18 20	[118]

Standard Reference Values

Grade	Condition	Test condition	*YS* [MPa]	*UTS* [MPa]	*El.* [%]	Ref.
304	Annealed–hot finished	RT	205	515	40	[130]
304L, 316L	Annealed–hot finished	RT	170	485	40	

Cond.—Condition; BD—Building Direction; YS—Yield Strength; UTS—Ultimate Tensile Strength; El.—Elongation; SD—Self-developed; NR—Non reported; RT—Room Temperature; HT—Heat-treated; H—Horizontal; V—Vertical.

Table 3 contains some fatigue resistance results, revealing the beneficial effect of machining operations, in particular on high cycle fatigue (which is more sensitive to surface conditions and eventual crack presence). Low cycle fatigue [131] showed little effect from reducing surface roughness.

Table 3. Fatigue endurance limits of L-PBF 316L samples, under different loading and surface conditions.

Alloy	Fatigue Endurance at 10^6 Cycles [MPa]	*R*	Surface Condition	*Ra* [μm]	Ref.
316L	130	−1	As built	13.29	[132]
316L	170	−1	Vibratory finished	1.74	
316L	240	−1	Turned	1.08	
316L	200	0.1	As built	10.0	[131]
316L	256	0.1	Machined	0.4	
316L	269	0.1	Polished	0.1	
316L	108	−1	As built	NR	[133]
316L	267	−1	Turned	NR	

Microstructural characterization performed by [127,134,135] on both 304 and 316L stainless steels showed common features:

- all samples feature columnar grains, see Figures 9–11, independently of the specific alloy;
- all samples feature columnar grains independently of scanning strategy or laser power, see Figures 9 and 10.
- grain size is basically independent of the selected scanning strategy, as long as the laser power is kept constant (see Figure 9);
- grain size decreases as the selected laser power decreases, as can be seen from Figure 10d: this effect is investigated in [104], with lower cooling rates being exhibited at higher laser power;
- as-built grains are characterized by needle-like structures with medium sizes of 500–800 nm and a high aspect ratio, oriented along different directions even in a single weld bead (see different growing orientations marked by red arrows in Figure 12);
- [127] also reported a top view microstructure, as shown in Figure 11b, exhibiting equiaxed grains all over the sample;
- two types of boundaries were observed by [135]: cell boundaries (formed by dislocations) and colony boundaries (prior high-angle austenite grain boundaries);
- in Figure 13, it can be seen how annealing heat treatment affects the starting microstructure (in Figure 13a): cell boundaries were unchanged after heat treatment at 800 °C, but they were not present after heat treatment at 900 °C. In contrast, colony boundaries were unmodified, meaning that no recrystallization phenomena had taken place.

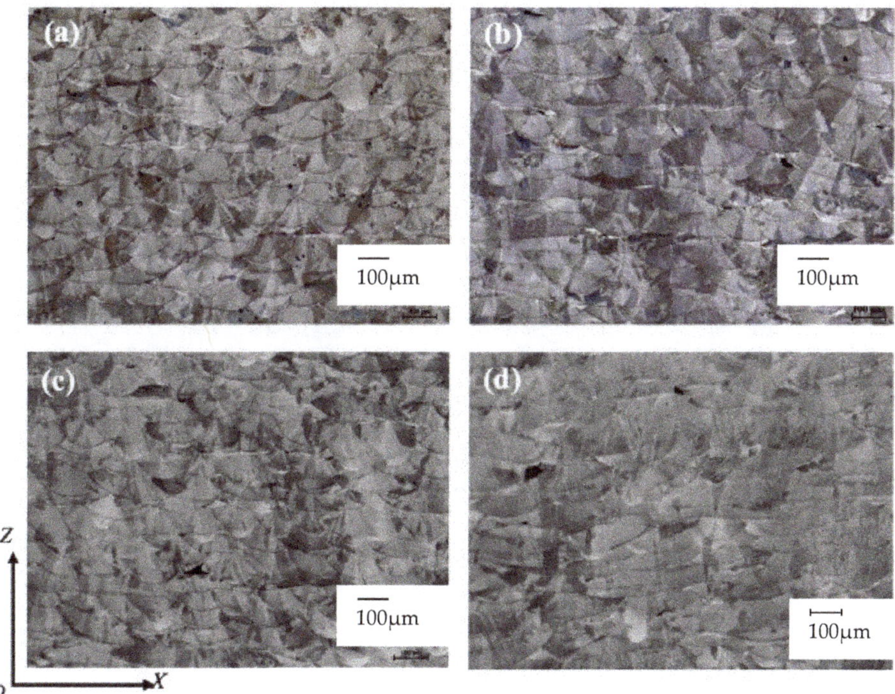

Figure 9. OM micrographs showing the microstructure of 316L samples fabricated at 200 W laser power, using different scanning strategies: (**a**) Meander; (**b**) Stripe; (**c**) Chess with 5 mm × 5 mm islands; (**d**) Chessboard with 1 mm × 1 mm islands. Z specifies the building direction [134].

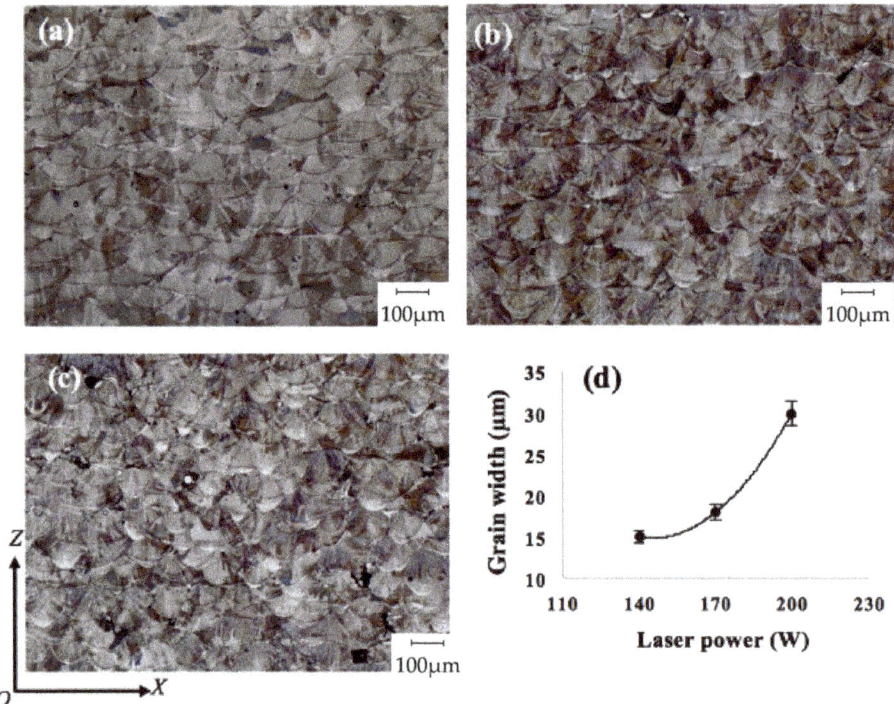

Figure 10. OM micrographs of L-PBF 316L microstructure produced at different laser powers, (**a**) 200 W; (**b**) 170 W; (**c**) 140 W; (**d**) grain width correlation to laser power values. Z specifies the building direction [134].

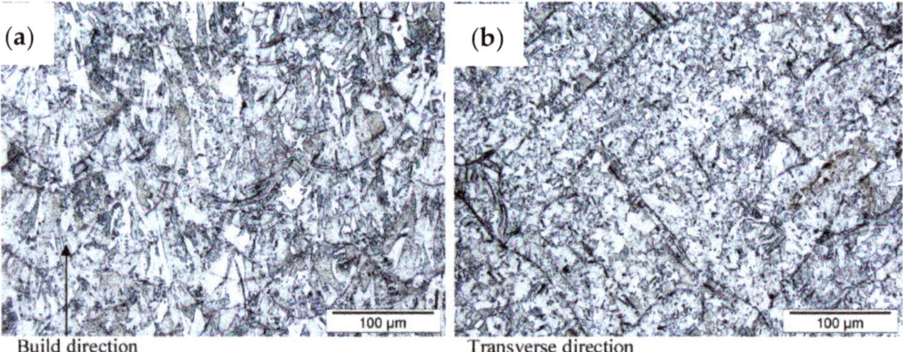

Figure 11. OM showing L-PBF 304 stainless steel microstructure: (**a**) along the vertical cross-section—the heat flow effect is evident in the build direction, and (**b**) nearly equiaxed grains in the planar direction, coincident to the layer top view. From [127], with permission from Elsevier, 2019.

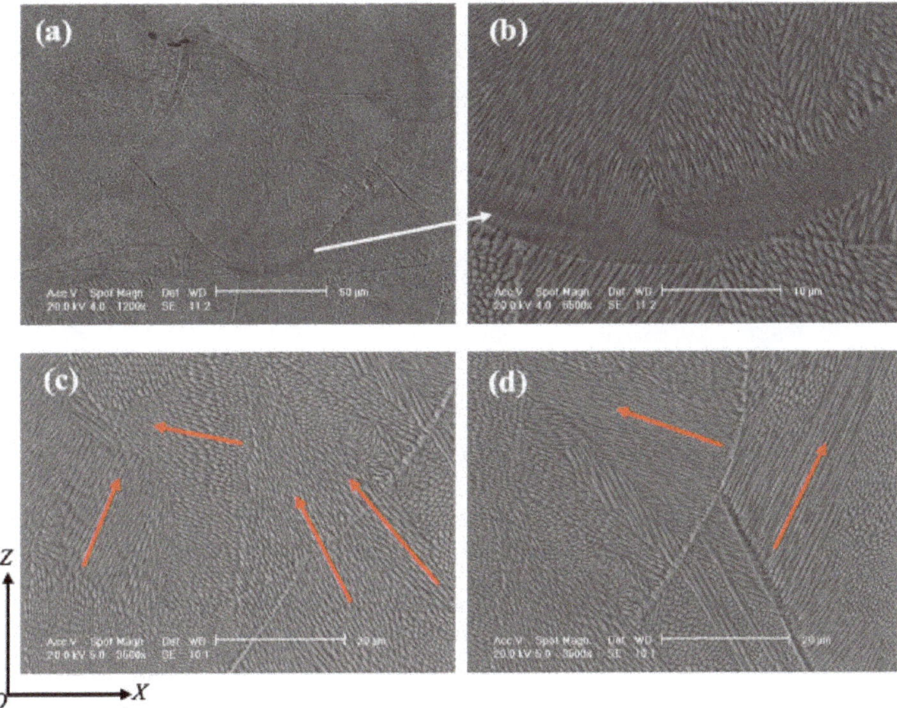

Figure 12. SEM micrographs showing the microstructure of 316L samples (laser power at 200 W): (**a**) within a weld bead; (**b**) at the bottom region of (**a**); (**c**) grain structure beyond two layers; (**d**) grains in two adjacent weld beads. Red arrows point out grain growth direction. Z specifies the building direction [134].

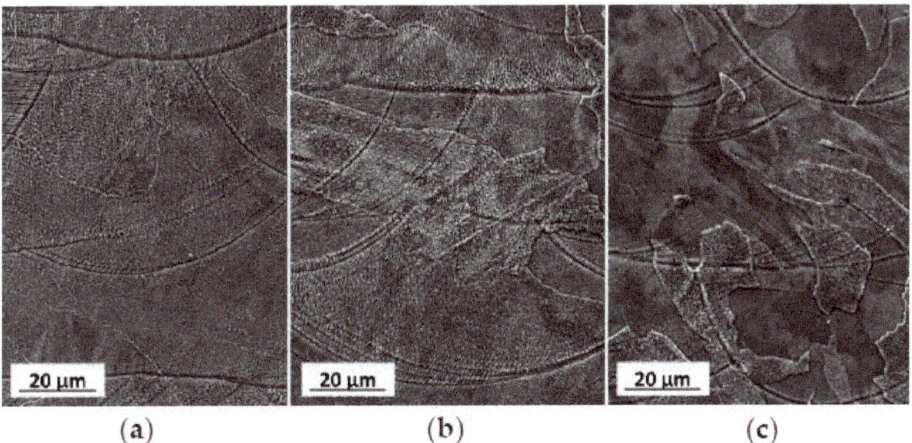

Figure 13. SEM images showing cross-section microstructures at melt-pool level of L-PBF 316 L steel in (**a**) as-built condition, and after annealing at (**b**) 800 °C/15 min and (**c**) 900 °C/15 min [135].

3.2. Precipitation Hardening Stainless Steel Grades

After austenitic grades, precipitation hardening stainless steels, and 15-5 PH and 17-4 PH alloys, particularly (mostly applied for aeronautic and oil and gas parts production), are the most widespread and characterized grades of L-PBF equipment.

Auguste et al., in [136], and Murr et al., in [137], both worked on L-PBF 17-4 PH stainless steel, studying the effects of different powder batches on the final microstructure; in [136], the main difference was the Niobium content of the feedstock, while [137] investigated the effect of powder processing atmosphere and working inert gas.

- 17-4 PH samples showed columnar grains oriented along the building direction, clearly visible in the samples in Figures 14a, 15 and 16. These observations are in accordance with those performed on austenitic stainless steels in the previous Section 3.1;
- As-built niobium-rich material, shown in Figure 14a, features mainly ferritic grains, with a minor content of martensite and residual austenite. Ferritic grains are characterized by a high aspect ratio, and the largest dimension reaches hundreds of micrometers. Ferritic microstructure is in contrast to the typical martensitic microstructure observed in this alloy;
- On the other hand, the as-built material shown in Figure 14b shows an overall martensitic microstructure, with a grain size in the range of 1–20 μm;
- Ferrite-rich samples do not show microstructural evolution after tempering at 480 °C (Figure 15a) and 550 °C (Figure 15b) with respect to Figure 14a. Martensitic microstructure and desired mechanical properties were achieved only after full material homogenization (Figure 15d and Table 3). This observation is in line with the thermal stability already underlined in austenitic 316L samples;
- Ar-atomized and N2-atomized powders produced completely martensitic phase materials when fabricated in an Ar environment (Figure 16 refers to a sample obtained from Ar-processed powders in Ar L-PBF atmosphere). Conversely, Ar-atomized powder and N_2-atomized powder showed different behavior after N_2 L-PBF processing gas environment; Ar-atomized powders fused in N2 produced fully martensitic components, while the N_2-atomized powder fabricated in a N_2-gas environment produced austenitic components containing roughly 15% martensite (Figure 17);
- Different microstructural observations obtained comparing Figures 16 and 17 were confirmed by hardness measurements, with the results being reported in Figure 18; the austenitic sample hardness is lower than the martensitic one, and little variation can be appreciated after aging treatment, in contrast to the martensitic one, which underwent second-phase precipitation.

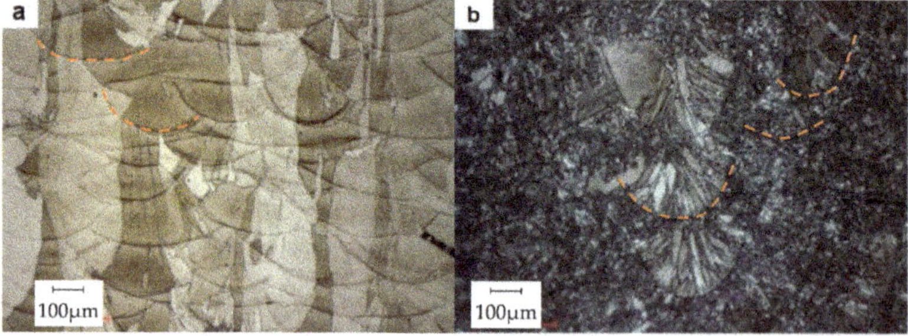

Figure 14. (**a**,**b**) OM cross-section observation of two 17-4 PH samples obtained from L-PBF, starting from different powder batches [136].

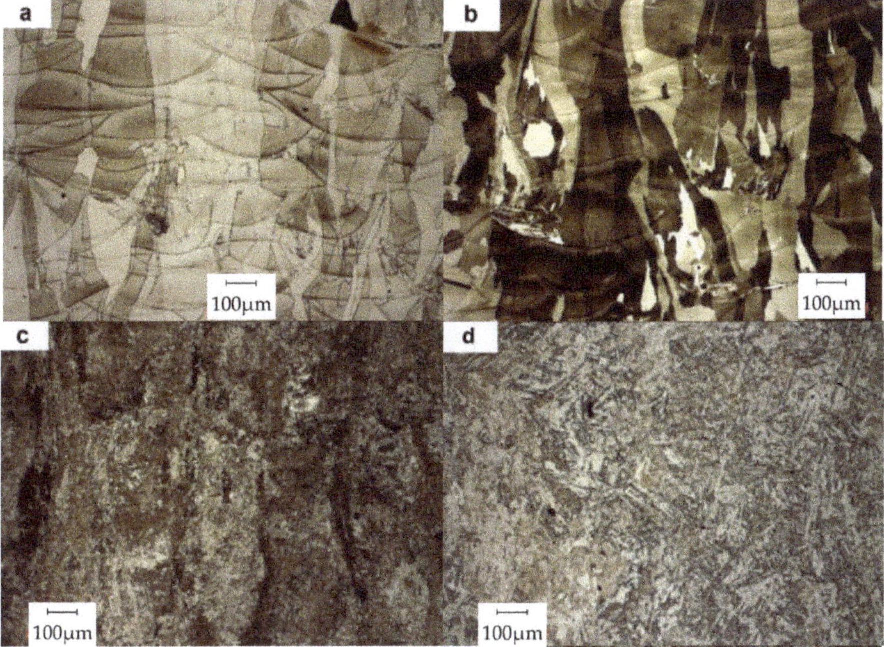

Figure 15. OM cross-section observation of L-PBF 17-4 PH sample (seen in Figure 14a) after heat treatment: (**a**) tempered at 480 °C/1 h; (**b**) tempered at 550 °C/4 h; (**c**) solution heat-treated (1040 °C/1.5 h), quenched and tempered at 480 °C/1 h; (**d**) homogenized (1190 °C/2 h), solution heat-treated (1040 °C/1.5 h) and tempered at 480 °C/1 h [136].

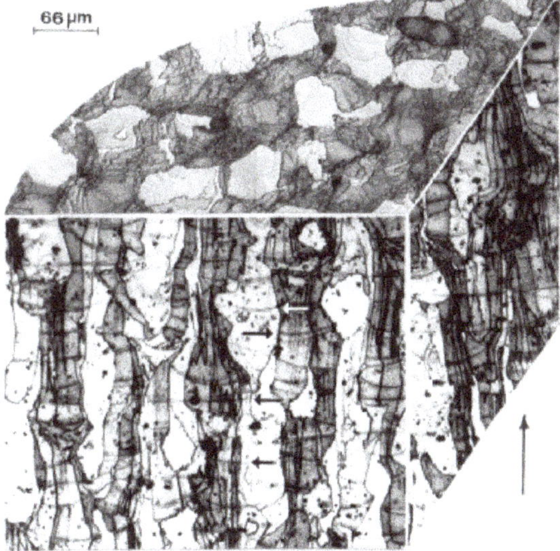

Figure 16. OM-3D reconstruction of 17-4 PH sample, effective building direction shown by vertical arrow in the lower right corner: the microstructure is fully martensitic and comes from Ar-atomized powders fabricated in an Ar-rich environment (Ar/Ar) [137].

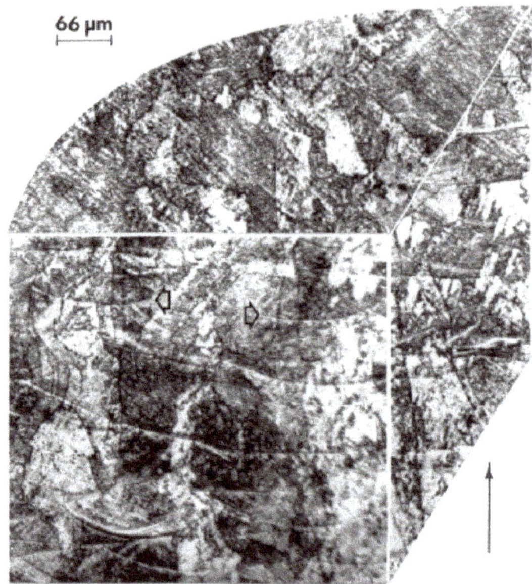

Figure 17. OM-3D reconstruction of 17-4 PH sample, effective building direction shown by vertical arrow in the lower right corner: the microstructure is mainly austenitic, with a minor volume fraction of martensite (15%) and comes from N_2-atomized powders fabricated in a N_2-rich environment (N_2/N_2) [137].

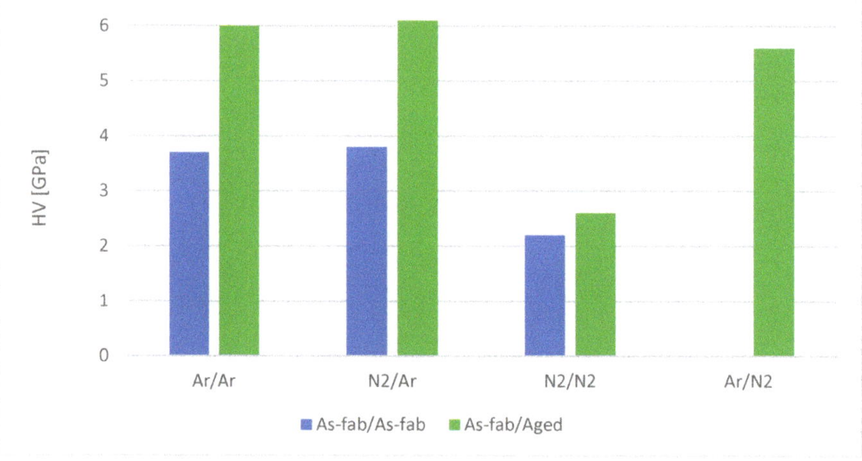

Figure 18. Experimental results from [137]: Vickers microhardness (HV) on as-manufactured samples (blue piles) and after aging at 482 °C for 1 h and air-cooled standard H900 heat treatment (green piles). Ar/Ar and N_2/Ar refer respectively to Ar- and N_2-atomized powders, both L-PBF processed in Ar atmosphere; similarly, Ar/N_2 and N_2/N_2 refer to Ar- and N_2-atomized powders, L-PBF fabricated in N_2 environment.

The tensile performances exhibited by L-PBF 17-4 PH specimens, summarized in Table 4, are characterized by a wide dispersion between different referenced works. In particular, the as-built alloy does not reach minimum standard requirements, resulting in the need to apply (and, eventually, develop) specifically tailored heat treatments. Yadollahi et al. [138] underlined the need for heat treating L-PBF 17-4 PH in order to reach standard tensile requirements and increase the low cycle fatigue life. On the other hand, standard solubilization and aging treatments have been found to lower high cycle fatigue life. Mahmoudi et al. in [139] found that vertically built samples were characterized by lower strength than those built horizontally, claiming that interlayer bonding was insufficient.

Table 4. Tensile strength of precipitation hardening stainless steel grades obtained from L-PBF, compared to standard reference values.

Grade	Equipment	Relative Density [%]	Condition	BD	Test Cond.	*YS* [MPa]	*UTS* [MPa]	*El.* [%]	Ref.
17-4 PH	EOS M270	NR	As built	H	RT	523	1028	-	[140]
				V		494	979	-	
			HT—650 °C/1 h	H		436	1295	-	
				V		483	1298	-	
17-4 PH	EOS M290	NR	As built	V	RT	835	1169	48.42	[141]
			HT—1050 °C/0.5 h + 552 °C/4 h			1176	1170	32.7	
17-4 PH	SLM Solutions 280HL	>99 average 99.6	As built	H	RT	850	890	13	[136]
				V		760	785	2.5	
			HT—480 °C/1 h	H		-	780	-	
				V		-	560	-	
			HT—550 °C/4 h	H		1210	1220	0.5	
				V		-	550	-	
			HT—1040 °C/1.5 h + quenching + 480 °C/1 h	H		785	990	4.6	
				V		590	680	1	
			HT—1190 °C/2 h + 1040 °C/1.5 h + quenching + 480 °C/1 h	H		1400	1295	3	
				V		1240	1305	1	
17-4 PH	EOS M290	NR	As built	H	RT	-	710	6.7–7.2	[142]
17-4 PH	3D Systems ProX 100	NR	As-built	H	RT	650	1050	9.8	[139]
				V		600–720	950–1050	3.5–6.4	
			HT—1038 °C/0.5 h + 482 °C/1 h	H		910	1220	7.8	
				V		730–950	970–1120	2.5–3.5	
15-5 PH	EOS M270	NR	As built	H	RT	1297	1450	12.53	[143]
				V		1100	1467	14.92	

Standard Reference Values

Grade	Condition	Test Condition	*YS* [MPa]	*UTS* [MPa]	*El.* [%]	Ref.
17-4 PH	H900 aging—482 °C/1 h	RT	>1170	>1310	>10	[144]
15-5 PH	H900 aging—482 °C/1 h	RT	>1170	>1310	>6 (transv.) >10 (long.)	

BD—Building Direction; YS—Yield Strength; UTS—Ultimate Tensile Strength; El.—Elongation; NR—Non reported; HT—Heat-treated; H—Horizontal; V—Vertical; RT—Room Temperature.

This observation is in contrast to what is expected from metallurgical features. Comparing results from similar specimens (same as-built condition and building direction) shows that mechanical properties registered high scattering; this could be correlated with unsuitable processing parameters, post-processing operations not being performed well (for example, sample machining), or samples being tested without machining. In the latter case, the results would be diminished by surface discontinuities.

3.3. Other Stainless Steel Grades

Literature research made it possible to highlight works inherent to the characterization of a martensitic stainless steel alloy, i.e., 420 (in Table 5), and two duplex stainless steel grades, whose results are summarized in Figure 19 and Table 6.

Saeidi et al. in [118] revealed that as-built L-PBF 420 alloy is characterized by a cellular microstructure of micron-sized martensitic cells (as can be seen in [118]); this feature leads to high values of ultimate tensile strength (according to Table 5).

Table 5. Tensile strength of martensitic 420 alloy obtained from L-PBF, compared to standard reference values.

Grade	Equipment	Relative Density [%]	Cond.	BD	Test Cond.	*YS* [MPa]	*UTS* [MPa]	*El.* [%]	Ref.
420	NR	NR	As built		RT	800	1800	5	[118]
420	SD	NR	As built	H V	RT	- -	505 1045	- -	[88]

Standard Reference Values

Grade	Condition	Test Condition	*YS* [MPa]	*UTS* [MPa]	*El.* [%]	Ref.
420	Annealed—holding T: 745–825 °C + air cooling	RT	-	<760	-	[145]
	QT800—quench at 950–1050 °C + oil or air cooling + tempering at 600–700 °C		>600	800–950	>12	

Cond.—Condition; BD—Building Direction; YS—Yield Strength; UTS—Ultimate Tensile Strength; El.—Elongation; SD—Self-developed; NR—Non reported; RT—Room Temperature; HT—Heat-treated; H—Horizontal; V—Vertical; QT—Quench and Tempered.

Duplex samples processed in L-PBF systems were mainly studied by Hengsbach et al. [146] and Saeidi et al. in [118,147]. The most relevant observations were:

- The as-built SAF 2507 alloy is characterized by an almost completely ferritic microstructure (Figure 19a,e), strongly different from the desired microstructure, caused by rapid cooling rates typical of L-PBF. The cooling rates experienced cause the alloy's solidification in delta ferrite, suppressing the austenite field;
- Proper heat treating conditions can restore an acceptable ferrite/austenite ratio; in particular, sample solubilization at 1000 °C (Figure 19c,g) made it possible to achieve a 34% austenite content;
- The performed tension tests (shown in Table 5) showed that L-PBF SAF2205 properly heat-treated satisfies the standard minimum requirements, while SAF2507 needs to be further optimized through subsequent heat treatment in order to achieve proper mechanical performances.

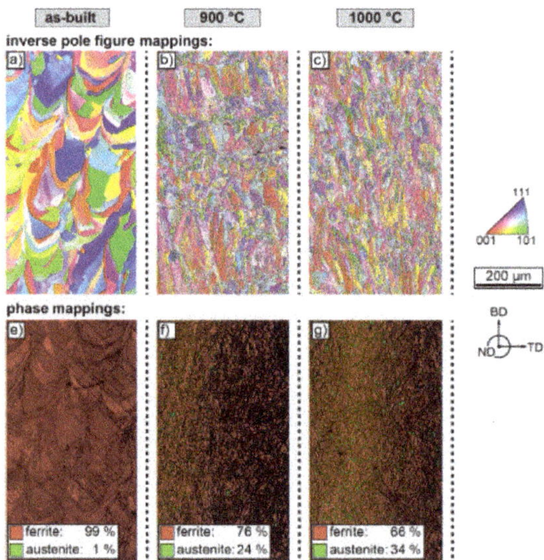

Figure 19. EBSD maps of duplex 2205 showing (**a**–**c**) inverse pole maps and (**e**–**g**) corresponding phase fraction, in (**a**,**e**) as-built condition, (**b**,**f**) after annealing at 900 °C, and (**c**,**g**) after annealing at 1000 °C [146], with permission from Elsevier, 2019.

Table 6. Tensile strength of duplex stainless steel alloys obtained from L-PBF, compared to standard reference values.

Grade	Equipment	Relative Density [%]	Condition	BD	Test Cond.	*YS* [MPa]	*UTS* [MPa]	*El.* [%]	Ref.
SAF2205	SLM Solutions 280HL	99.7–99.85	As built HT—1000 °C/0.083 h	V	RT	- -	940 770	12 28	[146]
SAF2507	NR	NR	As built	-	RT 1200 °C	1214 -	1321 500	- 30	[118]
SAF2507	EOS M270	99	HT—1200 °C/0.083 h	-	RT 1200 °C	- -	920 400	1.8 20	[147]

Standard Reference Values

Grade	Condition	Test Cond.	*YS* [MPa]	*UTS* [MPa]	*El.* [%]	Ref.
SAF2205	HT—1020–1100 °C + rapid cooling	RT	>450	>620	>25	[148]
SAF2507	HT—1025–1125 °C + rapid cooling	RT	>550	>800	>15	

BD—Building Direction; YS—Yield Strength; UTS—Ultimate Tensile Strength; El.—Elongation; NR—Non reported; RT—Room Temperature; HT—Heat-treated; H—Horizontal; V—Vertical; QT—Quench and Tempered.

4. Conclusions

The capability of L-PBF systems to process stainless steel alloys is reported in detail in this review, which therefore could be considered as a reference for all researchers and technologists involved in activities concerning stainless steel grades and L-PBF manufacturing. In fact, since research in this field has been receiving a strong impulse in recent years, it could be quite time-consuming for those people facing such topic for the first time.

The exploitation of the L-PBF technique makes it possible to achieve significant benefits when the following issues are required:

- Artefact weight reduction (made possible, for example, through topology optimization and/or lattice structures);
- Easy customization;
- Complex internal features manufacturing (e.g., inclined ducts).

The above-listed advantages, which are known to be relevant for high-cost alloys (such as superalloys or titanium alloys) and aerospace applications, are also significant for stainless steel grades, that, since their discovery, have found application in almost every field of application. In this context, the adoption of L-PBF reflects huge material savings and efficiency enhancement in applications ranging from biomedical devices to power production plants. L-PBF is particularly suited to additive manufacturing of stainless steels, since it makes it possible to produce complex features, in a wide range of cross-section thicknesses, with good compromise in terms of cost and time, while guaranteeing low oxides contamination. A variety of stainless steel grades coming from different classes (i.e., austenitic, martensitic, precipitation hardening and duplex) have been satisfactorily processed, meaning that it could be possible to widen the list of alloys in order to widen the field of applications.

This review shows that, to date, stainless steel alloys have been satisfactorily processed by L-PBF. Furthermore, the achieved mechanical properties make stainless steels fit the requirements of several applications. High mechanical properties are targeted, since the porosity level achieved by L-PBF is quite low and is comparable to conventionally processed materials. This statement is supported by the activities of many researchers, which has been demonstrated to satisfactorily achieve 99.9% relative density. Mechanical performances are even higher than the standard requirements, even if a wide range of values can be found in the literature, meaning that it is necessary to unify processing and post-processing routes. For example, standards defining the proper orientation to build and test samples, together with proper machine allowance and testing surface condition, still do not exist.

The analysis of the listed tested alloys, relative metallurgical microstructures and tensile strengths reveals a fundamental lack of alloys to be worked, standardization, and development of proper heat treatment. This inadequacy defines the path for further research and exploitation.

Funding: This research did not receive any specific grant from funding agencies in the public, commercial, or not-for-profit sectors.

Conflicts of Interest: The authors declare no conflict of interest.

References

1. Zhai, Y.; Lados, D.A.; LaGoy, J.L. Additive Manufacturing: Making Imagination the Major Limitation. *JOM* **2014**, *66*, 808–816. [CrossRef]
2. European Powder Metallurgy Association (EPMA)—Introduction to Additive Manufacturing Technology (Brochure). Available online: https://www.epma.com/epma-free-publications/product/introduction-to-additive-manufacturing-brochure (accessed on 3 August 2018).
3. Orme, M.; Madera, I.; Gschweitl, M.; Ferrari, M. Topology Optimization for Additive Manufacturing as an Enabler for Light Weight Flight Hardware. *Designs* **2018**, *2*, 51. [CrossRef]
4. Horn, T.J.; Harrysson, O.L.A. Overview of Current Additive Manufacturing Technologies and Selected Applications. *Sci. Prog.* **2012**, *95*, 255–282. [CrossRef] [PubMed]
5. Shukla, M.; Todorov, I.; Kapletia, D. Application of additive manufacturing for mass customisation: Understanding the interaction of critical barriers. *Prod. Plan. Control* **2018**, *29*, 814–825. [CrossRef]
6. Spallek, J.; Krause, D. Process Types of Customisation and Personalisation in Design for Additive Manufacturing Applied to Vascular Models. *Procedia CIRP* **2016**, *50*, 281–286. [CrossRef]
7. Glasschroeder, J.; Prager, E.; Zaeh, M.F. Powder-bed-based 3D-printing of function integrated parts. *Rapid Prototyp. J.* **2015**, *21*, 207–215. [CrossRef]
8. Sanín Pérez, P. *A Study of Additive Manufacturing Applied to the Design and Production of LED Luminaires*; Politecnico di Milano: Milano, Italy, 2013.
9. Tang, Y.; Yang, S.; Zhao, Y.F. Sustainable Design for Additive Manufacturing through Functionality Integration and Part Consolidation. In *Handbook of Sustainability in Additive Manufacturing*; Muthu, S.S., Savalani, M.M., Eds.; Springer: Singapore, 2016; pp. 101–144. ISBN 978-981-10-0547-3.
10. DebRoy, T.; Wei, H.L.; Zuback, J.S.; Mukherjee, T.; Elmer, J.W.; Milewski, J.O.; Beese, A.M.; Wilson-Heid, A.; De, A.; Zhang, W. Additive manufacturing of metallic components—Process, structure and properties. *Prog. Mater. Sci.* **2018**, *92*, 112–224. [CrossRef]
11. Poprawe, R.; Hinke, C.; Meiners, W.; Schrage, J.; Bremen, S.; Merkt, S. SLM Production Systems: Recent Developments in Process Development, Machine Concepts and Component Design. In *Advances in Production Technology*; Brecher, C., Ed.; Springer International Publishing: Cham, Switzerland, 2015; pp. 49–65. ISBN 978-3-319-12303-5.
12. Kellner, T. How 3D Printing Will Change Manufacturing. Available online: https://www.ge.com/reports/epiphany-disruption-ge-additive-chief-explains-3d-printing-will-upend-manufacturing/ (accessed on 15 April 2019).
13. Wimpenny, D.I.; Pandey, P.M.; Kumar, L.J. (Eds.) *Advances in 3D Printing & Additive Manufacturing Technologies*; Springer: Singapore, 2017; ISBN 978-981-10-0811-5.
14. Yang, L.; Hsu, K.; Baughman, B.; Godfrey, D.; Medina, F.; Menon, M.; Wiener, S. *Additive Manufacturing of Metals: The Technology, Materials, Design and Production*; Springer Series in Advanced Manufacturing; Springer International Publishing: Cham, Switzerland, 2017; ISBN 978-3-319-55127-2.
15. Wohlers, T.; Gornet, T. History of Additive Manufacturing. *Wohlers Rep.* **2014**, *24*, 118.
16. Gibson, I.; Rosen, D.; Stucker, B. *Additive Manufacturing Technologies*; Springer: New York, NY, USA, 2015; ISBN 978-1-4939-2112-6.
17. Milewski, J.O. *Additive Manufacturing of Metals*; Springer Series in Materials Science; Springer International Publishing: Cham, Switzerland, 2017; Volume 258, ISBN 978-3-319-58204-7.

18. Your Expert in Additive Manufacturing Since 25 Years. Available online: https://www.eos.info/about_eos/history (accessed on 19 April 2019).
19. Deckard, C.R. Apparatus for Producing Parts by Selective Sintering. U.S. Patent 5,597,589, 28 January 1997.
20. Park, J.H.; Kang, Y. Inclusions in Stainless Steels—A Review. *Steel Res. Int.* **2017**, *88*, 1700130. [CrossRef]
21. Mapelli, C.; Nolli, P. Formation Mechanism of Non-Metallic Inclusions in Different Stainless Steel Grades. *ISIJ Int.* **2003**, *43*, 1191–1199. [CrossRef]
22. Gokuldoss, P.K.; Kolla, S.; Eckert, J. Additive Manufacturing Processes: Selective Laser Melting, Electron Beam Melting and Binder Jetting—Selection Guidelines. *Materials* **2017**, *10*, 672. [CrossRef] [PubMed]
23. Zhang, B.; Li, Y.; Bai, Q. Defect Formation Mechanisms in Selective Laser Melting: A Review. *Chin. J. Mech. Eng.* **2017**, *30*, 515–527. [CrossRef]
24. Malekipour, E.; El-Mounayri, H. Common defects and contributing parameters in powder bed fusion AM process and their classification for online monitoring and control: A review. *Int. J. Adv. Manuf. Technol.* **2018**, *95*, 527–550. [CrossRef]
25. Zadpoor, A. Frontiers of Additively Manufactured Metallic Materials. *Materials* **2018**, *11*, 1566. [CrossRef] [PubMed]
26. Nilsson, J.O. Can mankind survive without stainless steel? *Stainl. Steel World.* 2014. 1–4. Available online: https://pdfs.semanticscholar.org/7bcf/2041cf93f4b5c3db7c7fe27b1d06d92f3602.pdf (accessed on 20 May 2019).
27. Rufini, R.; Di Pietro, O.; Di Schino, A. Predictive Simulation of Plastic Processing of Welded Stainless Steel Pipes. *Metals* **2018**, *8*, 519. [CrossRef]
28. Saha Podder, A.; Bhanja, A. Applications of Stainless Steel in Automobile Industry. *Adv. Mater. Res.* **2013**, *794*, 731–740. [CrossRef]
29. Baddoo, N.R. Stainless steel in construction: A review of research, applications, challenges and opportunities. *J. Constr. Steel Res.* **2008**, *64*, 1199–1206. [CrossRef]
30. Corradi, M.; Di Schino, A.; Borri, A.; Rufini, R. A review of the use of stainless steel for masonry repair and reinforcement. *Constr. Build. Mater.* **2018**, *181*, 335–346. [CrossRef]
31. Di Schino, A. Analysis of heat treatment effect on microstructural features evolution in a micro-alloyed martensitic steel. *Acta Metall. Slov.* **2017**, *22*, 266–270. [CrossRef]
32. Di Schino, A.; Di Nunzio, P.E. Metallurgical aspects related to contact fatigue phenomena in steels for back-up rolls. *Acta Metall. Slov.* **2017**, *23*, 62–71. [CrossRef]
33. Di Schino, A.; Porcu, G. Metallurgical design and development of C125 grade for mild sour service application. In *Proceedings of the Corrosion*; NACE Corrosion Paper: San Diego, CA, USA, 2006; pp. 1–14.
34. Kumar Sharma, D.; Filipponi, M.; Di Schino, A.; Rossi, F.; Castaldi, J. Corrosion behaviour of high temperature fuel cells: Issues for materials selection. *Metalurgija* **2019**, *58*, 347–351.
35. Di Schino, A.; Di Nunzio, P.E.; Turconi, G.L. Microstructure evolution during tempering of martenite in e medium C steel. *Mater. Sci. Forum* **2007**, *558*, 1435–1441. [CrossRef]
36. Cianetti, F.; Ciotti, M.; Palmieri, M.; Zucca, G. On the Evaluation of Surface Fatigue Strength of a Stainless-Steel Aeronautical Component. *Metals* **2019**, *9*, 455. [CrossRef]
37. Talha, M.; Behera, C.K.; Sinha, O.P. A review on nickel-free nitrogen containing austenitic stainless steels for biomedical applications. *Mater. Sci. Eng. C* **2013**, *33*, 3563–3575. [CrossRef] [PubMed]
38. Boulané-Petermann, L. Processes of bioadhesion on stainless steel surfaces and cleanability: A review with special reference to the food industry. *Biofouling* **1996**, *10*, 275–300. [CrossRef]
39. Bregliozzi, G.; Ahmed, S.I.-U.; Di Schino, A.; Kenny, J.M.; Haefke, H. Friction and Wear Behavior of Austenitic Stainless Steel: Influence of Atmospheric Humidity, Load Range, and Grain Size. *Tribol. Lett.* **2004**, *17*, 697–704. [CrossRef]
40. Di Schino, A.; Valentini, L.; Kenny, J.M.; Gerbig, Y.; Ahmed, I.; Haefke, H. Wear resistance of a high-nitrogen austenitic stainless steel coated with nitrogenated amorphous carbon films. *Surf. Coat. Technol.* **2002**, *161*, 224–231. [CrossRef]
41. Di Schino, A.; Kenny, J.M.; Abbruzzese, G. Analysis of the recrystallization and grain growth processes in AISI 316 stainless steel. *J. Mater. Sci.* **2002**, *37*, 5291–5298. [CrossRef]
42. Uriondo, A.; Esperon-Miguez, M.; Perinpanayagam, S. The present and future of additive manufacturing in the aerospace sector: A review of important aspects. *Proc. Inst. Mech. Eng. Part G J. Aerosp. Eng.* **2015**, *229*, 2132–2147. [CrossRef]

43. Leach, R. Metrology for Additive Manufacturing. Available online: http://eprints.nottingham.ac.uk/33924/1/AM%20metrology%20for%20MC.pdf (accessed on 20 May 2019).
44. Jurrens, K. Measurement Science and Standards for Metals-Based Additive Manufacturing. Available online: https://www.nrc.gov/docs/ML1815/ML18150A368.pdf (accessed on 20 May 2019).
45. Slotwinski, J.A.; Garboczi, E.J. Metrology Needs for Metal Additive Manufacturing Powders. *JOM* **2015**, *67*, 538–543. [CrossRef]
46. Mani, M.; Lane, B.M.; Donmez, M.A.; Feng, S.C.; Moylan, S.P. A review on measurement science needs for real-time control of additive manufacturing metal powder bed fusion processes. *Int. J. Prod. Res.* **2017**, *55*, 1400–1418. [CrossRef]
47. Everton, S.K.; Hirsch, M.; Stravroulakis, P.; Leach, R.K.; Clare, A.T. Review of in-situ process monitoring and in-situ metrology for metal additive manufacturing. *Mater. Des.* **2016**, *95*, 431–445. [CrossRef]
48. ASTM and ISO Additive Manufacturing Committees Approve Joint Standards under Partner Standards Developing Organization Agreement|www.astm.org. Available online: https://www.astm.org/cms/drupal-7.51/newsroom/astm-and-iso-additive-manufacturing-committees-approve-joint-standards-under-partner (accessed on 23 May 2019).
49. Committee F42 on Additive Manufacturing Technologies. Available online: https://www.astm.org/COMMITTEE/F42.htm (accessed on 23 May 2019).
50. Clayton, J.; Deffley, R. Optimising Metal Powders for Additive Manufacturing. *Met. Powder Rep.* **2014**, *69*, 14–17. [CrossRef]
51. He, X.; DebRoy, T.; Fuerschbach, P.W. Alloying element vaporization during laser spot welding of stainless steel. *J. Phys. Appl. Phys.* **2003**, *36*, 3079–3088. [CrossRef]
52. DebRoy, T.; David, S.A. Physical processes in fusion welding. *Rev. Mod. Phys.* **1995**, *67*, 85–112. [CrossRef]
53. Kou, S. *Welding Metallurgy*, 2nd ed.; Wiley-Interscience: Hoboken, NJ, USA, 2003; ISBN 978-0-471-43491-7.
54. Walton, O.R. Review of Adhesion Fundamentals for Micron-Scale Particles. *KONA Powder Part. J.* **2008**, *26*, 129–141. [CrossRef]
55. Khairallah, S.A.; Anderson, A.T.; Rubenchik, A.; King, W.E. Laser powder-bed fusion additive manufacturing: Physics of complex melt flow and formation mechanisms of pores, spatter, and denudation zones. *Acta Mater.* **2016**, *108*, 36–45. [CrossRef]
56. Kruth, J.P.; Wang, X.; Laoui, T.; Froyen, L. Lasers and materials in selective laser sintering. *Assem. Autom.* **2003**, *23*, 357–371. [CrossRef]
57. Bidare, P.; Bitharas, I.; Ward, R.M.; Attallah, M.M.; Moore, A.J. Fluid and particle dynamics in laser powder bed fusion. *Acta Mater.* **2018**, *142*, 107–120. [CrossRef]
58. Özel, T.; Arısoy, Y.M.; Criales, L.E. Computational Simulation of Thermal and Spattering Phenomena and Microstructure in Selective Laser Melting of Inconel 625. *Phys. Procedia* **2016**, *83*, 1435–1443. [CrossRef]
59. Megahed, M.; Mindt, H.-W.; N'Dri, N.; Duan, H.; Desmaison, O. Metal additive-manufacturing process and residual stress modeling. *Integr. Mater. Manuf. Innov.* **2016**, *5*, 61–93. [CrossRef]
60. Körner, C.; Attar, E.; Heinl, P. Mesoscopic simulation of selective beam melting processes. *J. Mater. Process. Technol.* **2011**, *211*, 978–987. [CrossRef]
61. Conti, P.; Cianetti, F.; Pilerci, P. Parametric Finite Elements Model of SLM Additive Manufacturing process. *Procedia Struct. Integr.* **2018**, *8*, 410–421. [CrossRef]
62. Manvatkar, V.; De, A.; DebRoy, T. Spatial variation of melt pool geometry, peak temperature and solidification parameters during laser assisted additive manufacturing process. *Mater. Sci. Technol.* **2015**, *31*, 924–930. [CrossRef]
63. Gürtler, F.-J.; Karg, M.; Leitz, K.-H.; Schmidt, M. Simulation of Laser Beam Melting of Steel Powders using the Three-Dimensional Volume of Fluid Method. *Phys. Procedia* **2013**, *41*, 881–886. [CrossRef]
64. King, W.E.; Anderson, A.T.; Ferencz, R.M.; Hodge, N.E.; Kamath, C.; Khairallah, S.A.; Rubenchik, A.M. Laser powder bed fusion additive manufacturing of metals; physics, computational, and materials challenges. *Appl. Phys. Rev.* **2015**, *2*, 041304. [CrossRef]
65. Zhang, Z.; Huang, Y.; Rani Kasinathan, A.; Imani Shahabad, S.; Ali, U.; Mahmoodkhani, Y.; Toyserkani, E. 3-Dimensional heat transfer modeling for laser powder-bed fusion additive manufacturing with volumetric heat sources based on varied thermal conductivity and absorptivity. *Opt. Laser Technol.* **2019**, *109*, 297–312. [CrossRef]

66. Foteinopoulos, P.; Papacharalampopoulos, A.; Stavropoulos, P. On thermal modeling of Additive Manufacturing processes. *CIRP J. Manuf. Sci. Technol.* **2018**, *20*, 66–83. [CrossRef]
67. Rodgers, T.M.; Madison, J.D.; Tikare, V. Simulation of metal additive manufacturing microstructures using kinetic Monte Carlo. *Comput. Mater. Sci.* **2017**, *135*, 78–89. [CrossRef]
68. Alimardani, M.; Toyserkani, E.; Huissoon, J.P. A 3D dynamic numerical approach for temperature and thermal stress distributions in multilayer laser solid freeform fabrication process. *Opt. Lasers Eng.* **2007**, *45*, 1115–1130. [CrossRef]
69. Ganeriwala, R.; Zohdi, T.I. Multiphysics Modeling and Simulation of Selective Laser Sintering Manufacturing Processes. *Procedia CIRP* **2014**, *14*, 299–304. [CrossRef]
70. Ren, K.; Chew, Y.; Fuh, J.Y.H.; Zhang, Y.F.; Bi, G.J. Thermo-mechanical analyses for optimized path planning in laser aided additive manufacturing processes. *Mater. Des.* **2019**, *162*, 80–93. [CrossRef]
71. Alimardani, M.; Toyserkani, E.; Huissoon, J.P.; Paul, C.P. On the delamination and crack formation in a thin wall fabricated using laser solid freeform fabrication process: An experimental–numerical investigation. *Opt. Lasers Eng.* **2009**, *47*, 1160–1168. [CrossRef]
72. Qiu, C.; Panwisawas, C.; Ward, M.; Basoalto, H.C.; Brooks, J.W.; Attallah, M.M. On the role of melt flow into the surface structure and porosity development during selective laser melting. *Acta Mater.* **2015**, *96*, 72–79. [CrossRef]
73. Scime, L.; Beuth, J. Using machine learning to identify in-situ melt pool signatures indicative of flaw formation in a laser powder bed fusion additive manufacturing process. *Addit. Manuf.* **2019**, *25*, 151–165. [CrossRef]
74. Coeck, S.; Bisht, M.; Plas, J.; Verbist, F. Prediction of lack of fusion porosity in selective laser melting based on melt pool monitoring data. *Addit. Manuf.* **2019**, *25*, 347–356. [CrossRef]
75. Zhou, Y.H.; Zhang, Z.H.; Wang, Y.P.; Liu, G.; Zhou, S.Y.; Li, Y.L.; Shen, J.; Yan, M. Selective laser melting of typical metallic materials: An effective process prediction model developed by energy absorption and consumption analysis. *Addit. Manuf.* **2019**, *25*, 204–217. [CrossRef]
76. Spierings, A.B.; Voegtlin, M.; Bauer, T.; Wegener, K. Powder flowability characterisation methodology for powder-bed-based metal additive manufacturing. *Prog. Addit. Manuf.* **2016**, *1*, 9–20. [CrossRef]
77. Parry, L.A.; Ashcroft, I.A.; Wildman, R.D. Geometrical effects on residual stress in selective laser melting. *Addit. Manuf.* **2019**, *25*, 166–175. [CrossRef]
78. Hitzler, L.; Hirsch, J.; Heine, B.; Merkel, M.; Hall, W.; Öchsner, A. On the Anisotropic Mechanical Properties of Selective Laser-Melted Stainless Steel. *Materials* **2017**, *10*, 1136. [CrossRef]
79. Sames, W.J.; List, F.A.; Pannala, S.; Dehoff, R.R.; Babu, S.S. The metallurgy and processing science of metal additive manufacturing. *Int. Mater. Rev.* **2016**, *61*, 315–360. [CrossRef]
80. Zhang, Y.; Wu, L.; Guo, X.; Kane, S.; Deng, Y.; Jung, Y.-G.; Lee, J.-H.; Zhang, J. Additive Manufacturing of Metallic Materials: A Review. *J. Mater. Eng. Perform.* **2018**, *27*, 1–13. [CrossRef]
81. Herzog, D.; Seyda, V.; Wycisk, E.; Emmelmann, C. Additive manufacturing of metals. *Acta Mater.* **2016**, *117*, 371–392. [CrossRef]
82. Concept Laser X LINE 2000R Metal Laser Melting System. Available online: https://www.concept-laser.de/fileadmin/Machine_brochures/CL_X_LINE_2000R_DS_EN_US_4_v1.pdf (accessed on 29 April 2019).
83. Machine Search|Senvol. Available online: http://senvol.com/machine-search/ (accessed on 5 May 2019).
84. MYSINT100, Stampante 3D a Fusione Laser Selettiva di Polvere Metallica. *SISMA*. Available online: https://www.sisma.com/prodotti/mysint100/ (accessed on 29 April 2019).
85. MYSINT 300, Stampante 3D a Fusione Laser Selettiva di Polvere Metallica. *SISMA*. Available online: https://www.sisma.com/prodotti/mysint300/ (accessed on 29 April 2019).
86. Ngo, T.D.; Kashani, A.; Imbalzano, G.; Nguyen, K.T.Q.; Hui, D. Additive manufacturing (3D printing): A review of materials, methods, applications and challenges. *Compos. Part B Eng.* **2018**, *143*, 172–196. [CrossRef]
87. Cordova, L.; Campos, M.; Tinga, T. Assessment of Moisture Content and Its Influence on Laser Beam Melting Feedstock. In Proceedings of the Euro PM2017 Congress & Exhibition: European Annual Powder Metallurgy congress and exhibition, Milan, Italy, 1–5 October 2017.
88. Zhao, X.; Song, B.; Zhang, Y.; Zhu, X.; Wei, Q.; Shi, Y. Decarburization of stainless steel during selective laser melting and its influence on Young's modulus, hardness and tensile strength. *Mater. Sci. Eng. A* **2015**, *647*, 58–61. [CrossRef]

89. Liverani, E.; Toschi, S.; Ceschini, L.; Fortunato, A. Effect of selective laser melting (SLM) process parameters on microstructure and mechanical properties of 316L austenitic stainless steel. *J. Mater. Process. Technol.* **2017**, *249*, 255–263. [CrossRef]
90. Stugelmayer, E. *Characterization of Process Induced Effects in Laser Powder Bed Fusion Processed AlSi10Mg alloy*; Master of Science Degree, Montana Tech: Butte, MT, USA, 2018.
91. Mercelis, P.; Kruth, J. Residual stresses in selective laser sintering and selective laser melting. *Rapid Prototyp. J.* **2006**, *12*, 254–265. [CrossRef]
92. Li, C.; Liu, Z.Y.; Fang, X.Y.; Guo, Y.B. Residual Stress in Metal Additive Manufacturing. *Procedia CIRP* **2018**, *71*, 348–353. [CrossRef]
93. Robinson, J.; Ashton, I.; Fox, P.; Jones, E.; Sutcliffe, C. Determination of the effect of scan strategy on residual stress in laser powder bed fusion additive manufacturing. *Addit. Manuf.* **2018**, *23*, 13–24. [CrossRef]
94. Liu, B.; Wildman, R.; Tuck, C.R.; Ashcroft, I.; Hague, R.J.M. *Investigaztion the Effect of Particle Size Distribution on Processing Parameters Optimisation in Selective Laser Melting Process*; Additive Manufacturing Research Group, Loughborough University: Loughborough, UK, 2011.
95. Clayton, J.; Millington-Smith, D.; Armstrong, B. The Application of Powder Rheology in Additive Manufacturing. *JOM* **2015**, *67*, 544–548. [CrossRef]
96. Rausch, A.M.; Markl, M.; Körner, C. Predictive simulation of process windows for powder bed fusion additive manufacturing: Influence of the powder size distribution. *Comput. Math. Appl.* **2018**, S0898122118303535. [CrossRef]
97. Spierings, A.B.; Herres, N.; Levy, G. Influence of the particle size distribution on surface quality and mechanical properties in AM steel parts. *Rapid Prototyp. J.* **2011**, *17*, 195–202. [CrossRef]
98. Dowson, G. *Introduction to Powder Metallurgy—The Process and its Products*; EPMA: Mannheim, Germany, 2008.
99. Sames, W.J.; Medina, F.; Peter, W.H.; Babu, S.S.; Dehoff, R.R. Effect of Process Control and Powder Quality on Inconel 718 Produced Using Electron Beam Melting. In *8th International Symposium on Superalloy 718 and Derivatives*; Ott, E., Banik, A., Andersson, J., Dempster, I., Gabb, T., Groh, J., Heck, K., Helmink, R., Liu, X., Wusatowska-Sarnek, A., Eds.; John Wiley & Sons, Inc.: Hoboken, NJ, USA, 2014; pp. 409–423. ISBN 978-1-119-01685-4.
100. Popovich, A.; Sufiiarov, V.; Polozov, I.; Borisov, E.; Masaylo, D. Producing hip implants of titanium alloys by additive manufacturing. *Int. J. Bioprinting* **2016**, *2*, 78–84. [CrossRef]
101. Choo, H.; Sham, K.-L.; Bohling, J.; Ngo, A.; Xiao, X.; Ren, Y.; Depond, P.J.; Matthews, M.J.; Garlea, E. Effect of laser power on defect, texture, and microstructure of a laser powder bed fusion processed 316L stainless steel. *Mater. Des.* **2019**, *164*, 107534. [CrossRef]
102. Vrancken, B. Study of Residual Stresses in Selective Laser Melting. Ph.D. Thesis, KU Leuven—Faculty of Engineering Science, Leuven, Belgium, June 2016.
103. Kamath, C.; El-dasher, B.; Gallegos, G.F.; King, W.E.; Sisto, A. Density of additively-manufactured, 316L SS parts using laser powder-bed fusion at powers up to 400 W. *Int. J. Adv. Manuf. Technol.* **2014**, *74*, 65–78. [CrossRef]
104. Keshavarzkermani, A.; Marzbanrad, E.; Esmaeilizadeh, R.; Mahmoodkhani, Y.; Ali, U.; Enrique, P.D.; Zhou, N.Y.; Bonakdar, A.; Toyserkani, E. An investigation into the effect of process parameters on melt pool geometry, cell spacing, and grain refinement during laser powder bed fusion. *Opt. Laser Technol.* **2019**, *116*, 83–91. [CrossRef]
105. Guan, K.; Wang, Z.; Gao, M.; Li, X.; Zeng, X. Effects of processing parameters on tensile properties of selective laser melted 304 stainless steel. *Mater. Des.* **2013**, *50*, 581–586. [CrossRef]
106. Jeon, T.; Hwang, T.; Yun, H.; VanTyne, C.; Moon, Y. Control of Porosity in Parts Produced by a Direct Laser Melting Process. *Appl. Sci.* **2018**, *8*, 2573. [CrossRef]
107. Wang, D.; Liu, Y.; Yang, Y.; Xiao, D. Theoretical and experimental study on surface roughness of 316L stainless steel metal parts obtained through selective laser melting. *Rapid Prototyp. J.* **2016**, *22*, 706–716. [CrossRef]
108. Yadollahi, A.; Mahtabi, M.J.; Khalili, A.; Doude, H.R.; Newman, J.C. Fatigue life prediction of additively manufactured material: Effects of surface roughness, defect size, and shape. *Fatigue Fract. Eng. Mater. Struct.* **2018**, *41*, 1602–1614. [CrossRef]
109. Fayazfar, H.; Salarian, M.; Rogalsky, A.; Sarker, D.; Russo, P.; Paserin, V.; Toyserkani, E. A critical review of powder-based additive manufacturing of ferrous alloys: Process parameters, microstructure and mechanical properties. *Mater. Des.* **2018**, *144*, 98–128. [CrossRef]

110. Gu, D.; Shen, Y. Balling phenomena in direct laser sintering of stainless steel powder: Metallurgical mechanisms and control methods. *Mater. Des.* **2009**, *30*, 2903–2910. [CrossRef]
111. Metelkova, J.; Kinds, Y.; Kempen, K.; de Formanoir, C.; Witvrouw, A.; Van Hooreweder, B. On the influence of laser defocusing in Selective Laser Melting of 316L. *Addit. Manuf.* **2018**, *23*, 161–169. [CrossRef]
112. Bassoli, E.; Sola, A.; Celesti, M.; Calcagnile, S.; Cavallini, C. Development of Laser-Based Powder Bed Fusion Process Parameters and Scanning Strategy for New Metal Alloy Grades: A Holistic Method Formulation. *Materials* **2018**, *11*, 2356. [CrossRef] [PubMed]
113. Mishurova, T.; Artzt, K.; Haubrich, J.; Requena, G.; Bruno, G. New aspects about the search for the most relevant parameters optimizing SLM materials. *Addit. Manuf.* **2019**, *25*, 325–334. [CrossRef]
114. Bruna-Rosso, C.; Demir, A.G.; Previtali, B. Selective laser melting finite element modeling: Validation with high-speed imaging and lack of fusion defects prediction. *Mater. Des.* **2018**, *156*, 143–153. [CrossRef]
115. Huang, Z.; Dantan, J.-Y.; Etienne, A.; Rivette, M.; Bonnet, N. Geometrical deviation identification and prediction method for additive manufacturing. *Rapid Prototyp. J.* **2018**, *24*, 1524–1538. [CrossRef]
116. Hussain, B.; El-Gizawy, A.S. Development of 3D Finite Element Model for Predicting Process-Induced Defects in Additive Manufacturing by Selective Laser Melting (SLM). In *Proceedings of the Volume 2: Advanced Manufacturing*; ASME: Phoenix, ZA, USA, 2016; p. V002T02A065.
117. Tang, H.P.; Qian, M.; Liu, N.; Zhang, X.Z.; Yang, G.Y.; Wang, J. Effect of Powder Reuse Times on Additive Manufacturing of Ti-6Al-4V by Selective Electron Beam Melting. *JOM* **2015**, *67*, 555–563. [CrossRef]
118. Saeidi, K.; Akhtar, F. Microstructure-Tailored Stainless Steels with High Mechanical Performance at Elevated Temperature. In *Stainless Steels and Alloys*; Duriagina, Z., Ed.; IntechOpen: London, UK, 2019; ISBN 978-1-78985-369-8.
119. Hooper, P.A. Melt pool temperature and cooling rates in laser powder bed fusion. *Addit. Manuf.* **2018**, *22*, 548–559. [CrossRef]
120. Molinari, A. La metastabilità strutturale delle leghe metalliche ottenute per Selective Laser Melting. *Metall. Ital.* **2017**, 21–27.
121. Murr, L.E.; Gaytan, S.M.; Ramirez, D.A.; Martinez, E.; Hernandez, J.; Amato, K.N.; Shindo, P.W.; Medina, F.R.; Wicker, R.B. Metal Fabrication by Additive Manufacturing Using Laser and Electron Beam Melting Technologies. *J. Mater. Sci. Technol.* **2012**, *28*, 1–14. [CrossRef]
122. EOS Metal Materials for Additive Manufacturing. Available online: https://www.eos.info/material-m (accessed on 20 April 2019).
123. Renishaw Renishaw: Metal Powders Supply. Available online: http://www.renishaw.com/en/metal-powders-supply--31458 (accessed on 5 May 2019).
124. SLM Solutions Group AG: SLM®Metal Powder. Available online: https://www.slm-solutions.com/en/products/accessories-consumables/slmr-metal-powder/ (accessed on 5 May 2019).
125. Metal Materials. Available online: https://it.3dsystems.com/materials/metal (accessed on 21 April 2019).
126. Materials Laser Melting—Concept Laser. Available online: https://www.concept-laser.de/en/products/materials.html (accessed on 30 March 2019).
127. Nguyen, Q.B.; Zhu, Z.; Ng, F.L.; Chua, B.W.; Nai, S.M.L.; Wei, J. High mechanical strengths and ductility of stainless steel 304L fabricated using selective laser melting. *J. Mater. Sci. Technol.* **2019**, *35*, 388–394. [CrossRef]
128. Marbury, F. *Characterization of SLM Printed 316L Stainless Steel and Investigation of Micro Lattice Geometry*; California Polytechnic State University: San Luis Obispo, CA, USA, 2017.
129. Casati, R.; Lemke, J.; Vedani, M. Microstructure and Fracture Behavior of 316L Austenitic Stainless Steel Produced by Selective Laser Melting. *J. Mater. Sci. Technol.* **2016**, *32*, 738–744. [CrossRef]
130. *ASTM A276/A276M—Standard Specification for Stainless Steel Bars and Shapes*; ASTM International: West Conshohocken, PA, USA, 2017.
131. Spierings, A.B.; Starr, T.L.; Wegener, K. Fatigue performance of additive manufactured metallic parts. *Rapid Prototyp. J.* **2013**, *19*, 88–94. [CrossRef]
132. Uhlmann, E.; Fleck, C.; Gerlitzky, G.; Faltin, F. Dynamical fatigue behavior of additive manufactured products for a fundamental life cycle approach. *Procedia CIRP* **2017**, *61*, 588–593. [CrossRef]
133. Riemer, A.; Leuders, S.; Thöne, M.; Richard, H.A.; Tröster, T.; Niendorf, T. On the fatigue crack growth behavior in 316L stainless steel manufactured by selective laser melting. *Eng. Fract. Mech.* **2014**, *120*, 15–25. [CrossRef]

134. Qiu, C.; Kindi, M.A.; Aladawi, A.S.; Hatmi, I.A. A comprehensive study on microstructure and tensile behaviour of a selectively laser melted stainless steel. *Sci. Rep.* **2018**, *8*, 7785. [CrossRef]
135. Krakhmalev, P.; Fredriksson, G.; Svensson, K.; Yadroitsev, I.; Yadroitsava, I.; Thuvander, M.; Peng, R. Microstructure, Solidification Texture, and Thermal Stability of 316 L Stainless Steel Manufactured by Laser Powder Bed Fusion. *Metals* **2018**, *8*, 643. [CrossRef]
136. Auguste, P.; Mauduit, A.; Fouquet, L.; Pillot, S. Study on 17-4 PH stainless steel produced by selective laser melting. *UPB Sci. Bull. Ser. B-chem. Mater. Sci.* **2018**, *80*, 197–210.
137. Murr, L.E.; Martinez, E.; Hernandez, J.; Collins, S.; Amato, K.N.; Gaytan, S.M.; Shindo, P.W. Microstructures and Properties of 17-4 PH Stainless Steel Fabricated by Selective Laser Melting. *J. Mater. Res. Technol.* **2012**, *1*, 167–177. [CrossRef]
138. Yadollahi, A.; Shamsaei, N.; Thompson, S.M.; Elwany, A.; Bian, L. Effects of building orientation and heat treatment on fatigue behavior of selective laser melted 17-4 PH stainless steel. *Int. J. Fatigue* **2017**, *94*, 218–235. [CrossRef]
139. Mahmoudi, M.; Elwany, A.; Yadollahi, A.; Thompson, S.M.; Bian, L.; Shamsaei, N. Mechanical properties and microstructural characterization of selective laser melted 17-4 PH stainless steel. *Rapid Prototyp. J.* **2017**, *23*, 280–294. [CrossRef]
140. Luecke, W.E.; Slotwinski, J.A. Mechanical Properties of Austenitic Stainless Steel Made by Additive Manufacturing. *J. Res. Natl. Inst. Stand. Technol.* **2014**, *119*, 398. [CrossRef] [PubMed]
141. Nezhadfar, P.D.; Masoomi, M.; Thompson, S.; Pham, N.; Shamsaei, N. Mechanical Properties of 17-4 Ph Stainless Steel Additively Manufactured Under Ar and N2 Shielding Gas. In Proceedings of the 29th Annual International Solid Freeform Fabrication, Austin, TX, USA, 13–15 August 2018; p. 10.
142. Ali, U.; Esmaeilizadeh, R.; Ahmed, F.; Sarker, D.; Muhammad, W.; Keshavarzkermani, A.; Mahmoodkhani, Y.; Marzbanrad, E.; Toyserkani, E. Identification and characterization of spatter particles and their effect on surface roughness, density and mechanical response of 17-4 PH stainless steel laser powder-bed fusion parts. *Mater. Sci. Eng. A* **2019**, *756*, 98–107. [CrossRef]
143. Rafi, H.K.; Starr, T.L.; Stucker, B.E. A comparison of the tensile, fatigue, and fracture behavior of Ti–6Al–4V and 15-5 PH stainless steel parts made by selective laser melting. *Int. J. Adv. Manuf. Technol.* **2013**, *69*, 1299–1309. [CrossRef]
144. *ASTM A564/A564M—Standard Specification for Hot-Rolled and Cold-Finished Age-Hardening Stainless Steel Bars and Shapes*; ASTM International: West Conshohocken, PA, USA, 2013.
145. *EN 10088-3: Stainless Steels—Part 3: Technical Delivery Conditions for Semi-Finished Products, Bars, Rods, Wire, Sections and Bright Products of Corrosion Resisting Steels for General Purposes*; CEN-CENELEC Management Centre: Brussels, Belgium, 2014.
146. Hengsbach, F.; Koppa, P.; Duschik, K.; Holzweissig, M.J.; Burns, M.; Nellesen, J.; Tillmann, W.; Tröster, T.; Hoyer, K.-P.; Schaper, M. Duplex stainless steel fabricated by selective laser melting—Microstructural and mechanical properties. *Mater. Des.* **2017**, *133*, 136–142. [CrossRef]
147. Saeidi, K.; Alvi, S.; Lofaj, F.; Petkov, V.I.; Akhtar, F. Advanced Mechanical Strength in Post Heat Treated SLM 2507 at Room and High Temperature Promoted by Hard/Ductile Sigma Precipitates. *Metals* **2019**, *9*, 199. [CrossRef]
148. *ASTM A789/A789M—Standard Specification for Seamless and Welded Ferritic/Austenitic Stainless Steel Tubing for General Service*; ASTM International: West Conshohocken, PA, USA, 2017.

© 2019 by the authors. Licensee MDPI, Basel, Switzerland. This article is an open access article distributed under the terms and conditions of the Creative Commons Attribution (CC BY) license (http://creativecommons.org/licenses/by/4.0/).

Review

Processing and Properties of Reversion-Treated Austenitic Stainless Steels

Antti Järvenpää [1,*], Matias Jaskari [1], Anna Kisko [2] and Pentti Karjalainen [2]

1 Kerttu Saalasti Institute, University of Oulu, FI-85500 Nivala, Finland; matias.jaskari@oulu.fi
2 Centre for Advanced Steels Research, University of Oulu, FI-90014 Oulu, Finland; anna.kisko@oulu.fi (A.K.); pentti.karjalainen@oulu.fi (P.K.)
* Correspondence: antti.jarvenpaa@oulu.fi; Tel.: +35-844-555-1633

Received: 6 December 2019; Accepted: 18 February 2020; Published: 21 February 2020

Abstract: Strength properties of annealed austenitic stainless steels are relatively low and therefore improvements are desired for constructional applications. The reversion of deformation induced martensite to fine-grained austenite has been found to be an efficient method to increase significantly the yield strength of metastable austenitic stainless steels without impairing much their ductility. Research has been conducted during thirty years in many research groups so that the features of the reversion process and enhanced properties are reported in numerous papers. This review covers the main variables and phenomena during the reversion processing and lists the static and dynamic mechanical properties obtained in laboratory experiments, highlighting them to exceed those of temper rolled sheets. Moreover, formability, weldability and corrosion resistant aspects are discussed and finally the advantage of refined grain structure for medical applications is stated. The reversion process has been utilized industrially in a very limited extent, but apparently, it could provide a feasible processing route for strengthened austenitic stainless steels.

Keywords: metastable austenitic stainless steel; cold rolling; annealing; reversion; grain size; mechanical properties; fatigue; corrosion; medical applications

---

1. Background

Austenitic stainless steels (ASSs) are potential candidates for structural parts in transportation and construction industries due to their excellent formability, work hardening capability and weldability, together with good corrosion resistance. However, their yield strength (YS) is generally low, for instance for the EN 1.4301/AISI 304 around 210–230 MPa and 330–350 MPa for EN 1.4318/AISI 301LN grade [1], which limits their use for structural applications. (For the ASSs grades, AISI codes will be used hereafter as commonly in the literature). Many grades are, however, metastable at room temperature, which means that the austenite phase transforms to martensite during cold working, so that their work hardening capability and ductility can be utilized to improve the strength by cold deformation (the temper rolling process). As an example, the strengthened condition +C1000 with the minimum proof strength ($R_{p0.2}$) of 700 MPa [2] can be achieved by about a 20% cold rolling thickness reduction for 304 and 301LN ASSs, with some loss of elongation. However, the disadvantage of strengthening by temper rolling is the formation of anisotropy in mechanical properties, the strength being different in different directions relative to the rolling direction [3,4]. Moreover, lowest values are obtained in compression tests parallel to the rolling direction [3]. The cold rolling tends also to impair the corrosion resistance of ASSs [5–7].

The strength of ASSs can also be increased by nitrogen alloying, and the processing and properties of high-nitrogen steels have been investigated extensively, e.g., [8,9]. However, in conventional processing of Cr-Ni alloys, the limit of nitrogen solubility is low and problems with the hot ductility may also appear with increased nitrogen content [4]. Hence, other strengthening methods are desirable.

Grain size (GS) refinement is an effective method for increasing the static strength properties of metals and alloys (the Hall–Petch relationship) and also their fatigue performance, especially in the high-cycle fatigue (HCF) regime. The GS of commercial ASSs is typically larger than 10 μm. Even though the impact of GS on strength is not so high in ASSs as in ferritic steels, it has been shown in a large number of papers that the refinement of GS can provide significant improvement in the YS of austenite, e.g., [10–29]. The traditional hot rolling process or cold rolling and recrystallization annealing are not effective for refining the GS of the austenite phase, although GS of 2 μm has been obtained by using warm rolling and annealing [30]. The role of dynamic recrystallization in GS refinement in ASSs has been reviewed by Zhao and Jiang [31], but it is not very efficient due to high temperatures required. However, the reversion treatment, in which deformation-induced α'-martensite (DIM) transforms back to austenite has been shown to refine the austenite GS to submicron size, resulting in excellent combinations of YS and tensile elongation (TE), as reported in numerous studies during the last 30 years.

The martensite to austenite reversion process in austenitic Cr-Ni steels was already studied in the 1970s and 1980s [32–37], but it continued comprehensively in Japan in early 1990s, e.g., [38–40] and later in many groups in various countries, e.g., [10–21,30,41–79]. The studies in Finland, starting in 2004, as supported by a Finnish stainless-steel company and in collaboration with two research groups in USA and one in the Czech Republic, may be mentioned separately [22–29,80–103]. In laboratory studies, the reversion process has been applied to several commercial Cr-Mn and Cr-Ni steel grades such as 201, 201L, 204Cu, 301, 301LN, 304, 304L (see later in Section 5.1).

As the reversion process is carried out simply using cold rolling and annealing, it seems more appropriate for bulk production of large-sized sheets and has more potential for actual applications than numerous other severe plastic deformation techniques applied to GS refinement. However, despite extensive academic research work, as far as the present authors know, there only exist a couple of industrial companies utilizing the reversion-treatment for ASSs. In Japan, Nano grains Co. Ltd (Komatsuseiki Kosakusho Co., Ltd.) produces grain-refined 304, 316 and 301 foils (thickness range 80–300 μm), with the GS finer than 1 μm, using repeated reversion treatment [104]. Especially the enhanced properties of micro-scale cutting and hole piercing are utilized in the manufacturing of orifices for electronic fuel injection [105–107]. Nippon Steel & Sumitomo Metal company lists in its product catalogue fine-grained 304 (SUS304 BA19) and 301L (NSSMC-NAR-301L BA1) grades: thin sheets, strips and foils (0.08–0.6 mm) having both high strength, ductility and formability and smooth formed surface due to refinement of GS [108,109]. The feasibility of the process for an industrial manufacturing using a continuous annealing line has been demonstrated in one laboratory study [12]. In recent studies by Järvenpää et al. [87], a pilot induction heating line has been employed to simulate industrial conditions in reversion annealing of the 301LN grade.

In order to highlight the advantages of grain-refined ASSs, in the present paper, the main process variables and their influences in the reversion-treatment are introduced, and the state of the art of mechanical properties of reversion-treated, fine-grained ASSs are reviewed. The formability, weldability, corrosion properties and medical applications are also included in this survey. A short overview of the state of the research on the reversion process has recently been published as a conference paper [103], but the present review is much more comprehensive than that.

2. Cold Rolling Stage

The first stage of the reversion treatment is cold rolling of an ASS sheet to obtain DIM, which can be reversed to fine-grained austenite in the subsequent annealing stage. In the following, we point out the different stability of the austenite in different ASSs towards transforming to martensite, depending mainly on the chemical composition of the steel. Some commercial ASS grades will be compared. The degree of cold rolling reduction is a factor affecting the fraction of DIM created, and in industrial rolling, it cannot very high. Therefore, only partial transformation of austenite to DIM can happen affecting the microstructure obtained in the annealing stage. Further, the DIM forming gradually will be inevitably deformed to different degrees, which also has an influence on microstructure heterogeneity and thereby on the mechanical properties.

The reversion heat treatment for ASSs consisting of the cold rolling and annealing stages is schematically presented in Figure 1. A metastable ASS must first be cold worked to transform the austenite to DIM and deform it. There is often some hexagonal ε-martensite formed as well, but its fraction and contribution are minor in the reversion process, so that much attention has not been paid on that phase. Highly-deformed cell-type DIM is preferable for a source of a large number of nucleation sites for new austenite grains to attain the desired highly-refined GS in the subsequent reversion annealing [38–40]. If the total cold-rolling reduction is small, the transformation of austenite to DIM tends to remain partial and coarse-grained deformed austenite (DA) grains are retained in the structure. Furthermore, the lath-type, slightly deformed DIM reverses into austenite with coarser GS. Accordingly, very high cold rolling reductions of 90% to 95% were recommended originally and applied in numerous studies, e.g., [14,15,17,19,20,39,40], which can, however, be impractical in industry.

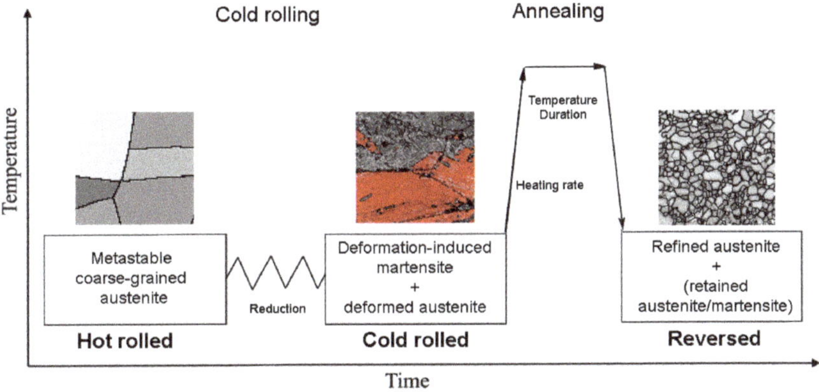

Figure 1. Reversion treatment of a metastable austenitic stainless-steel consisting of cold rolling and annealing stages. Important variables have been designated.

The austenite thermal stability has been discussed by Lo et al. [110] in their extensive review on ASSs, listing numerous Equations for the chemical compositional dependence of M_s temperature, for instance. The mechanical stability, the susceptibility of austenite to DIM formation, is mainly affected by the chemical composition of the steel and its GS, but also GS distribution, grain morphology and orientation as well as deformation conditions (stress state, temperature, pass strains, strain rate, etc.) have their influence [111,112]. Das et al. [113] have collected a very extensive literature data and developed a neural network model to predict the amount of DIM formation with its influencing parameters in a variety of ASSs. Mechanical driving force (stress) and temperature were found to have high significance, while concerning uniaxial tensile testing conditions. Regarding GS refinement, highly metastable grades are favored, and according to Tomimura et al. [38], the Ni equivalent (Ni + 0.35Cr) ≤ 16.0% (all compositions are in mass %) is required to transform more than 90% of austenite to DIM during 90% cold rolling at room temperature (RT).

The Nohara equation (Equation (1)), defining the temperature at which the amount of 50% DIM is transformed from austenite through cold deformation of 0.30 true strain, is commonly used for evaluation of the mechanical stability of austenite in rolling [114]. It shows that all alloying elements decrease the stability, though with varying power.

$$M_{d30}\ (^\circ\mathrm{C}) = 552 - 462(\mathrm{C} + \mathrm{N}) - 9.2\mathrm{Si} - 8.1\mathrm{Mn} - 13.7\mathrm{Cr} - 29(\mathrm{Ni} + \mathrm{Cu}) - 18.5\mathrm{Mo} - 68\mathrm{Nb} - 1.42(\mathrm{GS} - 8) \quad (1)$$

Here the alloying elements are in mass % and GS is in ASTM unit. The M_{d30} temperature is around 20–40 °C for 301LN, whereas lower for 304 (examples in Table 1). From Equation (1), it is also seen that during deformation Ni stabilizes austenite more than Mn does, which is contrary to that concerning thermodynamic austenite stability [115]. Refining the GS also increases the stability, as reported in numerous studies, e.g., [46,75,114]. In repeated reversion treatments, to be discussed shortly later, this means that in subsequent cold rolling stages, the DIM formation becomes reduced [21]. The relationship between austenite GS and its stability under tensile and fatigue loading will be discussed later in Section 5.

In addition to alloying adjustments, another method to enhance the DIM transformation is rolling at lowered temperatures, the technique which has been applied to 304/304L and 316L grades in particular [10,41,66,116–120].

To give examples, the DIM fractions formed during cold rolling for some Cr-Mn and Cr-Ni ASSs (the exact chemical compositions are listed in Table 1; concentrations are always in mass %) are plotted as a function of the cold rolling thickness reduction in Figure 2. They show that in the 304/304L grade, only a small fraction of DIM can be obtained at RT [20,121]. After a 90% cold rolling reduction, Di Schino et al. [11] obtained 35% and Ravi Kumar et al. [54] 43% of DIM. However, at 0 °C, in experiments performed by Forouzan et al. [14], about 55% DIM was formed by a 30% cold rolling reduction in a 304 ASS (M_{d30} = 13 °C), and even a completely martensitic structure could be obtained after 60% reduction. In a 301LN ASS ($M_{d30} \approx$ 23–27 °C) in order to achieve a fully martensitic structure, a cold rolling reduction of about 60% is required [23,29,102], but for 201 and 201L grades, the reduction of 40% can be enough for that [15,16,67]. However, it must be noted that for this higher metastability of 201/201L ASSs, the nitrogen content must be low, for in experiments performed by Kisko et al. [28,83,101] and Hamada et al. [80] using the 201 grade with 0.052%C and 0.245%N, the DIM fraction obtained was only 30% after 60% cold rolling reduction. Hence, the exact chemical composition can greatly affect the mechanical stability which can vary within an ASS grade, requiring attention. As seen from Figure 2, if single pass reductions are high, adiabatic heating tends to reduce markedly the extent of DIM formed [4,77].

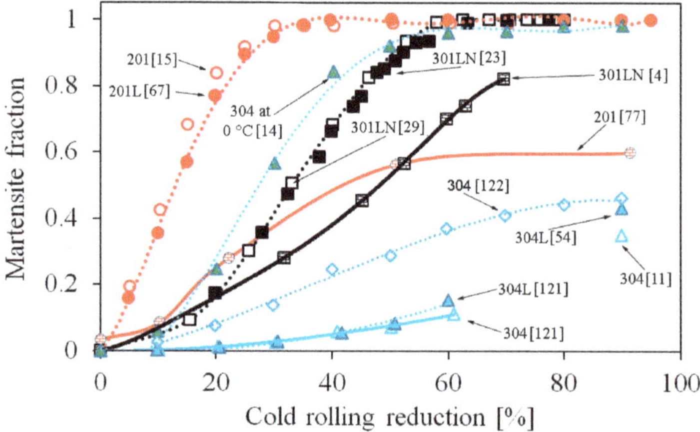

Figure 2. Examples of increase of the deformation-induced martensite fraction during cold rolling deformation in 201/201L, 304/304L and 301LN grades. The chemical compositions of the steels are in Table 1.

Mirzadeh and Najafizadeh [122] have modeled by means of an artificial neural network the effect of the cold working temperature, amount of deformation, strain rate and initial austenite GS on the volume fraction of DIM in a 301LN ASS. The appropriate grain refining zone can be determined using this model. The martensite content is increased when the degree of deformation is high or the deformation temperature is low. Moreover, by increasing the strain rate and the ASTM GS number (i.e., refining the GS), the amount of DIM is reduced.

Table 1. Chemical compositions and Md_{30} temperatures for Austenitic stainless steels (ASSs) in Figure 2.

Grade	C	Si	Mn	Cr	Ni	Mo	N	M_{d30}	Ref.
201	0.080	0.54	5.90	16.60	3.70	0.11	0.040		[15]
201	0.024	0.38	7.02	17.06	4.07	0.04	0.164		[77]
201L	0.027	0.51	5.91	16.20	3.88	0.08	0.040		[67]
301LN	0.017	0.52	1.29	17.30	6.50	0.15	0.150	27	[23]
301LN	0.025	0.53	1.25	17.50	6.50	0.09	0.150	23	[29]
301LN	0.024	0.54	1.22	17.80	6.40		0.133	28	[4]
304	0.040	0.34	1.15	18.06	8.33	0.05	0.048	6.2	[123]
304	0.060	0.02	0.33	18.40	8.60	0.06	0.024	9.6	[11]
304	0.065	0.48	1.35	18.20	8.37		0.046	−6.7	[121]
304L	0.021	0.56	1.37	18.44	9.38		0.038	−16	[121]
304L	0.027	0.43	1.58	18.20	8.22	0.35		13	[14]
304L	0.020	0.30	1.50	18.60	10.10		0.020	−29	[54]

Concerning the actual deformation degree experienced by DIM during cold rolling, it is important to realize that the DIM formation proceeds gradually, as seen in Figure 2. This means that at low rolling reductions, a significant fraction of the formed DIM remains inevitably only "slightly-deformed" and the resultant lath martensite plays a significant role in the microstructure evolution. In industrial practice, however, heavy cold rolling or complex processing routes are not desired. Promisingly, for instance for a 301LN ASS, cold rolling reductions as low as 35% to 52% seem to result in excellent strength-ductility combinations (see Table 2 in Section 5.1), even though the GS refinement is not most efficient and the GS obtained is not uniform [23,29]. Even very low reductions such as 10% to 40% have been reported to reduce significantly the average GS in 301LN during complete reversion [60,64].

The stability of austenite under tensile deformation will be further discussed in Section 5. It can be mentioned here that Ahmedabadi et al. [124] have recently modelled the DIM formation ($f_{\alpha'}$) during cold rolling of a 304 ASS using a sigmoidal logistic equation, Equation (2):

$$f_{\alpha'} = \frac{f_s}{1 + exp[-\beta(\varepsilon - \varepsilon_m)]} \quad (2)$$

In the equation, the parameter f_s is the maximum value of a given "S"-curve, ε_m is the abscissa of the mid-point of a given sigmoid, and ε is deformation strain. The parameter β represents the steepness of the curve. The fitting parameters for the model can be correlated to physical parameters associated with the austenite to DIM transformation. This model eliminates the limitations present in the previous models, as discussed. Importantly it shows that it is not always possible to completely transform austenite into martensite, which means that the DA affects the microstructure evolution in reversion annealing.

3. Reversion Annealing

During the annealing stage of the cold-rolled ASS sheet containing DIM, the reversion of DIM back to austenite can take place, refining the GS and enhancing the mechanical properties. There are two reversion mechanisms, shear and diffusional ones, the type depending on the chemical composition of the steel, heating rate and annealing temperature but hardly on the degree of cold rolling reduction. In this chapter, we first discuss the type and kinetics of the reversion in various Cr-Ni type ASSs and the factors affecting them. This section is followed by the discussion on the temperature range suitable for the reversion treatment, accounting for the different reversion mechanism. It can be noticed that certain typical temperatures (600–1000 °C) have been applied in experiments for commercial Cr-Ni and Cr-Mn ASSs, and the duration of annealing can be selected to be very short (less than 1 s) or even hours, highlighting the flexibility of the process.

3.1. Reversion Mechanism

In metastable ASSs, DIM reverses back to austenite during continuous heating or isothermal annealing at an appropriate temperature regime. Reversion of martensite to austenite is a phenomenon also taking place during intercritical annealing in low carbon martensitic stainless steels [125] and medium-Mn steels, e.g., [126]. A variety of experimental methods have been employed to study both martensitic transformation and DIM reversion in ASSs, e.g., microscopic methods, dilatometry, calorimetry, X-ray diffraction (both postmortem and in situ), internal friction, various magnetic, positron annihilation or hardness/mechanical properties measurements [127–132]. A small amount of ε-martensite can form in some ASSs at small deformation degrees, and it reverses at much lower temperatures than α′-martensite does (see e.g., [63]). Singh [37] reported that the ε-martensite was stable up to 200 °C, and according to Santos and Andrade [128], it reverses in the temperature range 50–200 °C and between 150–400 °C according to Dryzek et al. [129]. Very recently a latent strengthening mechanism, bake hardening without interstitials, due to the reversion of ε-martensite, has been reported in a metastable FCC high entropy alloy by Wei et al. [133]. Annealing for 20 min at 200 °C was enough for complete reversion accomplished by a shear-assisted displacive mechanism.

The reversion of α′-martensite (i.e., DIM) to austenite can occur by two different mechanisms, a diffusionless shear or diffusion-controlled one, as reported already in 1967 by Guy et al. [36]; see also [39,45]. Guy et al. observed that austenite with mechanical twins formed first from martensite in 18Cr-8Ni and 18Cr-12Ni steels, which then recovered to a subgrain structure. Concerning the GS refinement, both reversion mechanisms can readily lead to a micron-scale GS, though in principle the diffusional reversion is more efficient [38]. In Fe-Cr-Ni ternary alloys, in the first stage, the shear phase reversion results in austenite which contains traces of prior α′-martensite morphology, the same grain boundaries as those of original austenite and a high density of defects. After the fast transformation, defect-free austenite subgrains are formed which coalesce to a structure resembling recrystallized structure [38,39]. An example of the formation of dislocation free grains from subgrains is shown in Figure 3a. On the contrary, the diffusional reversion is characterized by the nucleation and growth of randomly oriented equiaxed austenite grains; the result is shown in Figure 3b. The nucleation occurs on the cell or lath boundaries of the deformed DIM, and austenitic grains grow in size with time but can stay in a nanometer or submicron range. Secondary phase precipitates can also form in the course of the reversion, for instance nano-size chromium nitrides in 301LN [22,26,89] (Figure 3c,d) and carbides in 301 [130,134,135].

The chemical composition of the steel and annealing temperature are two important variables affecting the type of the reversion mechanism [23,39]. Tomimura and co-workers [39] reported that the reverse transformation of martensite to austenite occurs by diffusionless mechanism in metastable Fe–Cr–Ni ternary alloys with the high ratio of Ni to Cr (about ≥0.6), whereas the diffusive reverse transformation happens at the low ratio of Ni to Cr (such as ≤0.5). Somani et al. [23] proposed a model to predict the reversion type based on the chemical composition of the steel, adopting the Cr- and Ni-equivalents including some other elements in addition to Cr and Ni. Using this model, for instance, an AISI 301 reverses by the shear mechanisms above 670 °C, whereas a 301LN grade exhibits diffusion-driven transformation [23], controlled by the fast diffusion of interstitial nitrogen atoms [136]. This model has also been used by Misra et al. [88]. Shirdel et al. [70] adopted different equivalents including microalloy elements as predicting the reversion mechanism for a 304L steel. According to that prediction, the shear mechanism is possible above 783 °C. Based on microstructure features, the diffusion-controlled mechanism was the dominant one at the temperature range of 600–650 °C, and the shear reversion mechanism might be operative at temperatures higher than 750 °C. Sun et al. [137] reported diffusional reversion for a 304 ASS after a 85% cold rolling reduction and annealing at 550–650 °C at a low heating rate of 10 °C/s based on model predictions and observed time-dependence of the reversion at 650 °C. However, Cios et al. [132] showed by the transmission Kikuchi diffraction measurements that the reverse transformation proceeded in a 304 ASS through a diffusionless mechanism.

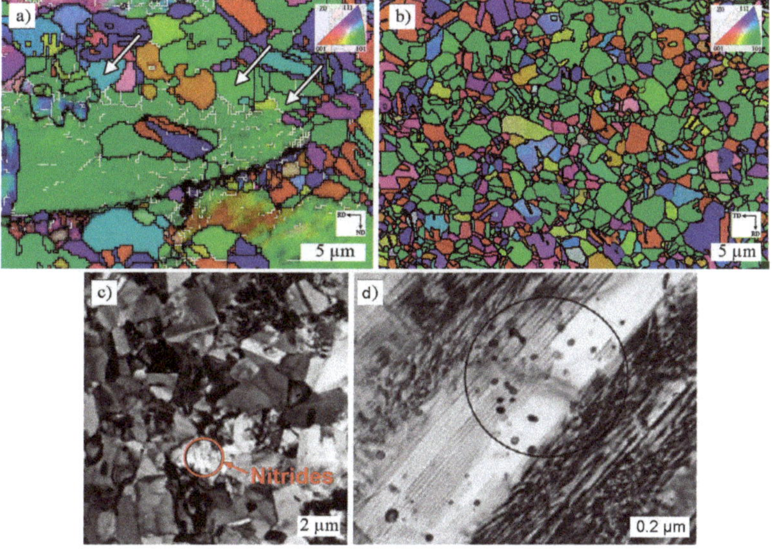

Figure 3. Reversion in 301LN ASS occurred by the shear reversion (**a**), where dislocation free grains are formed by continuous recrystallization (arrows pointing out such subgrains) [29], diffusional reversion (**b**) (reproduced from [84], with permission from Elsevier, 2017) and diffusional reversion with chromium nitride precipitation (**c**,**d**) (reproduced from [22], with permission from Springer Nature, 2007).

For a 301 ASS, microstructural observations of Johannsen et al. [135] indicated that it underwent a diffusional reversion from martensite to austenite in all annealed samples (90% cold rolled; heating rate 100 °C/min; annealed at 600–900 °C for 30 min) except when annealed at 800 °C and possibly at 850 °C while a shear reversion mechanism seemed to be active. Knutsson et al. [130] found that the diffusional reversion started at about 450 °C and the shear reversion at higher temperatures of 600 °C. In addition, carbo-nitride precipitation was observed for samples heat treated at these temperatures, which led to an increasing Ms-temperature and new α'-martensite formation upon cooling. They applied cold rolling reductions between 23% to 60%, but this did not affect the reversion rate.

Consistently with the above observations, Somani et al. [23] and Misra et al. [88] showed for a 301ASS that the shear transformation had occurred in the course of heating at a high rate of 200 °C/s or during annealing at temperatures around 800 and 900 °C. However, at 700 °C the reversion kinetics was clearly time dependent, i.e., the reversion was presumably diffusion controlled there (as mentioned above, the model developed predicted the start temperature of austenite formation As $\approx$ 670 °C for the shear transformation [23]). The continuous recrystallization of the shear-reversed austenite or the recrystallization of DA grains requires a higher temperature than the reversion itself does or it is slower than the martensitic reversion [121,138,139]. In a 301 grade, for the recrystallization that refines the GS following the shear reversion within few seconds, the temperature of 900 °C has been found feasible [23]. According to Sun et al. [140], in an 80% cold-rolled 304 ASS, the shear reversion took 1 min at 750 °C, while the recrystallization of DA grains required 15 min.

Yagodzinskyy et al. [134] investigated microstructural changes in 70% cold-rolled ASSs 301 and 301LN, containing 0.10% C and 0.16% N respectively, under slow linear heating at a rate of 15 °C/min and found the precipitation of carbides and nitrides respectively at martensite laths interfaces in the 301 steel and martensite/austenite interfaces in the 301LN steel. The start of reversion was at 550 and 500 °C in these steels, respectively. In the 301, the reversion was diffusion controlled at 600 °C, but there was also 30% retained austenite after the cold rolling, and stripes of α'-martensite formed recurrently in the retained austenite by a mechanism which seemed to be shear. The bimodal grain structure with the 1 µm average GS was formed after recrystallization at 700–800 °C. In contrast, the new austenite grains grew continuously in the 301LN under diffusion control and a uniform GS of 0.35 µm was obtained.

On the other hand, the start A_s and finish A_f temperatures for the diffusional reversion depend on both the heating rate and chemical composition of the steel. The effect of heating rate was already investigated in Fe-Ni-C alloys by Apple and Kraus in 1972, who observed the shear mechanism at high heating rates and diffusional at low heating rates [141]. The diffusional reversion proceeds more rapidly with increasing annealing temperature, and the A_f temperature depends on soaking time. The transformation kinetics from martensite to austenite has been modelled in a 301LN ASS by Rajasekhara and Ferreira and found to be controlled by the diffusion of nitrogen [136]. Sun et al. [142] reported diffusional reversion for the 304 ASS (Ni/Cr = 0.49) after a 93% cold rolling reduction and annealing at 650–700 °C at a low heating rate of 10 °C/s. Tomimura et al. [39] found that the reversion was diffusional at 652 °C in an 18Cr-9Ni alloy (Ni/Cr = 0.5) in spite of a high heating rate of 300 °C/s.

Somani et al. [23,80], Kisko et al. [27,28,81–83] and Järvenpää et al. [84–87] have used a high heating rate of 200 °C/s in their experiments. They found that in 301LN (Ni/Cr ≈ 0.4), the diffusion-controlled reversion started, depending on prior cold rolling reduction, around 650–700 °C and was completed within few seconds at 750–800 °C. Examples of the influence of cold rolling degree on the reversion kinetics in a 301LN ASS are illustrated in Figure 4, where also the data of Takaki et al. [40] for an 18Cr-8.5Ni steel is included. A higher reduction results in faster reversion. The diffusional reversion is fast at 750 °C and above in both steels being completed within 100 s even after 45% cold rolling reduction.

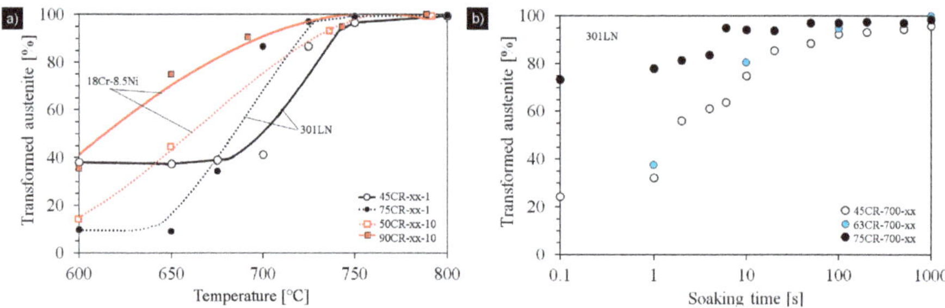

Figure 4. The effect of prior cold rolling reduction and annealing parameters on the reversion kinetics: Annealing temperature (**a**) and soaking time at 700 °C (**b**). Data collected from References [23,40] (**a**) and [23,26,88] (**b**). Legend: Reduction (%CR), annealing temperature (°C) and time (s).

However, under certain conditions, both the shear and diffusional reversion mechanisms seem to be active in a 301LN [84,102] and also in a 204Cu [82,101]. Guy et al. [35] found that in 18Cr-8Ni and 18Cr-12Ni ASSs the reverse transformation can proceed by both processes depending upon the heating time and temperature. Lee et al. [45] reported that in a 11Cr-9Ni-7Mn steel both mechanisms can occur sequentially, the diffusionless reverse transformation during continuous heating followed by the diffusive reverse transformation during the isothermal holding. Similarly, Knutsson et al. [130] pointed out that in a 301 ASS, the shear reversion only reverses 90% to 98% of the martensite, and a complete reversion can only be achieved by diffusional reversion for long holding times. According to Shakhova et al. [62], in S304H (18Cr-8Ni-2Cu-Nb) ASS after very severe caliber rolling to a total strain of 4, both reversion mechanisms were found to take place in annealing at 600–700 °C. The appearance of equiaxed austenite grains suggested diffusive reversion, while the elongated grains were signs of shear reversion. The latter was also indicated by the rather high dislocation density remained in the annealed microstructure.

3.2. Temperature Range for Reversion

The difference in the reversion mechanisms can be illustrated by the reversion-temperature-time diagram where the start and finish temperatures of the martensite reversion to austenite are drawn [39]. Moreover, from the diagram, it is possible to judge how the reversion process makes progress under certain conditions. In Figure 5, the respective temperatures are $A_{s'}$ and $A_{f'}$ for the shear reversion and A_s and A_f for the diffusional one. The martensitic shear reversion proceeds during heating in a narrow temperature range $A_{s'}$–$A_{f'}$. These temperatures depend on the chemical composition of steels being lowered by increasing the Ni/Cr ratio, but they are independent of the heating rate. The shear reversion rate is fast and independent of prior cold rolling reduction, and the reversed fraction is independent of isothermal holding time between these temperatures. The possible reversion treatment routes in Figure 5 are 1 and 2, leading to complete reversion, but the austenite grains can contain high dislocation density after short holding (route 1), or they can be dislocation-free and refined after sufficient holding (route 2).

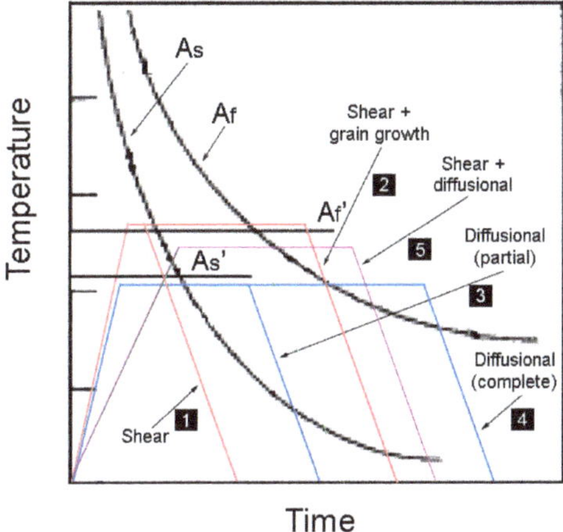

Figure 5. The temperature range for the reversion, dependent on the reversion mechanism. Various reversion treatments (numbered 1 ... 5) are possible. The original idea of the reversion-temperature-time diagram is from Reference [39].

On the other hand, the A_s and A_f temperatures for the diffusional reversion depend on the heating rate in addition to the chemical composition of the steel (Figure 5). The effect of heating rate was already investigated in Fe-Ni-C alloys by Apple and Krauss [141] in 1972. In isothermal annealing, the diffusional reversion proceeds more rapidly with increasing annealing temperature, and the A_f temperature depends on soaking time, as seen from the A_s and A_f curves. The reversion annealing routes 3, 4 and 5 are possible, and a short holding (route 3) below $A_{s'}$ would result in partial reversion and a sufficiently long holding (route 4) to complete reversion. A slow heating to the $A_{s'}$–$A_{f'}$ regime means the initiation of shear reversion which might be completed by the diffusional reversion (route 5) or some DIM remains stable.

There are several studies where the reversion range of 304 ASS has been investigated, but quite different values for A_s (or $A_{s'}$) and A_f (or $A_{f'}$) have been reported. As mentioned above, the mechanism seems to change at around 750 °C from diffusional to shear type with increasing heating or annealing temperature. Guy et al. [36] reported A_s = 540 °C for an 18Cr-8Ni steel. Tomimura et al. [39] found by the saturation magnetization measurements a narrow reversion range 575–625 °C for a 90% cold-rolled 16Cr-10Ni steel where DIM reverted to austenite by the shear mechanism during heating at a high rate of 300 °C/s. Mumtaz et al. [143] estimated from the changes in the saturation magnetization for a 304 ASS (40% to 55% cold rolled, rapid heating, though the rate not given, 5 min holding) the A_s temperature to be between 625–650 °C, and the A_f between 900–950 °C, hence a very wide range. At 700–800 °C, the reversion kinetics was dependent on the previous cold rolling reduction. Tavares et al. [144] obtained by thermomagnetic analysis the A_s and A_f temperatures during heating (heating rate was not given) as $A_s \approx$ 430–440 °C and $A_f \approx$ 610–616 °C. Cios et al. [63] determined A_s and A_f temperatures of 450 and 700 °C, respectively, from dilatometric curves. Shakhova et al. [62] data and ThermoCalc calculation for an Fe-0.1C-0.1N-18Cr-8Ni-2Cu-0.5Nb steel indicated the A_f temperature of 800 °C.

There are few studies on the reversion range itself in 301 and 301LN ASSs, although the reversion mechanisms have been investigated in numerous works. However, the A_f temperature for a 301LN ASS has not been clearly reported, and due to the diffusion-controlled reversion, it is dependent on cold rolling reduction and annealing duration, as evident from Figure 4. In that data, about 5%DIM is left after 10 s holding at 700 °C after the 75% reduction, while after the 63% reduction, Järvenpää et al. [84] reported 10% of DIM and Rajasekhara [26] 5% after 100 s and 20% after 10 s. Anyhow, in numerous studies, temperatures between 600 and 1000 °C have been used for reversion treatments of 301LN ASS [12,22–26,84–87].

For 201 and 201L grades, the temperature of 850 °C has been found to be an adequate reversion annealing temperature resulting in complete reversion within 30 s and in the finest GS [15,16]. Kisko et al. [27] obtained high tensile elongations (softened structure) in a 60% cold-rolled 201 ASS after 15 s at 800 °C. Sadeghpour et al. [69] found that in a 201L + 0.12Ti steel, the reversion rate depends on temperature and holding time, so that the reversion seemed to be diffusion controlled. The temperature of 900 °C was convenient for the reversion within a soaking time of 60 s.

4. Evolution of Reversion-Annealed Microstructures

In reversion annealing, a highly refined austenite GS is to be created. In this chapter the microstructure of reversion-treated ASS is discussed accounting for the fact that in practice the microstructure is not simple highly refined structure but is quite complex. Due to a limited cold-rolling reduction, the deformed structure consists of DA in addition to DIM, the latter deformed to various degrees. This results in non-homogeneous microstructure inherited from recrystallization of DA and reversion of DIM during the annealing stage. At low annealing temperatures, also recovered DA and retained DIM exist. In prolonged annealing, the grain growth of fine grains can be expected. The numerous studies to characterize and classify the microstructure are described with examples of GS and its distribution. Separately, a chance of repeating the reversion treatment or employing complex reversion routes have been mentioned, while looking for bimodal GS distributions.

4.1. Reversed Microstructures

The degree of cold rolling reduction prior the annealing influences on the reversed microstructure, e.g., [29,40,145]. After a high reduction, when the lath-martensitic structure is completely destroyed forming a cell structure, equiaxed austenite grains nucleate at random and grow in the recovered martensite matrix. In this case, small equiaxed austenite grains much below a micrometer can form during the full reversion. Instead, after a low cold rolling reduction, the DIM is still lath-martensitic, and reversed austenite nucleates on lath boundaries and forms a stratum structure of austenite laths and blocks. The reversed austenite just looks like a lath-martensitic structure. Guy et al. [36] found a twinned substructure in austenite. Finally, subboundaries in the shear-reversed austenite are gradually replaced in the continuous recrystallization process by high-angle grain boundaries forming thereby refined microstructure [29,38,40,54,146]. This happens especially after low cold rolling reductions [29].

However, if the microstructure contains both DIM and DA after cold rolling, different deformation states of DIM and DA tend to modify the microstructure, resulting in non-homogeneous GS distributions. In ASSs such as 304 and 316, grades where the austenite stability is high, low DIM fractions may be attained even after high cold rolling reductions (see Figure 2). Inhomogeneity in GS is a result of the reversion of DIM and the recrystallization of DA, the latter being often partial after low-temperature annealing. Similarly as in shear reversed austenite, the recrystallization of DA grains is sluggish in nature, and it also seems to take place by a gradual evolution of subgrains and their subsequent transformation into fine grains, i.e., by the continuous recrystallization type process, e.g., [29,39,40,45,54].

Järvenpää et al. [29,84] have investigated in detail the reversed microstructures created at low annealing temperatures after low cold rolling reductions and pointed out the complexity of the microstructure, consisting of fine-grained reversed austenite, DA with different recrystallization degree and retained DIM. They showed in reversed structures obtained at low temperatures (<800 °C), in addition to micron-scale reversed grains and coarse DA grains, the presence of medium-sized (GS range 3–10 µm) grains. Medium-size grains are formed from slightly deformed DIM so that their fraction depends on rolling reduction and annealing temperature. In Figure 6, examples of different microstructures created in a 301LN are displayed revealing a broad variation of the reversed austenite GS. In addition to the annealing temperature, the cold rolling reduction has a significant influence, whereas the fraction of slightly deformed austenite increases with decreasing rolling reduction [29,102].

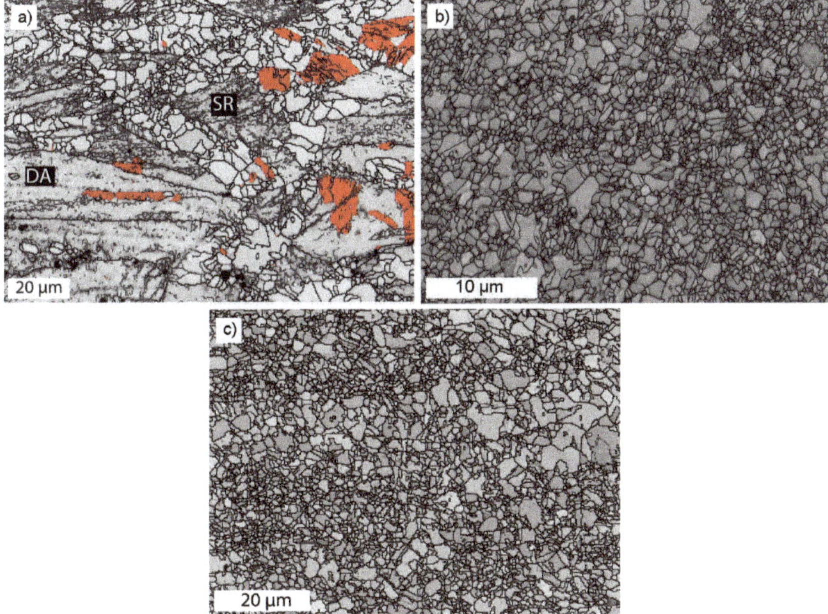

Figure 6. Examples of reversed microstructures in 301LN with various grain size. Partially reversed (32% cold rolling, 750 °C for 0.1 s) with shear reversed (SR) fine and medium-size grains, a large DA grain and retained DIM (red colored phase) (**a**) and diffusionally reversed ultrafine and medium-size grains in 63% cold-rolled steel, annealed at 800 °C for 1 s (**b**) and at 900 °C for 1 s (**c**).

A further factor resulting in non-homogeneous GS distribution is the elemental segregation present even in cold-rolled and annealed ASS sheets. Man et al. [100] have pointed out that the 301LN grade exhibits pronounced manganese banding, which is experienced in differences in the stability of austenite during cold rolling and consequently in GS after reversion annealing.

However, heterogeneous GS has been found to be beneficial especially for the ductility of ASSs (see more in Section 5), and therefore, in numerous studies, special processing routes have been applied intentionally to create such a structure. Ravi Kumar et al. [146] reported GS distribution varying widely between 100 nm and 2 µm after two-stage annealing of a 90% cold-rolled 316L ASS, resulting in a good combination of YS and ductility as compared to the coarse-grained counterpart after annealing at temperatures above 900 °C. Sun et al. [146] applied three cold rolling and annealing stages for a 304 ASS to obtain a GS of 150 nm providing good mechanical properties. In a more recent study,

Sun et al. [140] reported that in two-stage annealing of a 304 ASS containing 47% DIM and 53% DA after two-stage cold rolling (67% + 47%, respectively) at 800 and 850 °C, non-uniform grain structure was created, where the fraction of ultrafine grains (<1 μm) decreased and larger austenite grains (up to 3 μm) increased with prolonged annealing. Ravi Kumar et al. [18,146,147] and Ravi Kumar and Raabe [148] have obtained bulk ultrafine-grained ASS with mono- (maximum at GS ≈ 0.6 μm) and bimodal-type (minimum at GS ≈ 0.5 μm and maximum at GS ≈ 1.40–1.65 μm) GS distributions by 4-stage cyclic thermal processing in a 90% cold-rolled AISI 304L grade. This enhanced significantly the elongation. Roy et al. [117] also reported a bimodal GS distribution in a cryo-rolled 304L grade after annealing at 700–800 °C. The GS was very fine: one group between 50–100 nm and the other 200–300 nm from DA. A beneficial contribution to ductility was claimed. Poulon-Quintin et al. [30] obtained a bimodal grain structure consisting of ultrafine-grained austenite with average GS equal to 0.5 μm as a consequence of martensite reversion together with bands composed of recrystallized fine grains of 2.0 μm from strain-hardened austenitic bands in a 15Cr-9Ni-3Mo steel, which was 80% cold rolled forming 75%DIM and 25%DA.

In principle, the grain growth tendency of reversed very fine grains is high during the initial period of annealing (1–100 s) due to the high curvature of grain boundaries and consequently large driving forces for grain growth [26,89]. The value of the activation energy of grain growth was found by Rajasekhara et al. [26] to be equal to that of conventional ASSs (~205 kJ/mol). Kisko et al. [82] obtained much higher values for a Nb-microalloyed 204Cu grade, the value increasing from 363 to 458 kJ/mol with the increasing Nb content. Some experimental results for the growth of reversed fine grains in 201L, 204Cu, 301LN and 304 ASSs have been collected in Figure 7, revealing the trends at 800 and 900 °C. There is some difference between the data of 304L [17,70] and 304 [11] at 900 °C, but the processing of the structures was also different. Anyhow most importantly, the data indicate that grains do not grow much at 800 °C, if a GS of around 1 μm is concerned. The grain growth can also be retarded effectively by microalloying. The influence of Nb [19,82,83,101] and Ti [68] microalloying for restricting the grain growth has been determined. For instance, the grain growth was effectively retarded in a 204Cu ASS by 0.28 wt. % Nb alloying even at 1100 °C and by 0.11 wt. % Nb at 1000 °C [82].

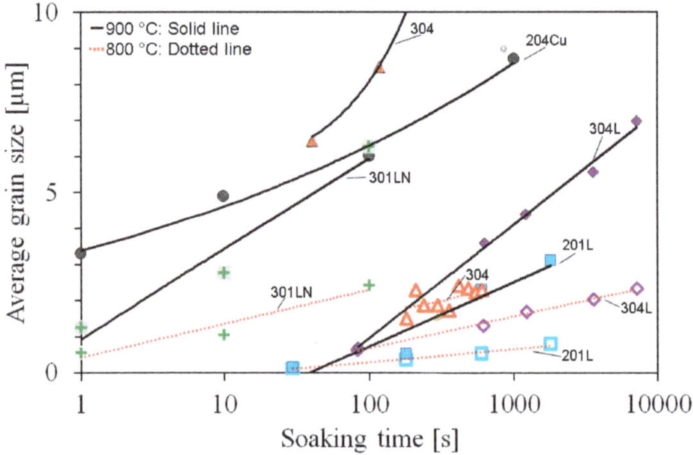

Figure 7. Grain growth of reversed grain size at 800 and 900 °C in 201L [67], 204Cu [82], 301LN [26], 304L [17,70] and 304 [11] ASSs.

Table 2. Examples of grain sizes and mechanical properties achieved in reversion treatments of Cr-Mn, Cr-Ni and some special austenitic stainless steels.

Grade	CR	T	t	GS	YS	TS	TE	Ref.	Grade	CR	T	t	GS	YS	TS	TE	Ref.
	[%]	[°C]	[s]	[μm]	[MPa]	[MPa]	[%]			[%]	[°C]	[s]	[μm]	[MPa]	[MPa]	[%]	
201	90	850	30	0.1	1370			[15]	301LN	80	700	1200	0.76	830	953	36	[47]
	60	800	2		770	1155	27	[28]		77	800	1	0.6	880		44	[91]
			15		686	1020	31	[28]		76	800	1		768	1060	43	[23]
			40		629	964	37	[28]			700	10		1004	1106	38	[23]
201L	95	850	30	0.07	1485	1780	33	[16]		63	800	1	0.54	720	1060	35	[22]
			180	0.14	1075	1710	37	[16]		60	800	1		711	995	50	[23]
		800	30	0.06	1300		33	[67]			700	100		901	1106	37	[23]
			180	0.35	1070		24	[67]		56	790	80	0.9	670	1030	55	[87]
201L-0.23N	60	800	1	0.5	1100	1260	30	[80]			690	60	0.7	815	1105	45	[87]
			10	1.5	800	1100	50	[80]			680	45	0.7	855	1135	44	[87]
			100		680	1020	58	[80]			660	30		985	1160	29	[87]
201-0.3Nb	90	900	60	0.09	1000	1500	35	[71]		50	800	160		798	1057	38	[12]
201Ti	90	900	60	0.05	1000	1330	42	[69]				107		773	1023	41	[12]
		850	60	0.05	1005		36	[68]				80		645	1036	41	[12]
		750	600	0.13	856		34	[68]			750	160		778	1057	32	[12]
204Cu	60	800	1		723	1041	47	[28]				107		682	1063	35	[12]
			1		606	1093	45	[83]				80		709	1042	35	[12]
		700	10		1185	1308	43	[83]		45	800	160		682	1000	46	[12]
			100		899	1345	37	[83]				107		662	1017	39	[12]
			1000		605	1180	36	[83]				80		644	1017	46	[12]

Table 2. *Cont.*

Grade	CR [%]	T [°C]	t [s]	GS [μm]	YS [MPa]	TS [MPa]	TE [%]	Ref.
204Cu-0.45Nb	60	800	1		997	1187	51	[83]
		700	10		1141	1297	40	[83]
			100		1124	1319	40	[83]
			1000		1015	1258	31	[83]
301	95	850	1	0.07	1970			[65]
	60	850	10		609	1122	36	[28]
		750	10		1001	1318	28	[28]
	52	900	1		584	1032	44	[23]
		800	1		1070	1290	28	[23]
			10	0.5	1000	1250	35	[88]
304	90	750	1200	1.02	550	840	41	[11]
	85	580	1800	0.15	1120	1440	12	[137]
	80	700	60	0.15	1020	1160	8	[50]
	70	750	600	1 / 0.2	950	1050	47	[148]
17.1Cr-11.3Mn-0.27N	80	850	100	0.85	900	1100	60	[72]
		750	10	0.5	1450	1500	19	[72]
17Cr-6Ni-2Cu	75	700-900		0.22	980	1100	30	[51]

Grade	CR [%]	T [°C]	t [s]	GS [μm]	YS [MPa]	TS [MPa]	TE [%]	Ref.
304L	32	750	160		809	1125	37	[12]
			107		785	1089	35	[12]
			80		748	1068		[12]
		760	85	1.1	590	1030	57	[87]
		695	65	0.8	750	1080	54	[87]
		675	45		780	1100	50	[87]
		820	130	1.6	520	1000	62	[87]
		690	70	1.2	785	1140	51	[87]
		660	40		915	1135	32	[87]
	20	750	600	2.9	749	1010	33	[60]
	0			20	331	815	62	[12]
					350	831	62	[12]
	90	900	120	2.2	994	1160	40	[20]
		700	1200	0.14	1000	1010	40	[14]
			18000	0.33	1000	1010	40	[14]
	65	850	60	0.62	885	1385	44	[70]
	67	550	150	0.27	1890	2050	6	[48]
	67	650	180	0.8	1170	1350	12	[48]
	0			35	220	1640	59	[48]

CR, Cold rolling reduction; T, annealing temperature; t, holding time; GS, grain size; YS, proof/yield stress; TS, tensile strength; TE, total elongation.

4.2. Repeated Reversion Treatments

Commonly, the prior austenite GS has an influence on its decomposed structure. Does the original GS affect the reversed structure? As a reversion treatment refines the GS, it is expected that by repeating the reversion treatment further refinement can be achieved. Ma et al. [21] repeated the conventional reversion treatment twice (75% reduction, annealing at 650 °C for 10 min + 50% reduction and annealing at 630 °C for 10 min) in an Fe-0.1C-10Cr-5Ni-8Mn alloy. After the first reversion, the structure consisted of nearly equiaxed austenite grains with the average size of about 300 nm, whereas after the second reversion, some grains are even smaller than 100 nm while others are a little larger than 200 nm. The YS increased from 708 MPa to 779 MPa by repeating the reversion, but the elongation decreased from 36% to 32%.

As mentioned in the previous section, in order to create the bimodal GS distribution, special reversion treatments with several rolling and annealing stages have been executed in processing of 304L steel. Sun et al. [137] have used three-stage processing and also two reversion stages [140]. Ravi Kumar et al. [147] and Ravi Kumar and Raabe [148] have employed complex rolling and annealing schedules, with iterative and isothermal annealing periods. The present authors have also applied various repeated schedules to 301LN ASS and obtained further GS refinement and higher uniformity of the grain structure. However, concerning industrial purposes complex reversion annealing treatments are hardly very interesting or practical, although a treatment repeated twice has been utilized [104,107].

5. Properties of the Reversion-Treated Structures

For steel users, the mechanical properties are important, and the reversion treatment is intended to enhance them, especially the YS of ASSs. In this extensive chapter, the tensile properties will first be discussed based on the vast experimental data collected from the literature, mainly for reversion-treated commercial Cr-Mn and Cr-Ni ASSs. The influence of the GS on YS has been analyzed. The enhanced YS-ductility combinations achieved are highlighted as a practical result while employing the reversion process. Further, the excellent tensile ductility being affected by the stability of the austenite has been discussed in detail. The effect of highly refined GS on the stability, a somewhat disputed issue, will be considered. Particularly, the influence of carbide/nitride precipitation occurring in many ASSs during reversion annealing reducing the stability, but neglected in many studies, will be addressed. The deformation mechanisms and the related strain hardening rate are also discussed, which is interesting in trying to understand the background of high ductility.

In a separate section, fatigue properties will be described showing distinct advantages attained from the reversion treatment if based on fatigue resistance under a given stress amplitude. Finally, the other properties, though more rarely investigated, such as the formability, weldability, corrosion resistance and surface properties for medical applications, will be considered, highlighting that they are comparable or better than the corresponding properties of conventional coarse-grained ASSs.

5.1. Tensile Strength Properties

A principal target of the reversion treatment is to increase YS of an ASS without impairing its high elongation. An example of typical results is illustrated in Figure 8, where a set of engineering stress–strain curves are plotted for a 301LN ASS showing how significantly YS is increased by refined GS as a result of reversion treatments. Moreover, the tensile strength (TS) is improved while TE is decreased. Certain typical changes, a yield point and a concave shape, appear in the flow curves, to be discussed later.

In order to give a view of the current state, some data concerning the mechanical properties of reversion treated ASSs were collected from the literature and listed in Table 2. In addition, the cold rolling reduction, annealing temperature and duration as well as the (average) GS are given in the list if available. The commercial steel grades 201/201L, 301/301LN and 304/304L mostly investigated are included. Some data are also for a 204Cu and Ti- or Nb-microalloyed 201and Nb-microalloyed 204Cu as well as for a couple of austenitic alloys.

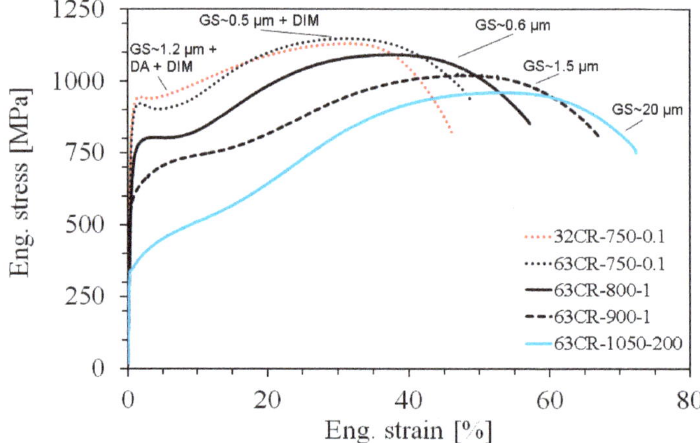

Figure 8. Examples of engineering tensile stress–strain curves for a 301LN ASS with refined grain size after various reversion treatments. Data collected from References [87,102,103]. Legend: Reduction (%CR), temperature (°C), duration (s).

From the data in Table 2, a general conclusion can be drawn that no distinct difference exists between the Cr-Ni and Cr-Mn type grades, and same temperatures between 650–900 °C and holding times 1–1800 s are feasible for the reversion. It seems that the YS (proof stress, $R_{p0.2}$) values reported for 201/201L ASSs can be well above 1000 MPa, even up to 1485 MPa [16], but in those instances severe cold rolling reductions of 90% to 95% have been applied to obtain the GS of an order of 60–100 nm. However, the 60% cold rolling of 201 with 0.23%N [80] similarly as 204Cu and 204Cu-0.45Nb [83] can result in YS of 1.1 GPa, while TE is still about 40%.

YS values of the reversion-treated 301/301LN ASSs achieved are typically over 700 MPa and up to 1 GPa, i.e., 2–3 times compared to the YS of respective commercial annealed sheets. Further, the reversed structures obtained at low temperatures (800 °C and below) exhibit significantly enhanced YS compared to the structures created at temperatures of 900 °C. Then, a slight upper yield point appears with discontinuous yielding (Lüders strain), as seen in Figure 8 [87,103].

Moreover, the 304 and 304L grades require a high cold rolling reduction at lowered temperatures, as pointed out in Section 2 (Figure 2), and then, comparable strength values have been obtained. It can be noted that exceptionally high strength values have been reported by Shen et al. [48] for a 304L ASS after 67% multipass warm rolled at 400 °C, which formed 73% of DIM. After annealing at 550 °C for 150 s, the reversed GS was 0.27 μm and the amount of retained DIM 32%. As listed in Table 2, the tensile properties reported were extremely high, YS 1890 MPa, TS 2050 MPa and TE 6%. Moreover, annealing at 650 °C resulted in high strength. YS values around 2 GPa with reasonable ductility have also been reported by Rasouli et al. [72] in interstitially alloyed Ni-free Cr-Mn grain-refined ASSs with a low C/N ratio. It can be mentioned that very recently, Xu et al. [149] have shown that the refined GS in an 18Cr-8.3Ni ASS exhibits the enhanced YS and YS-ductility combination in relation to its coarse-grained counterpart also at elevated temperatures, up to 600 °C.

Concerning the influence of the degree of cold rolling on YS, Somani et al. [23] and Misra et al. [24] reported that the YS of completely reversed structure in a 301LN steel, created at 800 °C–1s, increased from 800 to 950 MPa with increasing the reduction from 45% to 77%, while the TE remained practically unchanged (about 43%). This can be understood as a result from GS refinement due to the heavier cold rolling. However, the prior cold rolling reduction in the range of 32% to 63% did not affect markedly the mechanical properties of the low-temperature partially reversed structures, where a larger, non-uniform GS will be balanced by stronger retained phases, DIM and DA [87]. Experiments using induction heating indicate that the annealing temperature (660–820 °C) has an essential influence on YS, for it increased with decreasing the peak temperature, but any influence of the cold rolling reduction between 32% and 56% has not been realized (see Table 2) [87]. Regarding the YS of 201/201L ASS, it seems that the cold rolling reduction of 50% to 70% has resulted in considerably lower values (YS ≈ 400–750 MPa) than the reduction of 90% to 95% has done (YS above 1000 MPa), although the data are quite scarce. For instance, Rasouli et al. [72] have reported high YS values for a 17Cr-11Mn-0.275N ASS, YS 1450 MPa and TE 19%, although the 80% cold rolled structure contained 53% DIM only.

The evident fact is that the TS cannot be increased so significantly as the YS, only about 200 MPa to reach the level of 1100 MPa in most cases. This is due to somewhat lower strain hardening rate (SHR) owing to finer GS generally increasing the austenite stability and shorter elongation in reversed fine-grained structures (Figure 8).

The background mechanisms of high YS of reversion treated ASSs have been analyzed in some papers. For instance, in completely reversed structures of 301LN, the strengthening contributions of solid solution, precipitation, dislocations and GS have been evaluated, the values being approximately 200, 120, 40, 250 MPa respectively (annealing at 800 °C for 1–10 s; GS ≈ 0.54–1 μm) [22]. Thus, refined GS is an important strengthening factor. Moreover, Kisko et al. [83] estimated the amount of GS strengthening to be about 300 MPa in a 204Cu ASS as annealed at 900 °C for 10 s (GS 1.4–1.9 μm).

The Hall–Petch relation between YS and GS has been reported in numerous studies. Di Schino et al. [10] found that it holds down to 3 μm GS for a 301 ASS, and according to Huang et al. [47] at least down to 0.74 μm GS for 301LN and even to 0.135 μm GS for a 304L according to Forouzan et al. [14]. In structures reversed at low temperatures, the amount of precipitation in certain grades, the retained phases and their annealing state are additional contributors, so that it is obvious that any exact relationship between GS and YS cannot be expected for them. The several features affecting the Hall–Petch relationship for YS vs GS of reversed structures were discussed by Rajasekhara et al. [22]. Anyhow, several relationships have been proposed, and some examples are shown in Figure 9. Shakhova et al. [62] reported the slope 395 $\mu m^{-0.5}$ and the friction stress σ_o = 205 MPa for the data collected for Cr-Ni ASSs from the literature. Kisko [101] analyzed the literature data for Cr-Ni and Cr-Mn steels and realized a wide scatter, the slope varying in the range 256–377 $\mu m^{-0.5}$ and the σ_o between 241–428 MPa. Moreover, Järvenpää [102] has found a marked scatter especially in the slope, being in the range 261–576 $\mu m^{-0.5}$, while σ_o is 225–273 MPa. The GS is not uniform in many instances in the reversed structure, which is an apparent reason for the scatter in the values of the coefficients in the Hall–Petch relationship. A modified Hall–Petch relation has been suggested to account for the grain size distribution [150] but not applied to reversed structures.

As mentioned, Shen et al. [48] reported very high strength values close to 2 GPa for a 304L with very fine GS. Both DIM and mechanical twins were found to be effectively suppressed during tensile testing due to the refinement of GS so that the background of the strength seemed to be mainly the strengthening due to GS refinement. However, from the Hall–Petch relationship in Figure 9, it is expected that the GS of 270 nm (corresponding to 1.92 $\mu m^{-0.5}$) would result in YS of 1000 MPa, not close to 2 GPa. It can be concluded from the microstructural data that the structure annealed at 550 °C consisted of 41% reversed austenite (supposed to be formed by the diffusional mechanism), 32% retained recovered martensite and 27% DA. Therefore, it remains unclear, how the complex microstructure behaved to result in the GS of 270 nm and why this microstructure and GS provided the YS as high as 1.89 GPa. The dislocation structure must have a prominent contribution. As mentioned, the temper-rolled cold strengthened ASSs show directional anisotropy, so that the YS is lower in parallel than in transverse direction to the rolling direction [3,4]. However, the experiments have shown that this anisotropy disappears in completely reversed structure suggesting that it is related to the deformation in austenite [28,101]. This means that the reversed structure without any directional anisotropy can have the YS of about 650 MPa.

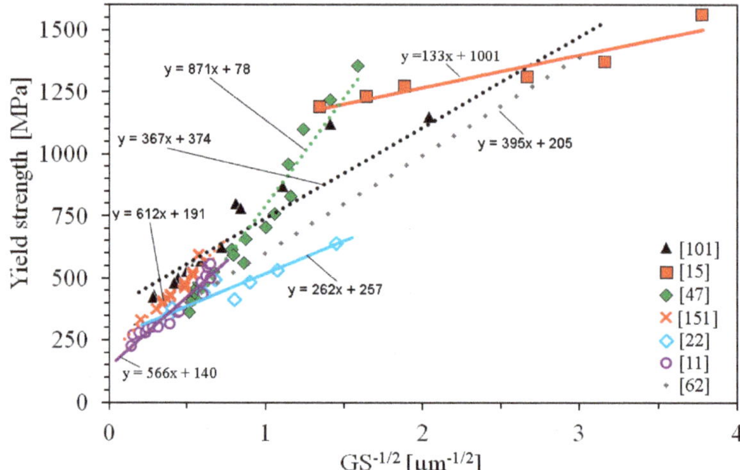

Figure 9. Examples of the Hall–Petch relations presented for reversed structures. Data for annealed coarse-grained 301LN is included from [151].

5.2. Tensile Ductility and Austenite Stability

5.2.1. Tensile Elongation

It is notable that the elongation values of reversed ASS structures are good in spite of high strength, though slightly decreasing with increasing YS, as seen in Figure 8. Generally, the TE is in the range of 30% to 40% even for the structures with the YS ≈ 1 GPa. It should, however, be noticed that many reported elongation values have been measured using non-standard tensile specimens with a short gauge length (15–25 mm instead of 80 mm), which may give too optimistic values.

In order to highlight the excellent strength-ductility combinations achieved with reversion-treated ASSs, YS vs TE combinations for 2XX and 3XX series, steels listed in Table 2 are plotted in Figure 10 together with some technical data for commercial temper-rolled steels and a trend line for the latter. The data though scattered indicate that many of the combinations achieved by reversion treatments are better than those obtained by temper-rolling. The combinations for 201/201L seem to distribute in two groups, well above the trend line and slightly below that, whereas for 301LN, they form one group above the trend line of the temper-rolled steels. From Table 2, it is seen that the higher values of 201/201L are obtained when very high cold rolling reductions are employed in the reversion treatment, whereas lower reductions are not so effective in GS refinement and increasing strength.

Furthermore, it has been found that the directional anisotropy, present in temper-rolled steels, is absent in completely reversed structures without DIM and DA [28]. The anisotropy appears in cold rolling of 5% reduction so that the texture cannot be the reason for that, but the dislocation structure. This means in practice that a steel with its YS order of 700 MPa can be manufactured without the directional anisotropy. Strength values in compression are not available so that they cannot be compared with tensile properties.

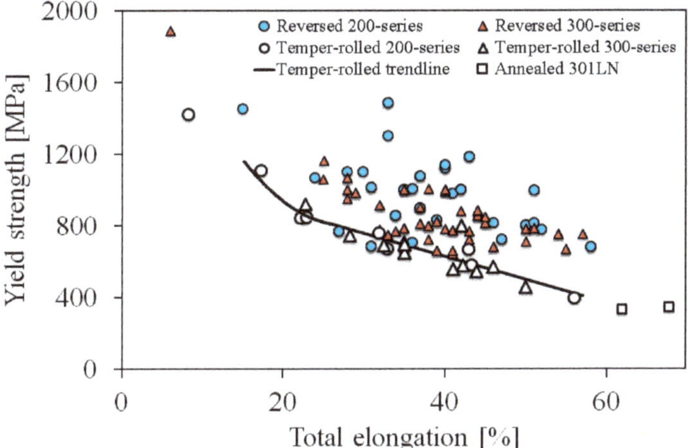

Figure 10. Yield strength-total elongation combinations for some reversion-treated steels listed in Table 2 compared to those of annealed 301LN and temper-rolled 301LN and 201L steel.

5.2.2. Elongation and Austenite Stability

In principle, the stability of austenite while transforming to martensite affects the strain hardening rate (SHR) and may influence on the ductility via the TRIP effect [152]. According to scarce data of Guo et al. [153], the tensile elongation of 301 ASS (M_{d30} = 39 °C) increases about 5%-units with increasing DIM fraction formed at different test temperatures, although the effect is quite marginal compared to the influence of martensite on SHR and TS. Cios et al. [63] measured the maximum elongation in a 304 ASS at −10 °C where about 50%DIM was formed. However, tensile tests at different temperatures by Weiss et al. [154] showed that the maximum uniform elongation for a 304L ASS is obtained at test temperatures of 20–40 °C, where only few per cents of DIM had formed, whereas the elongation decreased while more DIM was formed at lower test temperatures. They also refer to 11 investigations, which indicate that the maximum uniform elongation is obtained in conditions when about 20% of DIM was present. Talonen [155] also found that the elongation is highest at test temperatures where only a relatively low amount of DIM was formed in 304 and 301LN ASSs. In agreement, Hamada et al. [156] measured the maximum elongation at 50 °C for 201 and 201L ASSs, showing only 0% to 20% of DIM.

The rate of DIM formation in ASSs is classically described by the sigmoidal equation (OC-model) developed by Olson and Cohen [157]. The OC-model assumes that shear band intersections act as nuclei for martensite and that the nucleation and growth process of α′-martensite can be described by two parameters, α and β, as in the following equation:

$$f_{\alpha'} = 1 - exp\left[-\beta(1 - exp[-\alpha\varepsilon])^m\right] \tag{3}$$

where $f_{\alpha'}$ is the martensite fraction, ε is the true strain, α and β are parameters and m is the fixed exponent. The parameter α describes the rate of the shear band formation, assumed to be mainly dependent on the stacking fault energy (SFE) and strain rate. The parameter β is proportional to the probability that the α′-martensite is nucleated at a shear band intersection, being dependent on the chemical driving force and temperature. The exponent m describes the rate of the formation of the shear band intersections, often found to have the constant value of 4.5. In cold rolling, Forouzan et al. [14] reported α = 3.041 and β = 3.786 for AISI 304L (M_{d30} = 12.9 °C) and Somani et al. [23] α = 2.2 and β = 4.4 for 301LN (M_{d30} = 26.6 °C). The existing models used to predict the increase in the volume fraction of martensite with strain were examined and modified by Lichtenfeld et al. [158] to fit the experimental data for a metastable 304L and stable 309 ASSs as well as data for 304 and 301LN collected from the literature. They reported values α = 3.0 and β = 4.3 for 304L (M_{d30} = 2 °C) and α = 5.65 and β = 4.03 for 301LN (M_{d30} = 37 °C) in tensile tests at a low strain rate avoiding adiabatic heating. Talonen [155] observed that the OC-model fitted well with his tensile test data, obtained at low strain rates, with $\alpha \approx 11.7$ and $\beta \approx 6.1$ for a 301LN steel ($M_{d30} \approx$ 23°C). A recent logistic model was mentioned in Section 2, presented for 304 ASS cold-rolled to different reductions [124]. In that paper, also the limitations of the OC-model have been discussed.

In ultrafine-grained austenite, no shear bands are formed, but martensite nucleates on grain boundaries [75,81,102], so that the basis of the OC-model could be doubted. However, sigmodal shapes of DIM fraction vs strain have still been recorded (as seen in Figure 11). Marechal [75] even fitted his tensile data with the OC-model and determined the values of the parameters and their GS dependence, down to 0.5 µm GS. However, it is hard to draw any conclusions from the varying values of all the parameters.

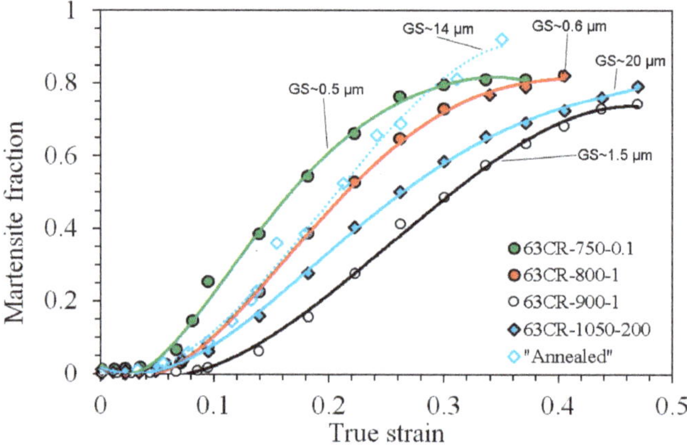

Figure 11. Examples of DIM evolution during tensile testing in various 301LN (M_{d30} = 27 °C) structures [82,85,87] and coarse-grained annealed 301LN with M_{d30} = 37 °C from Reference [155]. Legend: Reduction (%CR), temperature (°C), duration (s).

The austenite stability also depends on GS, as indicated by the Nohara equation (Equation (1)) for coarse grain sizes. Yoo et al. [159] found that DIM fraction after tensile fracture decreased from 60% to 25% with decreasing GS from 2 µm to 0.3 µm in a 10.30Cr–8.14Ni-7.47Mn steel. However, Matsuoka et al. [160] reported that the mechanical stability does not depend on GS in ASSs, because the martensite transformation is not necessarily multi-variant under tensile strain/stress, but the single-variant martensitic transformation is favored.

Marechal [75] and Järvenpää et al. [84] have extensively studied the stability of austenite in tensile testing in various reversed grain-refined structures of 301LN ASSs. Figure 11 shows examples of DIM vs strain in tensile tests determined by the latter authors. Generally, refining the GS down to about 1 µm was found to increase the stability, similarly to the GS change from 20 to 1.5 µm in the figure, but surprisingly, submicron-sized austenite was more unstable (GS of 0.6 and 0.5 µm) than the coarse-grained austenite [84]. Similarly, Kisko et al. [81] noticed the same trend in a 204Cu steel. Marechal [75] and Kisko et al. [81] explained the reduced stability of submicron grains by DIM nucleation on grain boundaries as a single variant (block), whereas the nucleation occurred at shear band intersections in the coarse-grained structure. Ravi Kumar and Gujral [161] and also Lee et al. [162] have reported DIM nucleation at the vicinity of grain boundaries in ultrafine-grained ASSs, although this was not related to a lower stability in their studies.

However, according to Järvenpää et al. [86], the real reason of the lower stability appearing in certain grain-refined structures in the 301LN ASS is the precipitation of chromium nitrides while annealed at low temperatures such as 650–800 °C (the curves for GS of 0.6 and 0.5 μm obtained at 800 and 750 °C, respectively, shown in Figure 11). A similar drop in the stability was observed by He et al. [163] in the low-temperature (800 °C–30 min) reversed structure of 321 (17Cr-8Ni-0.3Ti) ASS as comparing with the structure formed at 1000 °C. The reason for the low stability was suggested to be carbide precipitation. In consistence, Saenarjhan et al. [164] and Kim et al. [165] have reported the decrease of the austenite stability in a 15Cr-15Mn-4Ni steel due to $M_{23}C_6$ carbide precipitation during long-term annealing of the austenitic structure at around 800 °C. Johannsen et al. [135] noticed in a 90% cold-rolled 301 ASS, while annealing temperatures above 750 °C, an increase in the amount of martensite, which was formed during cooling. $(Fe,Cr,Mo)_{23}C_6$ carbides were formed within the grains and at grain boundaries, being an obvious reason for the reduced thermal stability of the reversed austenite. Moreover, Karimi et al. [166] observed new martensite forming during cooling in a 301 ASS, containing 0.11%C, after annealing at 900 °C for 100 min due to carbide precipitation.

Moreover, the results of Kisko et al. [83] reveal increasing DIM fraction after tensile testing of a 204Cu ASS while annealed at 700 °C for 1000 s, and the carbide or/and nitride precipitation could be the reason for that, though not confirmed. Lei et al. [51] found lower stability of a nano-ultrafine-grained (220 nm) structure of a 0.06C-17Cr-6Ni-2Cu steel compared to that of the coarse-grained (25 μm) structure, but they did not present any explanation. Huang et al. [47] have reported much lower stability of a ultrafine-grained 301 steel (GS 270 nm, deformed by ECAP and annealed for 1 h at 580–620 °C resulting in reversion) compared to a coarse-grained (GS 84 μm) counterpart during tensile straining. They suggested that micro-twins present in ultrafine grains could act as potential nucleation sites for martensitic transformation grains. However, the carbide precipitation might be the potential reason.

The annealing durations in most of the above-mentioned studies were quite long, 0.5–1 h. According to Kim et al. [165] the precipitation of carbides and Cr_2N in annealed austenite (Fe-0.2C-15Cr-15Mn-5Ni-0.2N) seems to be quite a slow process even at 700–800 °C. However, Järvenpää et al. [86] predicted that Cr_2N can start to precipitate in fine-grained austenite in a 301LN ASS within 10 s at 750 °C, and 0.15%N could be precipitated in 100 s. Evidently the precipitation can be very fast during the reversion treatment. Rajasekhara et al. [22,26] found by TEM nano-size CrxN precipitation within 1 s at 700–800 °C during reversion annealing. The precipitation took place presumably already in deformed martensite, where it is much faster than in annealed austenite. Hong et al. [167] observed that 30% tensile prestrain intensified precipitation in 347 type ASS, where carbides were formed in annealing at 650 °C around deformation bands. Knutssen et al. [130] found carbo-nitride precipitation in samples heat treated at 750 and 800 °C for 1 h after 61% cold rolling, but not after 23% reduction, indicating that high deformation was essential having the influence on diffusion rate and the rate of precipitation.

In 301LN ASS, the nitrogen is the element precipitating during reversion annealing. Lee et al. [162] reported only insignificant promoting effect of the depletion of solute atoms near high Cr-bearing particles of 10 nm in size and 3 nm Nb-bearing particles in a 70% cold-rolled Fe-0.1C-10Cr-5Ni-7.7Mn-0.3Nb steel, annealed at 664 °C for 10 min, on the total DIM fraction in tensile testing. However, Saenarjhan et al. [164] showed that the stability of austenite against martensitic transformation is enhanced with increasing C or N content, and N is a more effective element at an equivalent concentration. It is possible to estimate the potential effect of nitrogen precipitation in a 301LN ASS, considering the difference in the DIM formation between the data of Talonen et al. [155] in a 301LN with the M_{d30} = 37 °C and data reported by Järvenpää et al. [84,85,87,102] and Somani et al. [23] for another 301LN with the M_{d30} = 27 °C. These DIM vs strain curves are "Annealed" 14 μm GS and "63Cr-1050-200" 20 μm GS in Figure 11. For instance, at 0.2 true strain, the DIM fractions are about 45% and 32% in these steels, respectively. Thus, the difference of 10 °C in M_{d30} seems to result approximately in a 13% difference in the DIM fraction at 0.2 strain. Importantly, we can notice that the stability of the "Annealed 14 μm GS", and "63Cr-800-1 0.6 μm GS" steels is quite equal up to 0.25 strain.

In the Nohara equation (Equation (1)) for M_{d30}, the coefficient of N is 462, which means that the change of 0.022%N would cause the change of 10 °C in M_{d30}. However, during the reversion GS has been refined from 20 to 0.6 μm (ASTM8 to ASTM18). According to Equation (1), this GS refinement reduces M_{d30} by 1.42 × 10 = 14 °C (i.e., M_{d30} changes from 27 to 13 °C). Additionally, the M_{d30} difference of 10 °C between the steels based on the chemical compositions must be accounted, i.e., the total difference is 24 °C. With the power of 462 for N, the change of 0.052%N can cause this. This is about 1/3 of the total content of 0.15%N in the steel. Thus, we can conclude that binding of a part of N, which was available (0.15%) in the 301LN ASS studied by Järvenpää et al. [84,85,87,102], was good enough to result in the considerable stability drop, seen in Figure 11, observed between the low-temperature (precipitated) and high-temperature (no precipitates) annealed structures of this steel.

5.2.3. Elongation, Stacking Fault Energy and Deformation Mechanism

The M_{d30} temperature and SFE are relevant parameters concerning the mechanical stability of austenite towards the martensitic phase transformation. M_{d30} temperature was discussed in Section 2. It is commonly pointed out, that the SFE of a material determines the deformation mechanism and thereby affects the strain hardening behavior and ductility. The contribution of DIM formation to ductility was discussed earlier concluding that too a high fraction of DIM is not beneficial.

It is well known that a low SFE (≤20 mJ/m^2) favors the martensitic transformation (TRIP effect), whereas a higher SFE favors mechanical twinning (TWIP effect), e.g., [168]. Saeed-Akbari et al. [169] presented 20 mJ/m^2 as the upper limit for strain-induced α′-martensite formation in high-Mn steels. Allain et al. [168] suggested that with the SFE in the range 12–35 mJ/m^2, mechanical twinning would take place in addition to dislocation glide. It is well known that for instance in medium-Mn steels, the best combination of strength and ductility is obtained when both the TWIP and TRIP effects are activated in austenite grains, e.g., [170]. There are opinions that twinning plays a vital role in contributing to the excellent ductility of reversion-treated effectively grain-refined ASSs [90,149,171], where the DIM formation becomes restricted due to fine GS.

SFE of ASSs has been discussed for instance by Lo et al. [110]. Recently Noh et al. [115] considered the validity of several equations proposed in the literature for determining the SFE of ASSs and the influence of SFE on the thermal and mechanical stability of austenite. They found that the tendency for strain-induced martensite transformation was governed not by the thermodynamic stability but by the SFE, which was increased more effectively by Ni than by the equivalent amount of Mn. They developed a modified SFE equation (Equation (4)) and experimentally determined Ni equivalents which may provide a criterion for austenite stability under tensile deformation.

$$\text{SFE (mJ/m}^2) = 5.53 + 1.4[\text{Ni}] - 0.16[\text{Cr}] + 17.1[\text{N}] + 0.72[\text{Mn}], \quad (4)$$

Here the alloying elements are in mass %. Using Equation (4), the SFE of 20 mJ/m^2 was found to be the limit for stable austenite against DIM formation; hence, the same value as suggested to be the upper limit for martensite formation earlier [168,169]. For the 301LN steel used by Somani et al. [23] (see Table 1), the predicted SFE is 15.36 mJ/m^2, a value in fair agreement with that determined experimentally by Talonen (13–15 mJ/m^2) [155].

Equation (4) does not include the GS, but it is also known that in addition to temperature, fine GS increases the effective SFE [167,169]. The effective SFE is affected by grain boundaries tending to decrease the width of extended dislocations (see e.g., Mahato et al. [172]). It can be estimated from the work of Saeed-Akbari et al. [169]) that the submicron GS in reversed microstructures tends to increase the SFE of Cr-Mn and Cr-Ni ASSs by about 6 mJ/m^2. Hence, when the SFE of 301LN is 15 mJ/m^2, in a highly refined structure, the effective SFE could be 21 mJ/m^2. This is above the limit value of 20 mJ/m^2, and DIM formation might be restricted. Anyhow, we can conclude that refined GS reduces the DIM formation tendency and favors the activation of the TWIP effect instead of TRIP. This supports the arguments of Misra et al. [25,90,171] concerning the role of twinning in grain-refined ASSs.

Xu et al. [20] reported the occurrence of mechanical twinning instead of DIM in an 18Cr-8.3Ni ASS with the predicted SFE of 18.9 mJ/m^2 without accounting for any GS (≈2 μm) influence. We can conclude that the effective SFE has been about 25 mJ/m^2 for the grain-refined steel so that DIM formation is expected to be negligible. Lei et al. [51] found both TRIP and TWIP mechanisms in ultrafine-grained (GS 0.22 μm) 17Cr-6Ni-2Cu steel with the calculated SFE of 16.4 mJ/m^2 (using Noh's equation (Equation (4)) SFE is 12.6 mJ/m^2 without effects of GS and Cu, but both of them increase the SFE), so the existence of twinning is expectable. Hamada et al. [156] reported that the highest tensile elongation was achieved in coarse-grained 201 and 201L ASSs at 50 °C, where both the TRIP and TWIP effects were found to be active, apparently due to increase of the SFE by elevated deformation temperature. The SFE of the studied steels was estimated to be in the order of 20 mJ/m^2 at 50 °C.

In summary, it seems that the SFE of slightly below 20 mJ/m^2 is quite optimal for strain hardening, while twinning may contribute to high ductility with minor DIM formation in Cr-Ni and Cr-Mn ASSs. Thus, in principle the GS refinement affects beneficially the SFE of 301, 301LN and 304 steels while increasing it.

5.2.4. Strain Hardening Rate and Elongation

If we compared the SHR affected by DIM formation and the respective elongation in a tensile test, it is obvious that the elongation is not directly related to the maximum SHR or DIM fraction (Figure 12a). The latter fact was also pointed out earlier as an observation that the maximum elongation generally corresponds to DIM fractions below 20%. However, it seems that we can find a relationship between the tensile elongation (fracture strain) and the strain corresponding to the maximum SHR (called a peak strain), as demonstrated in Figure 12b. The figure includes data from tensile tests at different temperatures [154–156] and different grain sizes [83,85,102]. In both instances, the elongation increases with the increasing peak strain. From the relationship, it can be concluded that the maximum SHR should be reached as late as possible to delay necking (though before necking), and generally the DIM formation tends to occur too early, at too small strains. According to the data in Figure 12b, there are no cases where the austenite is too stable (concerning the ASS grades 201, 201LN, 204Cu, 301 and 301LN). Consistently, in medium-Mn steels with duplex structure, superior mechanical properties are obtained when the volume fraction and stability of austenite are maximized [170].

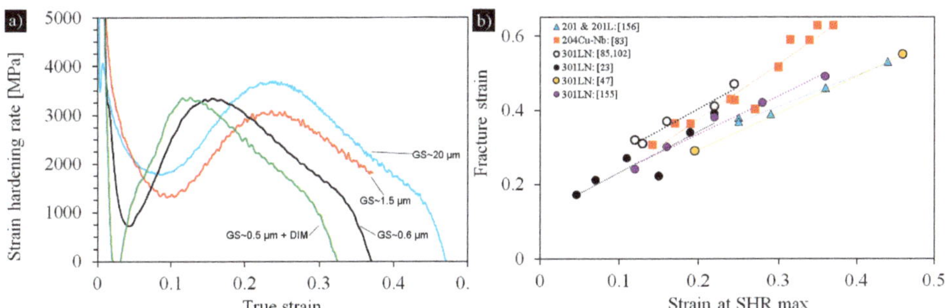

Figure 12. Strain hardening rate curves for selected AISI 301LN structures [85,87] (**a**) and the relation between fracture strain and the strain at the peak SHR (**b**). SHR = strain hardening rate.

Considering the reason for the SHR exhibiting a peak, according to Marechal [75], the kinetics of austenite to martensite phase transformation dominates at the strains before the peak, whereas at the peak and beyond, the apparent work hardening of the α′-martensite has the dominant effect on SHR. Thus, the rate of DIM formation should not be too fast but gradual.

5.2.5. Discontinuous Yielding

Discontinuous yielding, Lüders strain, has been frequently observed in ultrafine-grained ASSs [51,80,85,87,160,173]. Conventionally, the yield point is connected with a high density of mobile dislocations formed by unlocking pinned dislocations and rapid dislocation multiplication, when the material within the deformation band effectively softens and undergoes localized plastic deformation [174,175]. Lüders-type behavior has been detected in an ultrafine-grained Fe-C alloy with carbide particles [176] and Matsuoka et al. [160] reported it in 16Cr-10Ni ASS with 1 μm GS. Lei et al. [51] found a Lüders strain of 15% in an Fe-0.06C-17Cr-6Ni-2Cu with GS ≈ 220 nm. The appearance of a yield point with low subsequent strain hardening also happens in austenitic TWIP-type steels in connection with GS refinement down to a few microns [177].

Gao et al. [173] observed a very long Lüders strain of 0.4 in tensile deformation of an ultra-fine grained (bimodal GS, 1 μm and 0.2 μm) 304 ASS fabricated by a two-step cold rolling and annealing process. The authors explained the observation by intense formation of martensite, promoted by the incremental strain localization in the band region. After that, the strain localization region started to propagate to the undeformed region, leading to the Lüders-type deformation and also the high subsequent uniform elongation. Marechal [75] demonstrated the presence of 24% stress plateau in a stress–strain curve of a 301LN, cryorolled (100% DIM) and annealed at 750 °C for 30 min, forming a partially recrystallized fine grain structure without DIM.

Järvenpää et al. [29,87] showed that the discontinuous yielding in a 301LN ASS became more pronounced with decreasing GS obtained with decreasing annealing temperature so that the Lüders strain disappeared when the GS was larger than about 1 μm, and the annealing temperature increased to 900 °C. Typical stress–strain curves revealing that are displayed in Figure 13. At 690 °C the reversed GS is below 1 μm and Lüders strain of even more than 10% appears, though dependent on the presence of retained phases DIM and DA and the fraction of fine grains (affected by the prior cold rolling reduction).

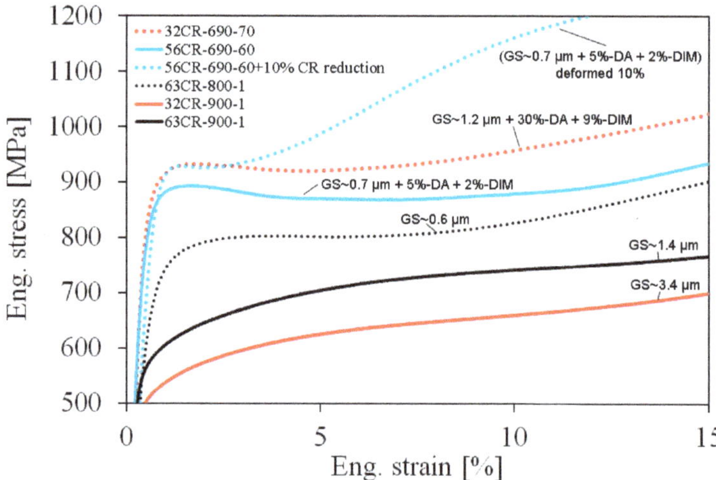

Figure 13. Initial parts of stress–strain curves of various reversed structures of a 30LN steel. Data from References [29,87]. Legend: Reduction (%CR), temperature (°C), duration (s).

Thus, the discontinuous yielding phenomenon in 301LN ASS could be connected with effective GS refinement or/and precipitation occurred at low annealing temperatures. Järvenpää [102] suggests that in addition to submicron-sized GS, the imposing softening and Lüders-type yielding are enhanced by the interaction between dislocations and fine precipitates, i.e., pinning effects. Marechal [75] explained and modelled the appearance of the discontinuous yielding with two factors: DIM formation at small strains (transformation strain contribution) and, as a key factor, the relative flow stress difference between work hardened fine-grained austenite and DIM, resulting in a negative strain hardening rate contribution balancing the influence of the work hardening of austenite. Accordingly, when the GS of austenite is sufficiently reduced to increase its strength and the transformation rate of DIM is also sufficient (affected possibly by the precipitation in a 301LN), though not particularly high, the strain localization will be induced.

At any rate, the discontinuous yielding of a partially reversed structure can be largely diminished by subsequent 10% cold rolling reduction, as demonstrated in Figure 13 [87]. Obviously, new mobile dislocations are created by this plastic deformation.

5.2.6. Tensile Properties of Temper-Rolled Reversed Structures

As mentioned, temper rolling is a common method to improve the strength of annealed ASSs. The influence of small cold rolling reductions on the tensile properties of reversed structures of 301LN grade has rarely been investigated. However, in a study [87], it was found that a 10% cold rolling reduction increased the YS of the low-temperature annealed (690 °C–60 s) structure only faintly (see Figure 13), although the DIM fraction grew by 10%. A 20% reduction increased the YS slightly more efficiently, but the improvement, in spite of 38% DIM, was still very modest in comparison with the effect of the same cold rolling reduction on a commercial steel containing 29% DIM. It is obvious that the influence of the DIM formation on the strength is more pronounced in a soft coarse-grained structure than in a strong reversed structure, but also the work-hardening of the soft coarse austenite grains must be much higher than the work-hardening of refined reversed grains.

Mao et al. [178] have recently reported that a good combination of strength and ductility can be obtained in a fine-grained (average GS 5 μm) 316L ASS by a cold rolling reduction of 30%. The steel sheet exhibited the YS and TS of 1045 and 1080 MPa, respectively, but with a short uniform elongation of 7%. Jung and Lee [179] applied warm rolling of 40% reduction at 500 °C to a 10.30Cr–8.14Ni-7.41Mn ASS, previously 75% cold rolled and annealed at 663 °C for 5 min to obtain a fine GS of 2 μm. This deformation increased the YS from 401 to 736 MPa (TS 1076 MPa, TE 25%). This observation on a significant improvement of the YS is somewhat different from those of Järvenpää et al. [87], but the rolling reduction used was also much higher.

5.3. Fatigue Behavior

5.3.1. Fatigue Strength

The effect of the GS refinement on fatigue strength has been studied extensively in nano-structured materials and also in metastable ASSs [43,60,64,74,87,180–196]. GS can be refined by various means in addition to the reversion, too. For instance, Ueno et al. [181] used equal channel angular pressing to obtain nanosized grain structure in a 316L ASS and observed a beneficial effect of GS refinement on fatigue strength. A drastic enhancing influence of nanocrystalline structure created by an ultrasonic attrition treatment has been reported for 316L and 301LN ASSs in bending fatigue cycling [185]. In this instance, in addition to the refined GS, compressive residual stresses and DIM formation during the attrition treatment can contribute to the fatigue strength improvement.

Examples of the results of the studies on the fatigue strength of reversion-treated ASSs are plotted in Figure 14, the strain amplitude–fatigue life curves in Figure 14a and the stress amplitude–fatigue life curves in Figure 14b. The influence of GS is not pronounced as based on strain amplitude (Figure 14a) compared to that if based on stress amplitude (Figure 14b). Di Schino et al. [43] investigated the high cycle fatigue (HCF) behavior of the N-alloyed (16.5Cr-1.0Ni-11.5Mn-0.3N) steel in axial load-controlled tests as a function of the GS, but they did not found any influence of grain refinement on the fatigue limit (2×10^6 cycles). This was explained in terms of the formation of slip bands on the steel surface promoted by nitrogen alloying and crack initiation in them. On the contrary, a slight beneficial effect of GS refinement was detected in the fatigue limit of a standard 304 ASS, as seen in Figure 14b.

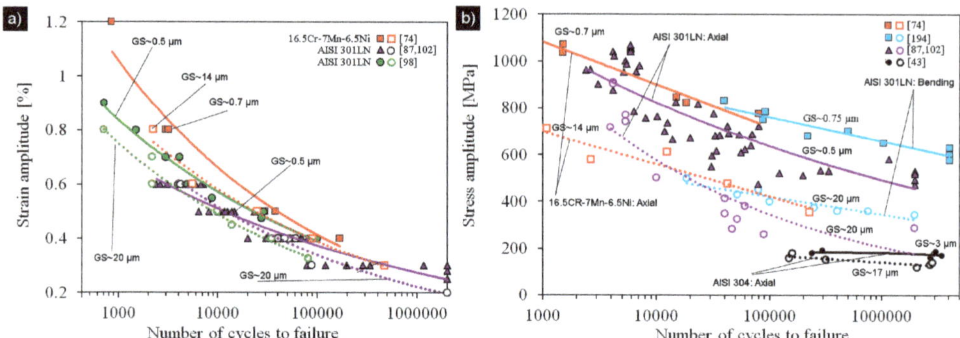

Figure 14. Fatigue life of reversion-treated 301LN in total strain-controlled tests compared to coarse-grained or annealed steel based on (**a**) strain amplitude and (**b**) mid-life stress amplitude. Data for 304 and 16.5Cr-7Mn-6.5N are from References [43] and [74], respectively.

Hamada et al. [194] noticed a significant improvement in the fatigue life of a reversion-treated 301LN steel (63% cold rolling reduction, annealing at 800 °C for 1 s; average GS ~0.75 μm) under bending fatigue loading compared that of the coarse-grained (GS ≈ 20 μm) counterpart (number of cycles to fracture more than 10^4 cycles). As evident from Figure 14b, the fatigue limit was significantly increased from 350 MPa to 630 MPa, reaching 59% of the TS. Planar dislocation structure typical to low SFE metals was found in fatigued coarse-grained steel, with intense formation of slip bands and crack propagation along these bands and grain boundaries. Contrary to that, fatigue loading created few deformation features, and fatigue damage was observed to occur by grain boundary cracking in the fine-grained counterpart. A small amount of DIM was formed in the fine-grained structure, whereas in the coarse-grained steel, appreciable amount of martensite was detected, and hardness increased about 47% during cycling. It should be noted that due to a high cycling frequency of 23 Hz, specimens tended to be heated during fatigue loading, which may have affected the austenite stability. Due to higher strength of the fine-grained steel, the adiabatic heating might be higher in fine-grained than in coarse-grained specimens.

Fargas et al. [64] carried out axial fatigue tests with a 301LN ASS under load control (R = 0.1) and found that GS refinement achieved by reversion treatments led to fatigue limits clearly higher than that of annealed coarse-grained counterpart. Interestingly, even a 20% cold rolling reduction providing 28% DIM was enough for this improvement. They also observed that after the fatigue tests, the amount of DIM was close to 44% for the coarse-grained (GS 11.7 ± 4.1 μm) steel, while the reverted samples (40% cold rolling reduction before annealing and GS 2.3 ± 1.5 μm) showed less than 23% of DIM. Hence, the stability of austenite increased with decreasing GS and less slip bands were also observed in fine-grained structures, both observations being consistent with those by Hamada et al. [194].

Liu et al. [182] used the tension-tension loading (R = 0.2 at 95 Hz) for an 18Cr-8Ni ASS (70% cold rolling; annealing 710 °C–10 min, 760 °C–5 min and 950 °C–5 min). They obtained the fatigue strengths of 811, 568 and 501 MPa for the reversion-treated structures with GS of 400 nm, 1.4 μm, and 12 μm, respectively, i.e., a distinct improvement in fatigue strength due to the GS refinement. The respective YS values were 878, 571 and 316 MPa so that the fatigue strength was lower than the YS for submicron-grained steel but higher for the coarse-grained one. After fatigue testing, distinct slip bands were formed on the surface of the coarse-grained sample, while shallower slip bands were present on fine-grained surfaces. The surface of the submicron-grained sample was smooth and without slip bands or other deformation marks. This sample also showed the smallest increase in hardness in fatigue, only 8%, whereas the hardness increased up to 80% in the coarse-grained counterpart. In the submicron-grained structure, a very small amount of martensite and dislocations were generated. Moreover, these observations on surface deformation and the higher stability of the refined structure are in agreement with those reported earlier by Hamada et al. [194] and Fargas et al. [64] for 301LN ASS.

In strain-controlled fatigue testing, nano- or ultrafine-grained materials usually exhibit longer fatigue lives in the HCF regime and shorter fatigue lives in the LCF regime compared to their coarse-grained counterparts due to their higher strength and lower ductility, respectively (see e.g., Höppel et al. [180]). However, as shown in Figure 14a, Man et al. [184,196] and Chlupova et al. [99] reported slightly higher fatigue resistance of the ultrafine-grained structure compared to that of coarse-grained counterpart of a 301LN ASS in the range 10^4–10^6 cycles. Moreover, Droste et al. [74] reported an improvement of the fatigue life due to the GS refinement (from 28 to 0.7 μm) both in LCF and HCF regimes in a metastable 16.5Cr-6.5Ni-7Mn ASS under total strain amplitude loading. They suggested that the small difference in hardness between austenite and DIM in the ultrafine-grained partially-reversed structure would lead to more homogeneous strain distribution during cycling, whereas in the coarse-grained structure, strain is localized in the austenite phase. The data of Järvenpää et al. [87,102] also inserted in Figure 14a, indicate hardly any influence of GS on the LCF life of the same steel. However, in a comparison based on the mid-life stress amplitudes, the reversed fine-grained structure was distinctly better than the coarse-grained counterpart (Figure 14b). In the HCF regime, the fatigue strength, i.e., the stress amplitude level at 10^6 cycles, was in the range of 460–570 MPa for the reversed structures (GS ≈ 0.6–3.4 μm), being even more than twice higher compared to corresponding value of the annealed coarse-grained (≈20 μm) structure.

Järvenpää et al. [29,87] have shown that low prior cold rolling reductions before the reversion annealing can lead to fatigue strength identical to that achieved after the 63% reduction. For the highest fatigue, strength of over 500 MPa was obtained in the partially reversed 32% and 56% cold-rolled structures created at lower temperatures with longer holding times. The fatigue strength was practically independent of the loading direction relative to the rolling direction. Mateo et al. [190] reported that even a cold rolling reduction as low as 20%, resulting in a very non-homogeneous reversed structure (the mean GS below 3 μm), led to 36% improvement (i.e., 100 MPa) of the fatigue limit compared to that of an annealed 301LN. However, the fatigue limit remained lower than that of a 20% temper-rolled sheet.

Similarly, as in the case of tensile properties, subsequent cold rolling deformation has been found to have only a slight effect on the fatigue strength of the partially reversed structure in a 301LN ASS, whereas the improvement is much more pronounced in the coarse-grained steel [87,102]. However, the fatigue limit (at 10^6 cycles) for the low-temperature reversed structure is about 600 MPa without any temper rolling, corresponding to the fatigue limit of 20% temper-rolled coarse-grained structure.

In summary, it can be noticed that the fatigue strength of reversion-treated ASSs has been extensively studied for Cr-Ni and few Cr-Ni-Mn grades. Research on Cr-Mn type steels seems to be lacking. The results indicate the fatigue strength is highly improved by the GS refinement, as compared on the basis of stress amplitude. As is typical, the comparison based on strain amplitude does not show significant influence of GS. The influence of GS can be seen in the formation of slip bands and crack initiation.

5.3.2. Cyclic Stability

In metastable ASSs DIM forms during tensile testing and also during fatigue loading. Hard martensite in the soft austenitic matrix is known to improve fatigue strength; e.g., [60,64,98,186–192]. The cyclic deformation-induced martensitic transformation depends on the temperature, total strain amplitude and the cumulated plastic strain [186], and significantly on GS as pointed out in the previous section.

It has been shown that an increase in the DIM fraction leads to higher fatigue strength in the HCF regime due to increased strength, but relatively, LCF strength is decreased due to impaired ductility. It has been argued that there is an upper limit to martensite fraction in improving the fatigue resistance in the HCF regime, but the exact value of this limit seems to be unclear. Topic et al. [191] showed that in a 304 ASS (coarse-grained condition), pre-formed DIM is beneficial for fatigue life if its fraction is below 20%. Mateo et al. [60,190] found that even a 28% DIM fraction did not impair fatigue strength and Fargas et al. [64] reported a clear improvement even with a 38% DIM fraction (obtained by 40% reduction of a coarse-grained annealed sample) in cold rolled specimens.

Initial cyclic softening followed by pronounced cyclic hardening has been observed in cyclic loading in a coarse-grained annealed 304 ASS. The cyclic hardening is connected with DIM formation during dynamic straining, e.g., [192]. Poulon et al. [195] observed in a 15.38Cr-9.06Ni-3.06Mo ASS, with a fine 2.5 μm and coarse 21 μm GS, a slight cyclic softening followed by a secondary cyclic hardening, whatever the GS. The secondary cyclic hardening was more intense and started earlier at high strain amplitudes, being promoted especially in the coarse-grained steel. During the hardening stage, in grains larger than 2 μm, the formation of DIM occurs inside austenite grains on dislocation structures with the presence of nuclei on cell walls or dipolar walls. However, in fine grains, where plasticity activity occurs at grain boundaries, an intragranular dislocation structure is not a prerequisite for martensite nucleation but nuclei can appear along grain boundaries.

Biermann et al. [186] and Droste et al. [74] divided cyclic behavior of a 16.5Cr-7Mn-6.5Ni ASS into three stages: a primary hardening followed by a stage of cyclic softening being related to an increasing dislocation density and their interactions in the beginning of cyclic deformation and a rearrangement and annihilation of dislocations in the stage of cyclic softening and, finally, a secondary hardening due to the martensitic transformation. The secondary hardening increases with increasing strain amplitudes, and the onset occurs at a lower number of cycles. Increasing temperature reduces the degree of secondary hardening. In a recent study, laser beam annealing has been used to reverse the DIM in this steel, and cyclic hardening behavior and fatigue properties investigated [193]. Earlier, laser annealing was used for the reversion in a 301LN steel [197].

In studies of Järvenpää et al. [84,87], Chlupova et al. [99] and Man et al. [196], a minimal softening was found in reversed structures of a 301LN ASS at a strain amplitude of 0.4%, as seen in in Figure 15a. The same has reported by Droste et al. [74] for a 16.5Cr-7Mn-6.5Ni ASS. The softening was followed by slight hardening of the coarse-grained structure. At a higher total strain amplitude of 0.6%, both the coarse-grained and fine-grained reversed structures experienced unstable behavior, characterized by a pronounced hardening period, as illustrated in Figure 15a. A slight softening took place before that in the fine-grained structure, but not in the coarse-grained counterpart. However, the fine-grained reversed structures, which were strong as shown by the stress amplitude level (initially about 750 MPa), exhibited much less cyclic hardening than the coarse-grained structure, which had a low initial stress amplitude (about 320 MPa). Therefore, the stress amplitudes were close to each other after about 5000 cycles. DIM transformation occurred over the course of cycling after certain incubation periods (Figure 15b), and obviously, this is connected with cyclic hardening. However, in spite of a very different amount of hardening between the structures with these different GSs, the amount of DIM formed in cycling was high but quite similar among the studied structures (about 80% DIM after 5000 cycles at 0.6%). Thus, the influence of GS on DIM formation was not pronounced, which is somewhat contrary to the observations from load-controlled tests mentioned in the previous section. However, the influence of a given DIM fraction on cyclic hardening is much more significant in soft coarse-grained austenitic structure than in strong fine-grained structure, similarly as reported by Droste et al. [74]. Järvenpää et al. [84] have discussed this.

In summary, after initial slight softening or constant stage, cyclic hardening takes place in grain-refined as well as coarse-grained ASSs, which can be attributed to DIM formation. The amount of DIM is independent of the GS, but the degree of cyclic hardening is more pronounced in soft coarse-grained austenite than in grain-refined austenite. The background of the phenomena is quite well understood.

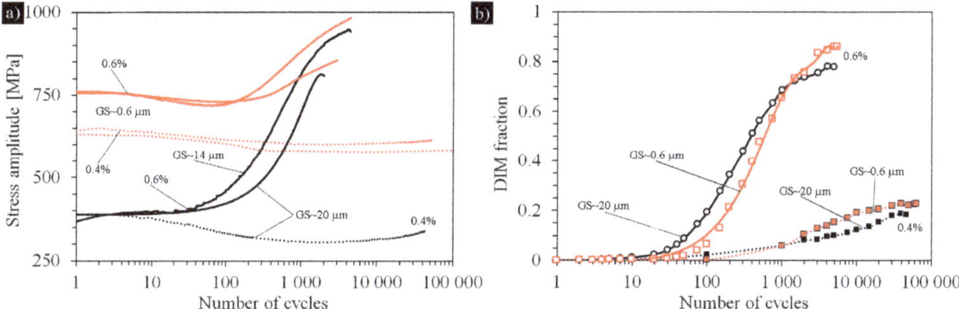

Figure 15. Cyclic medium-life stress amplitude (**a**) and DIM fraction (**b**) vs number of cycles curves of coarse-grained and reversed fine-grained 301LN at 0.4% and 0.6% strain amplitudes. Data from References [87,99,102].

5.4. Stretch Formability

The formability properties of reversion-treated structures are scarcely reported in the literature. Tensile ductility of ASSs is high, so the formability can be expected to remain on a good level in spite of the GS refinement and high strength. In one study, the stretch formability properties of the reversion treated 3 mm thick sheets of a 301LN ASS with the YS in the range 682–998 MPa, and temper-rolled structures of 301LN and 301ASSs, for reference, have been determined by the Erichsen cupping tests [198]. The results indicated that the stretch formability of the reversion-treated structures is identical or slightly better than that of the temper-rolled counterparts at a given YS level. Furthermore, the stretched cup surface of reversion-treated sheets is smoother than the surface of the cups of commercial temper-rolled sheets. This is in accordance with a technical data from Nippon Steel & Sumitomo Metal reporting excellent strength and formability for grain-refined thin sheets of AISI 304 (SUS304 BA1) and 301L (NSSMC-NAR-301L BA1) [108,109]. Further tests on the formability have been carried out by the authors but not published yet, and the study continues.

5.5. Weldability

Welding of steel sheets is often needed for constructive applications. In fusion welding processes, cold-strengthened ASSs tend to soften at the heat-affected zone (HAZ), and therefore, welding with restricted heat inputs must be applied. Weldability of reversion-treated ASSs has been investigated very scarcely. In a very recent study [199], it has been shown that autogenous laser welding of a reversion-treated ultrafine-grained (YS ~ 670 MPa) 301LN steel sheet (3 mm thick) created a weld metal with coarse columnar grains and a very narrow HAZ, similarly as happens in a temper-rolled (YS ~ 740 MPa) sheet [200–202]. Figure 16 displays an example of a laser welded joint and its microstructure and hardness distribution in a 3 mm 301LN steel. The weld metal has a low hardness (240 HV), naturally identical in the both sheets corresponding to the hardness of a hot-rolled sheet. The HAZ in the reversion-treated sheet exhibited grain coarsening, but in the temper-rolled counterpart, the GS is locally refined due to the recrystallization of the cold-rolled structure [201,202]. However, the grain structure of the HAZ was still finer in the reversion-treated sheet and the softened zone was slightly narrower. The YSs determined of the specimens cut across the weld seams were equal (629 MPa), but the elongation was clearly better in the seam of the reversion-treated sheet. The YS of the laser welded sheets was about double compared to that (330 MPa) of the hot rolled AISI 301LN sheet. However, the fatigue limits were quite equal in the hot rolled and laser-welded reversion-treated and laser-welded temper-rolled sheets.

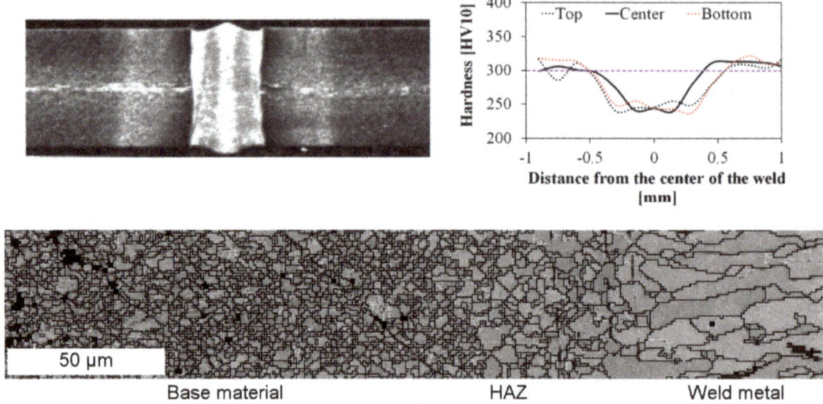

Figure 16. Cross-section of a laser weld on a reversion-treated 301LN sheet, hardness profile and grain size distribution. The hardness profile is from Reference [199].

5.6. Corrosion Resistance

High corrosion resistance is one of the most important properties of ASSs. Lv et al. [203] have investigated the effect of cold rolling temperature in the range from −196 °C to RT on corrosion resistance of a 304L ASS by potentiodynamic polarization tests in a borate buffer solution. In rolling, the austenitic microstructure was transformed completely to martensite. Results showed that samples cold rolled at −120 °C had an equiaxed GS of about 300 nm, which displayed the best corrosion performance. Moreover, the reverse transformation treatment was performed by annealing at 850 °C for 180–270 s, but no results were given concerning the reversed structures.

Han et al. [204] showed that a refinement of austenite GS of a 304 ASS strip by a thermomechanical treatment improved the TS and particularly the YS and also increased the intergranular corrosion resistance. This was explained as a result from the increased grain boundary area and thereby a decreased degree of the Cr depletion caused by carbide precipitation in the steel.

Di Schino et al. [41,42] have investigated the effect of refined GS on various properties of a low-Ni nitrogen alloyed Cr-Mn (Fe-0.037C-18.5Cr-1.07Ni-11.4Mn-0.37N) ASS. In the experiments, cold rolling of 75% reduction created about 45%DIM, and after the reversion annealing at 900 °C, GS varied from 2.4 μm after 10 s to 3 μm after 10 ks holding. The general corrosion rate of the ASS with 2.5 to 40 μm GSs was measured in 5% H_2SO_4 boiling solution for 10 h. It was found that the corrosion rate increases with decreasing GS. It was concluded that increasing the grain boundary surface area by grain refining causes destabilization of the surface passive film. The intergranular corrosion rate was estimated by immersing the samples in H_2SO_4 -$FeSO_4$ (Streicher solution) for 10 h. The intergranular corrosion rate was found to decrease with decreasing GS. As GS is refined, the grain boundary area per unit volume increases and the degree of the Cr depletion caused by carbide precipitation will decrease for a given carbon content. The pitting corrosion rate was measured in a 10% $FeCl_3$-$6H_2O$ solution at RT for 10 h. In contrast to the reduction of the general corrosion resistance, the grain refining led to improved pitting corrosion resistance in the low Ni-nitrogen alloyed steel as well as in a 304 ASS used for the reference. The pitting of ultrafine grained steel was found to initiate in several sites but with small individual pits, whereas in the large-grained steel pits were coarse and deep though lower in number. Moreover, the pitting potential was found to shift towards more noble potentials with decreasing GS.

Consistent with the result of Di Schino et al. [41,42], also Hamada et al. [205] reported that in a 301LN ASS the submicron-grained structure, formed upon reversion annealing at 800 °C for 1 s, exhibited a better pitting corrosion resistance than the coarse-grained structure. This was attributed to the preferential and faster pitting attack on grain boundaries in the coarse-grained steel, presumably due to more severe impurity segregation or chromium-nitride precipitation than in the ultra-fine-grained structure.

As discussed in Section 5.2.2, precipitation of tiny Cr_xN particles has been detected in 301LN ASS, occurring presumably in DIM before the reversion at 600–800 °C [22,26]. Hence, some concern can arise if the precipitation impairs the corrosion resistance, even though the improvement in intergranular corrosion resistance was reported [41,42], as mentioned above. The unpublished research on this issue performed at the University of Oulu has indicated that there exists no risk to intergranular corrosion in the instance that no DIM is present, i.e., after complete reversion. Electron microscopy studies have revealed that particles locate inside austenite grains, not along grain boundaries [22,26,86], obviously due to precipitation in deformed martensite before its transformation. However, the presence of more than 2.7% DIM seems to degrade the corrosion resistance of low-temperature reversed structures. According to the Strauss test, sensitization also in a 204Cu ASS has been found possible after reversion annealing at 700 °C for 100–200 s (the DIM content 2.5% to 3%).

In summary, the general corrosion resistance is impaired, but the pitting corrosion resistance is improved by refined GS. Retained DIM may be detrimental for corrosion resistance, and some concern exists regarding the sensitization of ASSs with nitrogen alloying or high carbon content. Further investigations might be performed to further clarify the issue.

5.7. Properties for Medical Applications

Additionally, metals or alloys with nano- or ultrafine-grained structures are attractive materials for biomedical applications. Higher strength is one property desired and can be increased by the reversion treatment, but also, their osteoblast activity seems to be improved by the GS. To enhance osseointegration of orthopedic implants, the surfaces of implants, which are in direct contact with bone tissue, are of importance. If the implant surfaces provide a better environment for bone cell functions, then the integration of the implant with the juxtaposed bone tissue can be improved. By modifying biomaterial surfaces to possess features having nanophase topography, implant surfaces mimic the feature size of natural tissues, providing a more realistic niche to promote cellular functions on implant surfaces. It seems that a steel surface with nano-/submicron-sized GS is this kind of advanced topography without the risk of delamination as is the case with anodized oxide films, for instance. The favorable enhancement of osteoblasts functions and cellular attachment on nano-grained surface is attributed to ultrafine GS, i.e., the availability of greater open lattice in the position of high angle grain boundaries and high hydrophilicity.

Professor Misra's group together with researchers at the University of Oulu has investigated reversion-treated 301LN ASS for medical purposes [206–210]. For instance, Venkatsurya et al. [207] compared osteoblast response of grain boundary grooved and planar nano-/ultrafine-grained surfaces. Nune et al. [208] studied the influence of GS on osteoblast differentiation and mineralization. The determining role of grain refined structure on osteoblasts functions was fundamentally defined so that surface roughness and grain structure are the contributing factors that strongly influence protein adsorption on substrates and consequently cell attachment; proliferation and expression level of actin, vinculin and fibronectin [209]. Recently, the corresponding behavior of a 304 ASS has been investigated in another research group [211], and similar features as for the 301LN ASS have been reported.

Tufan et al. [212] obtained by machining substantial grain refinement up to 98.2 nm on the 316L ASS surfaces. Biological experiments showed nearly 280% increase in the bone cell density on turned samples compared to the as-received 316L stainless steel at the 5th day of culture, which was explained with enhanced hydrophilicity of the turned sample as a result from the combined effect of surface nanocrystallization and rougher nanoscale surface topography.

It is well known that a Cu-alloyed 304 ASS has antibacterial properties when Cu precipitates in surface layers and release Cu ions, e.g., [213,214]. In annealed condition, the strength properties of this steel are relatively low (YS ≈ 250 MPa), but they can, evidently, be improved by refining the GS by the reversion treatment, as shown for a 304 ASS in numerous studies. The Cu precipitation in annealed austenite needed for creating the antibacterial property requires quite a long aging time, but it can take place at temperatures 650–800 °C [214]. At these temperatures, the grain growth in ASSs is negligible, as discussed in Section 4.1. Further, Cu precipitation in deformed DIM can be supposed to be a much faster process than in annealed austenite. Thus, the micron-scale reversed GS can be retained during the aging treatment. The study on this issue has just been finished, and the results will be published soon.

The integration of cellular and molecular biology with material science and engineering as described here provides a route to modulate cellular and molecular reactions in promoting osteoinductive signaling of surface adherent cells.

6. Conclusions

A large number of laboratory experiments have demonstrated that many commercial metastable Cr-Ni and Cr-Mn ASS grades can, even after a reasonably low cold rolling reduction (≈20% to 50%), be reversion treated in a temperature region of 600–900 °C for a desired short duration, and excellent yield strength-elongation combinations and high fatigue strength are obtained. The mechanical properties surpass those of temper-rolled sheets. Moreover, the formability, corrosion and biomedical properties are enhanced by the reversion treatment, although these properties have been much rarely investigated. Weldability issues are most scarcely covered yet. After this intense research, the topic of reversion phenomenon is quite matured and well understood, even though many details depend on the exact chemical composition of the steel and processing variables (cold rolling degree, annealing temperature, etc.) The highly refined grain size is the main factor for the enhanced properties, but also partly reversed structures with retained martensite and austenite can exhibit excellent strength-ductility combinations, although the grain size remains coarser and non-uniform. Numerous processing factors affect the properties so that they must be adjusted for the case sensitively and the robustness of the processing checked. However, studies demonstrate that the reversion processing route could be adopted industrially for manufacturing austenitic stainless steels with advanced properties.

Recently it has been reported that the reversion-treatment approach can be adopted also to high-Mn-TRIP steels. Various phenomena and features, found during the reversion treatment of ASSs and discussed in previous sections, can exist in high-Mn-TRIP steels (Mn > 15%), too. In cold rolling of certain Fe-Mn-C alloys, e-martensite can form and reverse in subsequent annealing [215,216]. In low-carbon high-Mn steels, two types of α'-martensite have been found to exist after cold rolling with retained austenite [217–220]. The reversion of α'-martensite takes place during annealing in two temperature ranges, together with a new phenomenon of Mn partition, not reported in the reversion stage in ASSs. Both shear and diffusional reversion mechanisms can occur, and GS refinement can be effective in austenite formed from cell-type martensite while ineffective in austenite reversed from lath-type martensite. During deformation, TRIP and also TWIP mechanisms can occur. Thus, this field is rich in details to be further investigated.

Reversion has been reported in high-entropy alloys [221], with the chemical composition metastable enough for deformation-induced martensite formation during cold rolling. This will expand the research field for studying reversion-affected phenomena and related properties.

Author Contributions: Conceptualization, data curation, draft preparation, A.J., M.J. and A.K.; visualization, A.J.; writing—review and editing, supervising, P.K. All authors have read and agreed to the published version of the manuscript.

Funding: This research was funded by the ACADEMY OF FINLAND, grant number 311934.

Acknowledgments: Outokumpu Stainless Oy is thanked for providing the commercial material and for long-lasting support for the reversion-related research at the University of Oulu.

Conflicts of Interest: The authors declare no conflict of interest.

Nomenclature

α	Parameter in Olson–Cohen model (Equation (3))
α'-martensite	Deformation-induced body-centered cubic martensite
β	Parameter in Equations (2) and (3)
ε-martensite	Hexagonal epsilon martensite
ε	Strain in Equations (2) and (3)
$\varepsilon_{m'}$	Abscissa of the mid-point of a sigmoidal curve (Equation (2))
σ_o	Friction stress in the Hall–Petch relation
A_f	Finish temperature of the martensite to austenite reversion
A_s	Start temperature of the martensite to austenite reversion
ASS	Austenitic stainless steel
CR	Cold rolling
DA	Deformed austenite, retained in cold rolling
DIM	Deformation-induced martensite (α'-martensite)
FCC	Face-centered cubic
$f_{\alpha'}$	Martensite fraction in Equation (2) and Olson–Cohen model (Equation (3))
f_s	Saturation martensite fraction in cold rolling deformation (Equation (2))
GS	Grain size
HAZ	Heat affected zone
HCF	High cycle fatigue
m	Exponent in Olson–Cohen model (Equation (3))
M_{d30}	Temperature where 50% a'-martensite is formed by 30% true strain
M_s	Starting temperature of martensite transformation
LCF	Low cycle fatigue
R	R ratio, minimum peak stress divided by the maximum peak stress in fatigue cycling
$R_{p0.2}$	Proof strength, yield strength
RT	Room temperature
SHR	Strain hardening rate
TE	Tensile elongation, fracture strain
TS	Tensile strength
YS	Yield strength

References

1. *Stainless Steel: Tables of Technical Properties*, 2nd ed.; Euro Inox: Brussels, Belgium, 2007; Volume 5, ISBN 9782879972428.
2. Mechanical behaviour and design values of properties. In *Design Manual for Structural Stainless Steel*; Euro Inox/The Steel Construction Institute: Brussels, Belgium; Chicago, IL, USA, 2006; p. 146. ISBN 2-87997-204-3.
3. Taulavuori, T.; Aspegren, P.; Säynäjäkangas, J.; Salmén, J.; Karjalainen, L.P. The anisotropic behaviour of the nitrogen alloyed stainless steel grade 1.4318. In Proceedings of the 7th High Nitrogen Steels Conference, Ostend, Denmark, 19–22 September 2004; pp. 405–412.
4. Karjalainen, L.P.; Taulavuori, T.; Sellman, M.; Kyröläinen, A. Some strengthening methods for austenitic stainless steels. *Steel Res. Int.* **2008**, *79*, 404–412. [CrossRef]
5. Mazza, B.; Pedeferri, P.; Sinigaglia, D.; Cigada, A.; Fumagalli, G.; Re, G. Electrochemical and corrosion behaviour of work-hardened commercial austenitic stainless steels in acid solutions. *Corros. Sci.* **1979**, *19*, 907–921. [CrossRef]
6. Barbucci, A.; Delucchi, M.; Panizza, M.; Sacco, M.; Cerisola, G. Electrochemical and corrosion behaviour of cold rolled AISI 301 in 1 M H_2SO_4. *J. Alloys Compd.* **2001**, *317–318*, 607–611. [CrossRef]
7. Mudali, U.K.; Shankar, P.; Ningshen, S.; Dayal, R.K.; Khatak, H.S.; Raj, B. On the pitting corrosion resistance of nitrogen alloyed cold worked austenitic stainless steels. *Corros. Sci.* **2002**, *44*, 2183–2198. [CrossRef]
8. Gavriljuk, V.G.; Berns, H.; Escher, C.; Glavatskaya, N.I.; Sozinov, A.; Petrov, Y. Grain boundary strengthening in austenitic nitrogen steels. *Mater. Sci. Eng. A* **1999**, *271*, 14–21. [CrossRef]
9. Simmons, J.W. Overview: High-nitrogen alloying of stainless steels. *Mater. Sci. Eng. A* **1996**, *207*, 159–169. [CrossRef]

10. Di Schino, A.; Barteri, M.; Kenny, J.M. Development of ultra fine grain structure by martensitic reversion in stainless steel. *J. Mater. Sci. Lett.* **2002**, *21*, 751–753. [CrossRef]
11. Di Schino, A.; Salvatori, I.; Kenny, J.M. Effects of martensite formation and austenite reversion on grain refining of AISI 304 stainless steel. *J. Mater. Sci.* **2002**, *37*, 4561–4565. [CrossRef]
12. Srikanth, S.; Saravanan, P.; Kumar, V.; Saravanan, D.; Sivakumar, L.; Sisodia, S.; Ravi, K.; Jha, B.K. Property Enhancement in metastable 301LN austenitic stainless steel through strain-induced martensitic transformation and its reversion (SIMTR) for metro coach manufacture. *Int. J. Metall. Eng.* **2013**, *2*, 203–213. [CrossRef]
13. Ravi Kumar, B.; Sharma, S.; Kashyap, B.P.; Prabhu, N. Ultrafine grained microstructure tailoring in austenitic stainless steel for enhanced plasticity. *Mater. Des.* **2015**, *68*, 63–71. [CrossRef]
14. Forouzan, F.; Najafizadeh, A.; Kermanpur, A.; Hedayati, A.; Surkialiabad, R. Production of nano/submicron grained AISI 304L stainless steel through the martensite reversion process. *Mater. Sci. Eng. A* **2010**, *527*, 7334–7339. [CrossRef]
15. Moallemi, M.; Najafizadeh, A.; Kermanpur, A.; Rezaee, A. Effect of reversion annealing on the formation of nano/ultrafine grained structure in 201 austenitic stainless steel. *Mater. Sci. Eng. A* **2011**, *530*, 378–381. [CrossRef]
16. Rezaee, A.; Kermanpur, A.; Najafizadeh, A.; Moallemi, M. Production of nano/ultrafine grained AISI 201L stainless steel through advanced thermo-mechanical treatment. *Mater. Sci. Eng. A* **2011**, *528*, 5025–5029. [CrossRef]
17. Sabooni, S.; Karimzadeh, F.; Enayati, M.H. Thermal stability study of ultrafine grained 304L stainless steel produced by martensitic process. *J. Mater. Eng. Perform.* **2014**, *23*, 1665–1672. [CrossRef]
18. Behjati, P.; Kermanpur, A.; Najafizadeh, A.; Baghbadorani, H.S. Effect of annealing temperature on nano/ultrafine grain of Ni-free austenitic stainless steel. *Mater. Sci. Eng. A* **2014**, *592*, 77–82. [CrossRef]
19. Baghbadorani, H.S.; Kermanpur, A.; Najafizadeh, A.; Behjati, P.; Rezaee, A.; Moallemi, M. An investigation on microstructure and mechanical propertiesof a Nb-microalloyed nano/ultrafine grained 201 austenitic stainless steel. *Mater. Sci. Eng. A* **2015**, *636*, 593–599. [CrossRef]
20. Xu, D.M.; Li, G.Q.; Wan, X.L.; Xiong, R.L.; Xu, G.; Wu, K.M.; Somani, M.C.; Misra, R.D.K. Deformation behavior of high yield strength—High ductility ultrafine-grained 316LN austenitic stainless steel. *Mater. Sci. Eng. A* **2017**, *688*, 407–415. [CrossRef]
21. Ma, Y.-Q.; Jin, J.-E.; Lee, Y.-K. A repetitive thermomechanical process to produce nano-crystalline in a metastable austenitic steel. *Scr. Mater.* **2005**, *52*, 1311. [CrossRef]
22. Rajasekhara, S.; Ferreira, P.J.; Karjalainen, L.P.; Kyröläinen, A. Hall-Petch behavior in ultra-fine-grained AISI 301LN stainless steel. *Metall. Mater. Trans. A* **2007**, *38*, 1202–1210. [CrossRef]
23. Somani, M.C.; Juntunen, P.; Karjalainen, L.P.; Misra, R.D.K.; Kyröläinen, A. Enhanced mechanical properties through reversion in metastable austenitic stainless steels. *Metall. Mater. Trans. A* **2009**, *40*, 729–744. [CrossRef]
24. Misra, R.D.K.; Nayak, S.; Mali, S.A.; Shah, J.S.; Somani, M.C.; Karjalainen, L.P. Microstructure and deformation behavior of phase-reversion-induced nanograined/ultrafine-grained austenitic stainless steel. *Metall. Mater. Trans. A* **2009**, *40*, 2498–2509. [CrossRef]
25. Misra, R.D.K.; Zhang, Z.; Jia, Z.; Somani, M.C.; Karjalainen, L.P. Probing deformation processes in near-defect free volume in high strength—high ductility nanograined/ultrafine-grained (NG/UFG) metastable austenitic stainless steels. *Scr. Mater.* **2010**, *63*, 1057–1060. [CrossRef]
26. Rajasekhara, S.; Karjalainen, L.P.; Kyröläinen, A.; Ferreira, P.J. Microstructure evolution in nano/submicron grained AISI 301LN stainless steel. *Mater. Sci. Eng. A* **2010**, *527*, 1986–1996. [CrossRef]
27. Kisko, A.; Hamada, A.; Karjalainen, L.P.; Talonen, J. Microstructure and mechanical properties of reversion treated high Mn austenitic 204Cu and 201 stainless steels, HMnS 2011. In Proceedings of the 1st International Conference on High Manganese Steels, Seoul, Korea, 15–18 May 2011; p. B-19.
28. Kisko, A.; Rovatti, L.; Miettunen, I.; Karjalainen, L.P.; Talonen, J. Microstructure and anisotropy of mechanical properties in reversion-treated high-Mn type 204Cu and 201 stainless steels. In Proceedings of the 7th European Stainless Steel Conference—Science and Market, Como, Italy, 21–23 September 2011.
29. Järvenpää, A.; Jaskari, M.; Karjalainen, L.P. Reversed microstructures and tensile properties after various cold rolling reductions in AISI 301LN steel. *Metals* **2018**, *8*, 109. [CrossRef]

30. Poulon-Quintin, A.; Brochet, S.; Vogt, J.-B.; Glez, J.-C.; Mithieux, J.-D. Fine grained austenitic stainless steels: The role of strain induced α′ martensite and the reversion mechanism limitations. *ISIJ Int.* **2009**, *49*, 293–301. [CrossRef]
31. Zhao, J.; Jiang, Z. Thermomechanical processing of advanced high strength steels. *Prog. Mater. Sci.* **2018**, *94*, 174–242. [CrossRef]
32. Smith, H.; West, D.R.F. The reversion of martensite to austenite in certain stainless steels. *J. Mater. Sci.* **1973**, *8*, 1413–1420. [CrossRef]
33. Smith, H.; West, D.R.F. Annealing of austenite formed by reversion from martensite in an Fe–16Cr–12Ni alloy. *Metals Technol.* **1974**, *1*, 37–40. [CrossRef]
34. Coleman, T.H.; West, D.R.F. Deformation-induced martensite and its reversion to austenite in an Fe–16Cr–12Ni alloy. *Met. Technol.* **1976**, *3*, 49–53. [CrossRef]
35. Guy, K.B.; Butler, E.P.; West, D.R.F. Reversion of bcc α′ martensite in Fe–Cr–Ni austenitic stainless steels. *Met. Sci.* **1983**, *17*, 167–176. [CrossRef]
36. Guy, K.; Butler, E.P.; West, D.R.F. Martensite formation and reversion in austenitic stainless steels. *J. Phys. Colloq.* **1982**, *43*, C4-575–C4-580. [CrossRef]
37. Singh, J. Influence of deformation on the transformation of austenitic stainless steels. *J. Mater. Sci.* **1985**, *20*, 3157–3166. [CrossRef]
38. Tomimura, K.; Takaki, S.; Tanimoto, S.; Tokunaga, Y. Optimal chemical composition in Fe-Cr-Ni Alloys for ultra grain refining by reversion from deformation induced martensite. *ISIJ Int.* **1991**, *31*, 721–727. [CrossRef]
39. Tomimura, K.; Takaki, S.; Tokunaga, Y. Reversion mechanism from deformation induced martensite to austenite in metastable austenitic stainless steels. *ISIJ Int.* **1991**, *31*, 1431–1437. [CrossRef]
40. Takaki, S.; Tomimura, K.; Ueda, S. Effect of pre-cold-working on diffusional reversion of deformation induced martensite in metastable austenitic stainless steel. *ISIJ Int.* **1994**, *34*, 522–527. [CrossRef]
41. Di Schino, A.; Barteri, M.; Kenny, J.M. Grain size dependence of mechanical, corrosion and tribological properties of high nitrogen stainless steels. *J. Mater. Sci.* **2003**, *38*, 3257–3262. [CrossRef]
42. Di Schino, A.; Barteri, M.; Kenny, J.M. Effects of grain size on the properties of a low nickel austenitic stainless steel. *J. Mater. Sci.* **2003**, *38*, 4725–4733. [CrossRef]
43. Di Schino, A.; Kenny, J.M. Grain size dependence of the fatigue behaviour of a ultrafine-grained AISI 304 stainless steel. *Mater. Lett.* **2003**, *57*, 3182–3185. [CrossRef]
44. Choi, J.-Y.; Jin, W. Strain induced martensite formation and its effect on strain hardening behavior in the cold drawn 304 austenitic stainless steels. *Scr. Mater.* **1997**, *36*, 99–104. [CrossRef]
45. Lee, S.-J.; Park, Y.-M.; Lee, Y.-K. Reverse transformation mechanism of martensite to austenite in a metastable austenitic alloy. *Mater. Sci. Eng. A* **2009**, *515*, 32–37. [CrossRef]
46. Jung, Y.-S.; Lee, Y.-K.; Matlock, D.K.; Mataya, M.C. Effect of grain size on strain-induced martensitic transformation start temperature in an ultrafine grained metastable austenitic steel. *Met. Mater. Int.* **2011**, *17*, 553–556. [CrossRef]
47. Huang, J.; Ye, X.; Gu, J.; Chen, X.; Xu, Z. Enhanced mechanical properties of type AISI 301LN austenitic stainless steel through advanced thermo mechanical process. *Mater. Sci. Eng. A* **2012**, *532*, 190–195. [CrossRef]
48. Shen, Y.F.; Jia, N.; Wang, Y.D.; Sun, X.; Zuo, L.; Raabe, D. Suppression of twinning and phase transformation in an ultrafine grained 2GPa strong metastable austenitic steel: Experiment and simulation. *Acta Mater.* **2015**, *97*, 305–315. [CrossRef]
49. Gong, N.; Wu, H.; Niu, G.; Zhang, D. Effects of annealing temperature on nano/ultrafine-grained structure in austenite stainless steel. *Mater. Sci. Technol.* **2017**, *33*, 1667–1672. [CrossRef]
50. Gong, N.; Wu, H.-B.; Yu, Z.-C.; Niu, G.; Zhang, D. Studying mechanical properties and micro deformation of ultrafine-grained structures in austenitic stainless steel. *Metals* **2017**, *7*, 188. [CrossRef]
51. Lei, C.; Li, X.; Deng, X.; Wang, Z.; Wang, G. Deformation mechanism and ductile fracture behavior in high strength high ductility nano/ultrafine grained Fe-17Cr-6Ni austenitic steel. *Mater. Sci. Eng. A* **2018**, *709*, 72–81. [CrossRef]
52. Xu, D.M.; Wan, X.L.; Yu, J.X.; Xu, G.; Li, G.Q. Effect of grain refinement on strain hardening and fracture in austenitic stainless steel. *Mater. Sci. Technol.* **2018**, *34*, 1344–1352. [CrossRef]
53. Ravi Kumar, B.; Das, S.K.; Mahato, B.; Ghosh, R.N. Role of strain-induced martensite on microstructural evolution during annealing of metastable austenitic stainless steel. *J. Mater. Sci.* **2010**, *45*, 911–918. [CrossRef]

54. Ravi Kumar, B.; Sharma, S. Recrystallization behavior of a heavily deformed austenitic stainless steel during iterative type annealing. *Metall. Mater. Trans. A* **2014**, *45*, 6027–6038. [CrossRef]
55. Ghosh, S.K.; Mallick, P.; Chattopadhyay, P.P. Effect of reversion of strain induced martensite on microstructure and mechanical properties in an austenitic stainless steel. *J. Mater. Sci.* **2011**, *46*, 3480–3487. [CrossRef]
56. Ghosh, S.K.; Jha, S.; Mallick, P.; Chattopadhyay, P.P. Influence of mechanical deformation and annealing on kinetics of martensite in a stainless steel. *Mater. Manuf. Process.* **2013**, *28*, 249–255. [CrossRef]
57. Mallick, P.; Tewary, N.K.; Ghosh, S.K.; Chattopadhyay, P.P. Microstructure-tensile property correlation in 304 stainless steel after cold deformation and austenite reversion. *Mater. Sci. Eng. A* **2017**, *707*, 488–500. [CrossRef]
58. Mallick, P.; Tewary, N.K.; Ghosh, S.K.; Chattopadhyay, P.P. Effect of TMCP on microstructure and mechanical properties of 304 stainless steel. *Steel Res. Int.* **2018**, *89*, 1800103. [CrossRef]
59. Mallick, P.K. Strain Induced Martensite and its Reversion in 304 Stainless Steel. Ph.D. Thesis, Indian Institute of Engineering Science and Technology, West Bengal, India, 2017.
60. Mateo, A.; Zapata, A.; Fargas, G. Improvement of mechanical properties on metastable stainless steels by reversion heat treatments. *IOP Conf. Ser. Mater. Sci. Eng.* **2013**, *48*, 12001. [CrossRef]
61. Odnobokova, M.; Belyakov, A.; Enikeev, N.; Molodov, D.A.; Kaibyshev, R. Annealing behavior of a 304L stainless steel processed by large strain cold and warm rolling. *Mater. Sci. Eng. A* **2017**, *689*, 370–383. [CrossRef]
62. Shakhova, I.; Dudko, V.; Belyakov, A.; Tsuzaki, K.; Kaibyshev, R. Effect of large strain cold rolling and subsequent annealing on microstructure and mechanical properties of an austenitic stainless steel. *Mater. Sci. Eng. A* **2012**, *545*, 176–186. [CrossRef]
63. Cios, G.; Tokarski, T.; Żywczak, A.; Dziurka, R.; Stępień, M.; Gondek, Ł.; Marciszko, M.; Pawłowski, B.; Wieczerzak, K.; Bała, P. The investigation of strain-induced martensite reverse transformation in AISI 304 austenitic stainless steel. *Metall. Mater. Trans. A* **2017**, *48*, 4999–5008. [CrossRef]
64. Fargas, G.; Zapata, A.; Roa, J.J.; Sapezanskaia, I.; Mateo, A. correlation between microstructure and mechanical properties before and after reversion of metastable austenitic stainless steels. *Metall. Mater. Trans. A* **2015**, *46*, 5697–5707. [CrossRef]
65. Eskandari, M.; Najafizadeh, A.; Kermanpur, A.; Karimi, M. Potential application of nanocrystalline 301 austenitic stainless steel in lightweight vehicle structures. *Mater. Des.* **2009**, *30*, 3869–3872. [CrossRef]
66. Hedayati, A.A.; Najafizadeh, A.; Kermanpur, A.; Forouzan, F. The effect of cold rolling regime on microstructure and mechanical properties of AISI 304L stainless steel. *Mater. Sci. Eng. A* **2010**, *210*, 1017–1023. [CrossRef]
67. Rezaee, A.; Najafizadeh, A.; Kermanpur, A.; Moallemi, M. The influence of reversion annealing behavior on the formation of nanograined structure in AISI 201L austenitic stainless steel through martensite treatment. *Mater. Des.* **2011**, *32*, 4437–4442. [CrossRef]
68. Sadeghpour, S.; Kermanpur, A.; Najafizadeh, A. Influence of Ti microalloying on the formation of nanocrystalline structure in the 201L austenitic stainless steel during martensite thermomechanical treatment. *Mater. Sci. Eng. A* **2013**, *584*, 177–183. [CrossRef]
69. Sadeghpour, S.; Kermanpur, A.; Najafizadeh, A. Formation of nano/ultrafine grain structure in a Ti-modified 201L stainless steel through martensite thermomechanical treatment. *ISIJ Int.* **2014**, *54*, 920–925. [CrossRef]
70. Shirdel, M.; Mirzadeh, H.; Parsa, M.H. Nano/ultrafine grained austenitic stainless steel through the formation and reversion of deformation-induced martensite: Mechanisms, microstructures, mechanical properties, and TRIP effect. *Mater. Charact.* **2015**, *103*, 150–161. [CrossRef]
71. Baghbadorani, H.S.; Kermanpur, A.; Najafizadeh, A.; Behjati, P.; Moallemi, M.; Rezaee, A. Influence of Nb-microalloying on the formation of nano/ultrafine-grained microstructure and mechanical properties during martensite reversion process in a 201-type austenitic stainless steel. *Metall. Mater. Trans. A* **2015**, *46*, 3406–3413. [CrossRef]
72. Rasouli, D.; Kermanpur, A.; Ghassemali, E.; Najafizadeh, A. On the reversion and recrystallization of austenite in the interstitially alloyed Ni-free nano/ultrafine grained austenitic stainless steels. *Met. Mater. Int.* **2019**, *25*, 846–859. [CrossRef]
73. Kheiri, S.; Mirzadeh, H.; Naghizadeh, M. Tailoring the microstructure and mechanical properties of AISI 316L austenitic stainless steel via cold rolling and reversion annealing. *Mater. Sci. Eng. A* **2019**, *759*, 90–96. [CrossRef]

74. Droste, M.; Ullrich, C.; Motylenko, M.; Fleischer, M.; Weidner, A.; Freudenberger, J.; Rafaja, D.; Biermann, H. Fatigue behavior of an ultrafine-grained metastable CrMnNi steel tested under total strain control. *Int. J. Fatigue* **2018**, *106*, 143–152. [CrossRef]
75. Maréchal, D. Linkage between Mechanical Properties and Phase Transformations in a 301LN Austenitic Stainless Steel. Ph.D. Thesis, University of British Columbia, Vancouver, BC, Canada, 2011.
76. Fava, J.; Spinosa, C.; Ruch, M.; Carabedo, F.; Landau, M.; Cosarinsky, G.; Savin, A.; Steigmann, R.; Craus, M.L. Characterization of reverse martensitic transformation in cold-rolled austenitic 316 stainless steel. *Rev. Mater.* **2018**, *23*. [CrossRef]
77. Souza Filho, I.R.; Zilnyk, K.D.; Sandim, M.J.R.; Bolmaro, R.E.; Sandim, H.R.Z. Strain partitioning and texture evolution during cold rolling of AISI 201 austenitic stainless steel. *Mater. Sci. Eng. A* **2017**, *702*, 161–172. [CrossRef]
78. Souza Filho, I.R.; Sandim, M.J.R.; Cohen, R.; Nagamine, L.C.C.M.; Hoffmann, J.; Bolmaro, R.E.; Sandim, H.R.Z. Effects of strain-induced martensite and its reversion on the magnetic properties of AISI 201 austenitic stainless steel. *J. Magn. Magn. Mater.* **2016**, *419*, 156–165. [CrossRef]
79. Souza Filho, I.R.; Junior, D.R.A.; Gauss, C.; Sandim, M.J.R.; Suzuki, P.A.; Sandim, H.R.Z. Austenite reversion in AISI 201 austenitic stainless steel evaluated via in situ synchrotron X-ray diffraction during slow continuous annealing. *Mater. Sci. Eng. A* **2019**, *755*, 267–277. [CrossRef]
80. Hamada, A.S.; Kisko, A.P.; Sahu, P.; Karjalainen, L.P. Enhancement of mechanical properties of a TRIP-aided austenitic stainless steel by controlled reversion annealing. *Mater. Sci. Eng. A* **2015**, *628*, 154–159. [CrossRef]
81. Kisko, A.; Misra, R.D.K.; Talonen, J.; Karjalainen, L.P. The influence of grain size on the strain-induced martensite formation in tensile straining of an austenitic 15Cr–9Mn–Ni–Cu stainless steel. *Mater. Sci. Eng. A* **2013**, *578*, 408–416. [CrossRef]
82. Kisko, A.; Talonen, J.; Porter, D.A.; Karjalainen, L.P. Effect of Nb microalloying on reversion and grain growth in a high-Mn 204Cu austenitic stainless steel. *ISIJ Int.* **2015**, *55*, 2217–2224. [CrossRef]
83. Kisko, A.; Hamada, A.S.; Talonen, J.; Porter, D.; Karjalainen, L.P. Effects of reversion and recrystallization on microstructure and mechanical properties of Nb-alloyed low-Ni high-Mn austenitic stainless steels. *Mater. Sci. Eng. A* **2016**, *657*, 359–370. [CrossRef]
84. Järvenpää, A.; Jaskari, M.; Man, J.; Karjalainen, L.P. Austenite stability in reversion-treated structures of a 301LN steel under tensile loading. *Mater. Charact.* **2017**, *127*, 12–26. [CrossRef]
85. Järvenpää, A.; Jaskari, M.; Man, J.; Karjalainen, L.P. Stability of grain-refined reversed structures in a 301LN austenitic stainless steel under cyclic loading. *Mater. Sci. Eng. A* **2017**, *703*, 280–292. [CrossRef]
86. Järvenpää, A.; Jaskari, M.; Juuti, T.; Karjalainen, P. Demonstrating the effect of precipitation on the mechanical stability of fine-grained austenite in reversion-treated 301LN stainless steel. *Metals* **2017**, *7*, 733. [CrossRef]
87. Järvenpää, A.; Jaskari, M.; Karjalainen, L.P. Properties of induction reversion-refined microstructures of AISI 301LN under monotonic, cyclic and rolling deformation. In *THERMEC 2018*; Trans Tech Publications Ltd.: Stafa-Zurich, Switzerland, 2019; pp. 601–607. [CrossRef]
88. Misra, R.D.K.; Nayak, S.; Venkatasurya, P.K.C.; Ramuni, V.; Somani, M.C.; Karjalainen, L.P. Nanograined/ultrafine-grained structure and tensile deformation behavior of shear phase reversion-induced 301 austenitic stainless steel. *Metall. Mater. Trans. A* **2010**, *41*, 2162–2174. [CrossRef]
89. Rajasekhara, S.; Ferreira, P.J.; Karjalainen, L.P.; Kyrolainen, A. Microstructure evolution in nano/sub-micron grained AISI 301 stainless steel. In Proceedings of the 6th European Congress Stainless Steel Science and Market, Helsinki, Finland, 10–13 June 2008.
90. Misra, R.D.K.; Kumar, B.R.; Somani, M.; Karjalainen, P. Deformation processes during tensile straining of ultrafine/nanograined structures formed by reversion in metastable austenitic steels. *Scr. Mater.* **2008**, *59*, 79–82. [CrossRef]
91. Misra, R.D.K.; Nayak, S.; Mali, S.A.; Shah, J.S.; Somani, M.C.; Karjalainen, L.P. On the significance of nature of strain-induced martensite on phase-reversion-induced nanograined/ultrafine-grained austenitic stainless steel. *Metall. Mater. Trans. A* **2010**, *41*, 3–10. [CrossRef]
92. Misra, R.D.K.; Zhang, Z.; Venkatasurya, P.K.C.; Somani, M.C.; Karjalainen, L.P. Martensite shear phase reversion-induced nanograined/ultrafine-grained Fe–16Cr–10Ni alloy: The effect of interstitial alloying elements and degree of austenite stability on phase reversion. *Mater. Sci. Eng. A* **2010**, *527*, 7779–7792. [CrossRef]

93. Misra, R.D.K.; Zhang, Z.; Jia, Z.; Surya, P.K.C.V.; Somani, M.C.; Karjalainen, L.P. Nanomechanical insights into the deformation behavior of austenitic alloys with different stacking fault energies and austenitic stability. *Mater. Sci. Eng. A* **2011**, *528*, 6958–6963. [CrossRef]
94. Challa, V.S.A.; Wan, X.L.; Somani, M.C.; Karjalainen, L.P.; Misra, R.D.K. Strain hardening behavior of phase reversion-induced nanograined/ultrafine-grained (NG/UFG) austenitic stainless steel and relationship with grain size and deformation mechanism. *Mater. Sci. Eng. A* **2014**, *613*, 60–70. [CrossRef]
95. Challa, V.S.A.; Wan, X.L.; Somani, M.C.; Karjalainen, L.P.; Misra, R.D.K. Significance of interplay between austenite stability and deformation mechanisms in governing three-stage work hardening behavior of phase-reversion induced nanograined/ultrafine-grained (NG/UFG) stainless steels with high strength-high ductility combination. *Scr. Mater.* **2014**, *86*, 60–63. [CrossRef]
96. Challa, V.S.A.; Misra, R.D.K.; Somani, M.C.; Wang, Z.D. Strain hardening behavior of nanograined/ultrafine-grained (NG/UFG) austenitic 16Cr–10Ni stainless steel and its relationship to austenite stability and deformation behavior. *Mater. Sci. Eng. A* **2016**, *649*, 153–157. [CrossRef]
97. Misra, R.D.K.; Injeti, V.S.Y.; Somani, M.C. The significance of deformation mechanisms on the fracture behavior of phase reversion-induced nanostructured austenitic stainless steel. *Sci. Rep.* **2018**, *8*, 7908. [CrossRef]
98. Chlupová, A.; Man, J.; Polák, J.; Karjalainen, L.P. Microstructural investigation and mechanical testing of an ultrafine-grained austenitic stainless steel. In *NANOCON 2013*; Tanger Ltd.: Ostrava, Czech Republic, 2013; pp. 733–738.
99. Chlupová, A.; Man, J.; Kuběna, I.; Polák, J.; Karjalainen, L.P. LCF behaviour of ultrafine grained 301LN stainless steel. *Procedia Eng.* **2014**, *74*, 147–150. [CrossRef]
100. Man, J.; Kuběna, I.; Smaga, M.; Man, O.; Järvenpää, A.; Weidner, A.; Chlup, Z.; Polák, J. Microstructural changes during deformation of AISI 300 grade austenitic stainless steels: Impact of chemical heterogeneity. *Procedia Struct. Integr.* **2016**, *2*, 2299–2306. [CrossRef]
101. Kisko, A. Microstructure and Properties of Reversion Treated Low-Ni High-Mn Austenitic Stainless Steels. Ph.D. Thesis, University of Oulu, Oulu, Finland, 2016; ISSN 1796-2226.
102. Järvenpää, A. Microstructures, Mechanical Stability and Strength of Low-Temperature AISILN Stainless steel Under Monotonic and Dynamic Loading. Ph.D. Thesis, University of Oulu, Oulu, Finland, 2019; ISSN 1796-2226.
103. Järvenpää, A.; Karjalainen, L.P. An overview of mechanical properties of today's grain-refined austenitic stainless steels. In Proceedings of the ESSC & DUPLEX, ASMET, Vienna, Austria, 31 September–2 October 2019; pp. 12–21.
104. Komatsuseiki Kosakusho Co., Ltd. Available online: https://www.komatsuseiki.co.jp/english/future/03.php (accessed on 9 August 2016).
105. Komatsu, T.; Matsumura, T.; Torizuka, S. Effect of grain size in stainless steel on cutting performance in micro-scale cutting. *Int. J. Autom. Technol.* **2011**, *5*, 334–341. [CrossRef]
106. Komatsu, T.; Kobayashi, H.; Torizuka, S.; Nagayama, S. Micro hole piercing for ultra fine grained steel. In *THERMEC 2013*; Trans Tech Publications Ltd.: Stafa-Zurich, Switzerland, 2014; Volume 783, pp. 2653–2658. [CrossRef]
107. Komatsu, T.; Yoshino, T.; Matsumura, T.; Torizuka, S. Effect of crystal grain size in stainless steel on cutting process in micromilling. *Procedia CIRP* **2012**, *1*, 150–155. [CrossRef]
108. Nippon Steel & Sumitomo Metal Product Catalog: SUS304 BA1. Available online: https://stainless.nipponsteel.com/product/grade/nssmc_series/product/sus304_ba1.php (accessed on 6 December 2019).
109. Nippon Steel & Sumitomo Metal Product Catalog: NSSMC-NAR-301L BA1. Available online: https://stainless.nipponsteel.com/product/grade/nssmc_series/product/nssmc-nar-301l_ba1.php (accessed on 6 December 2019).
110. Lo, K.H.; Shek, C.H.; Lai, J.K.L. Recent developments in stainless steels. *Mater. Sci. Eng. R* **2009**, *65*, 39–104. [CrossRef]
111. Pereloma, E.; Gazder, A.; Timokhina, I. Retained austenite: Transformation-induced plasticity. *Encycl. Iron Steel Alloys* **2016**, 3088–3103. [CrossRef]
112. Beese, A.M.; Mohr, D. Effect of stress triaxiality and Lode angle on the kinetics of strain-induced austenite-to-martensite transformation. *Acta Mater.* **2011**, *59*, 2589–2600. [CrossRef]
113. Das, A.; Tarafder, S.; Chakraborti, P.C. Estimation of deformation induced martensite in austenitic stainless steels. *Mater. Sci. Eng. A* **2011**, *529*, 9–20. [CrossRef]

114. Nohara, K.; Ono, Y.; Ohashi, N. Composition and grain size dependencies of strain-induced martensitic transformation in metastable austenitic stainless steels. *ISIJ Int.* **1977**, *63*, 772–782. [CrossRef]
115. Noh, H.-S.; Kang, J.-H.; Kim, K.-M.; Kim, S.-J. Different effects of Ni and Mn on thermodynamic and mechanical stabilities in Cr-Ni-Mn austenitic steels. *Metall. Mater. Trans. A* **2019**, *50*, 616–624. [CrossRef]
116. Furukane, S.; Torizuka, S. Effect of grain size and dislocation density on strain-induced martensitic transformation in austenitic stainless steels. *Tetsu Hagane* **2019**, *105*, 827–836. [CrossRef]
117. Roy, B.; Kumar, R.; Das, J. Effect of cryorolling on the microstructure and tensile properties of bulk nano-austenitic stainless steel. *Mater. Sci. Eng. A* **2015**, *631*, 241–247. [CrossRef]
118. Xiong, Y.; He, T.; Wang, J.; Lu, Y.; Chen, L.; Ren, F.; Liu, Y.; Volinsky, A.A. Cryorolling effect on microstructure and mechanical properties of Fe–25Cr–20Ni austenitic stainless steel. *Mater. Des.* **2015**, *88*, 398–405. [CrossRef]
119. Zheng, C.; Liu, C.; Ren, M.; Jiang, H.; Li, L. Microstructure and mechanical behavior of an AISI 304 austenitic stainless steel prepared by cold- or cryogenic-rolling and annealing. *Mater. Sci. Eng. A* **2018**, *724*, 260–268. [CrossRef]
120. Mallick, P.; Tewary, N.K.; Ghosh, S.K.; Chattopadhyay, P.P. Effect of cryogenic deformation on microstructure and mechanical properties of 304 austenitic stainless steel. *Mater. Charact.* **2017**, *133*, 77–86. [CrossRef]
121. Martins, L.F.M.; Plaut, R.L.; Padilha, A.F. Effect of carbon on the cold-worked state and annealing behavior of two 18wt%Cr-8wt%Ni austenitic stainless steels. *ISIJ Int.* **1998**, *38*, 572–579. [CrossRef]
122. Mirzadeh, H.; Najafizadeh, A. Correlation between processing parameters and strain-induced martensitic transformation in cold worked AISI 301 stainless steel. *Mater. Charact.* **2008**, *59*, 1650–1654. [CrossRef]
123. Xu, D.; Wan, X.; Yu, J.; Xu, G.; Li, G. Effect of cold deformation on microstructures and mechanical properties of austenitic stainless steel. *Metals* **2018**, *8*, 522. [CrossRef]
124. Ahmedabadi, P.M.; Kain, V.; Agrawal, A. Modelling kinetics of strain-induced martensite transformation during plastic deformation of austenitic stainless steel. *Mater. Des.* **2016**, *109*, 466–475. [CrossRef]
125. Niessen, F. Phase Transformations in Supermartensitic Stainless Steels. Ph.D. Thesis, Technical University of Denmark, Lyngby, Denmark, 2018.
126. Yang, D.P.; Wu, D.; Yi, H.L. Reverse transformation from martensite into austenite in a medium-Mn steel. *Scr. Mater.* **2019**, *161*, 1–5. [CrossRef]
127. Santos, T.F.A.; Andrade, M.S. Avaliação dilatométrica da reversão das martensitas induzidas por deformação em um aço inoxidável austenítico do tipo ABNT 304. *Matéria* **2008**, *13*, 587–596. [CrossRef]
128. Santos, T.F.A.; Andrade, M.S. Internal friction on AISI 304 stainless steels with low tensile deformations at temperatures between 50 and 20 °C. *Adv. Mater. Sci. Eng.* **2010**, *2010*, 326736. [CrossRef]
129. Dryzek, E.; Sarnek, M.; Wróbel, M. Reverse transformation of deformation-induced martensite in austenitic stainless steel studied by positron annihilation. *J. Mater. Sci.* **2014**, *49*, 8449–8458. [CrossRef]
130. Knutsson, A.; Hedström, P.; Odén, M. Reverse martensitic transformation and resulting microstructure in a cold rolled metastable austenitic stainless steel. *Steel Res. Int.* **2008**, *79*, 433–439. [CrossRef]
131. Talonen, J.; Aspegren, P.; Hänninen, H. Comparison of different methods for measuring strain induced α-martensite content in austenitic steels. *Mater. Sci. Technol.* **2004**, *20*, 1506–1512. [CrossRef]
132. Cios, G.; Tokarski, T.; Bała, P. Strain-induced martensite reversion in 18Cr–8Ni steel – transmission Kikuchi diffraction study. *Mater. Sci. Technol.* **2018**, *34*, 580–583. [CrossRef]
133. Wei, S.; Jiang, M.; Tasan, C.C. Interstitial-free bake hardening realized by epsilon martensite reverse transformation. *Metall. Mater. Trans. A* **2019**, *50*, 3985–3991. [CrossRef]
134. Yagodzinsky, Y.; Saukkonen, T.; Romu, J.; Hänninen, H. Comparative study of nitrogen and carbon effects on mechanism of reversion of a'-martensite to austenite in metastable AISI 301 steel grades. In Proceedings of the International Conference on High Nitrogen Steels, Jiuzhaigou Valley, China, 29–31 August 2006; pp. 59–66.
135. Johannsen, D.L.; Kyrolainen, A.; Ferreira, P.J. Influence of annealing treatment on the formation of nano/submicron grain size AISI 301 austenitic stainless steels. *Metall. Mater. Trans. A* **2006**, *37*, 2325–2338. [CrossRef]
136. Rajasekhara, S.; Ferreira, P.J. Martensite—Austenite phase transformation kinetics in an ultrafine-grained metastable austenitic stainless steel. *Acta Mater.* **2011**, *59*, 738–748. [CrossRef]

137. Sun, G.S.; Du, L.X.; Hu, J.; Xie, H.; Wu, H.Y.; Misra, R.D.K. Ultrahigh strength nano/ultrafine-grained 304 stainless steel through three-stage cold rolling and annealing treatment. *Mater. Charact.* **2015**, *110*, 228–235. [CrossRef]
138. Padilha, A.F.; Plaut, R.L.; Rios, P.R. Annealing of cold-worked austenitic stainless steels. *ISIJ Int.* **2003**, *43*, 135–143. [CrossRef]
139. Haessner, F.; Plaut, R.L.; Padilha, A.F. Separation of static recrystallization and reverse transformation of deformation-induced martensite in an austenitic stainless steel by calorimetric measurements. *ISIJ Int.* **2003**, *43*, 1472–1474. [CrossRef]
140. Sun, G.S.; Du, L.X.; Hu, J.; Misra, R.D.K. Microstructural evolution and recrystallization behavior of cold rolled austenitic stainless steel with dual phase microstructure during isothermal annealing. *Mater. Sci. Eng. A* **2018**, *709*, 254–264. [CrossRef]
141. Apple, C.A.; Krauss, G. The effect of heating rate on the martensite to austenite transformation in Fe-Ni-C alloys. *Acta Metall.* **1972**, *20*, 849–856. [CrossRef]
142. Sun, G.S.; Du, L.X.; Hu, J.; Xie, H.; Misra, R.D.K. Low temperature superplastic-like deformation and fracture behavior of nano/ultrafine-grained metastable austenitic stainless steel. *Mater. Des.* **2017**, *117*, 223–231. [CrossRef]
143. Mumtaz, K.; Takahashi, S.; Echigoya, J.; Kamada, Y.; Zhang, L.F.; Kikuchi, H.; Ara, K.; Sato, M. Magnetic measurements of the reverse martensite to austenite transformation in a rolled austenitic stainless steel. *J. Mater. Sci.* **2004**, *39*, 1997–2010. [CrossRef]
144. Tavares, S.S.M.; da Silva, M.R.; Neto, J.M.; Miraglia, S.; Fruchart, D. Ferromagnetic properties of cold rolled AISI 304L steel. *J. Magn. Magn. Mater.* **2002**, *242–245*, 1391–1394. [CrossRef]
145. Behjati, P.; Kermanpur, A.; Karjalainen, L.P.; Järvenpää, A.; Jaskari, M.; Samaei Baghbadorani, H.; Najafizadeh, A.; Hamada, A. Influence of prior cold rolling reduction on microstructure and mechanical properties of a reversion annealed high-Mn austenitic steel. *Mater. Sci. Eng. A* **2016**, *650*, 119–128. [CrossRef]
146. Ravi Kumar, B.; Sharma, S.; Mahato, B. Formation of ultrafine grained microstructure in the austenitic stainless steel and its impact on tensile properties. *Mater. Sci. Eng. A* **2011**, *528*, 2209–2216. [CrossRef]
147. Ravi Kumar, B.; Mahato, B.; Sharma, S.; Sahu, J.K. Effect of cyclic thermal process on ultrafine grain formation in AISI 304L austenitic stainless steel. *Metall. Mater. Trans. A* **2009**, *40*, 3226–3234. [CrossRef]
148. Ravi Kumar, B.; Raabe, D. Tensile deformation characteristics of bulk ultrafine-grained austenitic stainless steel produced by thermal cycling. *Scr. Mater.* **2012**, *66*, 634–637. [CrossRef]
149. Xu, D.M.; Li, G.Q.; Wan, X.L.; Misra, R.D.K.; Yu, J.X.; Xu, G. On the deformation mechanism of austenitic stainless steel at elevated temperatures: A critical analysis of fine-grained versus coarse-grained structure. *Mater. Sci. Eng. A* **2019**. [CrossRef]
150. Lehto, P.; Remes, H.; Saukkonen, T.; Hänninen, H.; Romanoff, J. Influence of grain size distribution on the Hall–Petch relationship of welded structural steel. *Mater. Sci. Eng. A* **2014**, *592*, 28–39. [CrossRef]
151. *Technical Data for Annealed 301LN*; Outokumpu Stainless Oy: Helsinki, Finland, 2016.
152. Bleck, W.; Guo, X.; Ma, Y. The TRIP effect and its application in cold formable sheet steels. *Steel Res. Int.* **2017**, *88*, 1700218. [CrossRef]
153. Guo, X.; Post, J.; Groen, M.; Bleck, W. Stress oriented delayed cracking induced by dynamic martensitic transformation in meta-stable austenitic stainless steels. *Steel Res. Int.* **2011**, *82*, 6–13. [CrossRef]
154. Weiß, A.; Gutte, H.; Scheller, P.R. Deformation induced martensite formation and its effect on transformation induced plasticity (TRIP). *Steel Res. Int.* **2006**, *77*, 727–732. [CrossRef]
155. Talonen, J. Effect of Strain-Induced α'-Martensite Transformation on Mechanical Properties of Metastable Austenitic Stainless Steels. Ph.D. Thesis, TKK Dissertations 71, Espoo, Finland, 2007.
156. Hamada, A.S.; Karjalainen, L.P.; Misra, R.D.K.; Talonen, J. Contribution of deformation mechanisms to strength and ductility in two Cr–Mn grade austenitic stainless steels. *Mater. Sci. Eng. A* **2013**, *559*, 336–344. [CrossRef]
157. Olson, G.B.; Cohen, M. Kinetics of strain-induced martensitic nucleation. *Metall. Trans. A* **1975**, *6*, 791. [CrossRef]
158. Lichtenfeld, J.A.; Van Tyne, C.J.; Mataya, M.C. Effect of strain rate on stress-strain behavior of alloy 309 and 304L austenitic stainless steel. *Metall. Mater. Trans. A* **2006**, *37*, 147–161. [CrossRef]
159. Yoo, C.-S.; Park, Y.-M.; Jung, Y.-S.; Lee, Y.-K. Effect of grain size on transformation-induced plasticity in an ultrafine-grained metastable austenitic steel. *Scr. Mater.* **2008**, *59*, 71–74. [CrossRef]

160. Matsuoka, Y.; Iwasaki, T.; Nakada, N.; Tsuchiyama, T.; Takaki, S. Effect of grain size on thermal and mechanical stability of austenite in metastable austenitic stainless steel. *ISIJ Int.* **2013**, *53*, 1224–1230. [CrossRef]
161. Ravi Kumar, B.; Gujral, A. Plastic deformation modes in mono- and bimodal-type ultrafine-grained austenitic stainless steel. *Metallogr. Microstruct. Anal.* **2014**, *3*, 397–407. [CrossRef]
162. Lee, Y.K.; Jin, J.E.; Ma, Y.Q. Transformation-induced extraordinary ductility in an ultrafine-grained alloy with nanosized precipitates. *Scr. Mater.* **2007**, *57*, 707–710. [CrossRef]
163. He, Y.M.; Wang, Y.H.; Guo, K.; Wang, T.S. Effect of carbide precipitation on strain-hardening behavior and deformation mechanism of metastable austenitic stainless steel after repetitive cold rolling and reversion annealing. *Mater. Sci. Eng. A* **2017**, *708*, 248–253. [CrossRef]
164. Saenarjhan, N.; Kang, J.-H.; Kim, S.-J. Effects of carbon and nitrogen on austenite stability and tensile deformation behavior of 15Cr-15Mn-4Ni based austenitic stainless steels. *Mater. Sci. Eng. A* **2019**, *742*, 608–616. [CrossRef]
165. Kim, K.-S.; Kang, J.-H.; Kim, S.-J. Effects of carbon and nitrogen on precipitation and tensile behavior in 15Cr-15Mn-4Ni austenitic stainless steels. *Mater. Sci. Eng. A* **2018**, *712*, 114–121. [CrossRef]
166. Karimi, M.; Najafizadeh, A.; Kermanpur, A.; Eskandari, M. Effect of martensite to austenite reversion on the formation of nano/submicron grained AISI 301 stainless steel. *Mater. Charact.* **2009**, *60*, 1220–1223. [CrossRef]
167. Hong, S.-M.; Kim, M.-Y.; Min, D.-J.; Lee, K.; Shim, J.-H.; Kim, D.-I.; Suh, J.-Y.; Jung, W.-S.; Choi, I.-S. Unraveling the origin of strain-induced precipitation of M23C6 in the plastically deformed 347 austenite stainless steel. *Mater. Charact.* **2014**, *94*, 7–13. [CrossRef]
168. Allain, S.; Chateau, J.-P.; Bouaziz, O.; Migot, S.; Guelton, N. Correlations between the calculated stacking fault energy and the plasticity mechanisms in Fe–Mn–C alloys. *Mater. Sci. Eng. A* **2004**, *387–389*, 158–162. [CrossRef]
169. Saeed-Akbari, A.; Imlau, J.; Prahl, U.; Bleck, W. Derivation and variation in composition-dependent stacking fault energy maps based on subregular solution model in high-manganese steels. *Metall. Mater. Trans. A* **2009**, *40*, 3076–3090. [CrossRef]
170. Lee, S.; Shin, S.; Kwon, M.; Lee, K.; De Cooman, B.C. Tensile properties of medium Mn steel with a bimodal UFG $\alpha + \gamma$ and coarse δ-ferrite microstructure. *Metall. Mater. Trans. A* **2017**, *48*, 1678–1700. [CrossRef]
171. Misra, R.D.K.; Challa, V.S.A.; Venkatsurya, P.K.C.; Shen, Y.F.; Somani, M.C.; Karjalainen, L.P. Interplay between grain structure, deformation mechanisms and austenite stability in phase-reversion-induced nanograined/ultrafine-grained austenitic ferrous alloy. *Acta Mater.* **2015**, *84*, 339–348. [CrossRef]
172. Mahato, B.; Shee, S.K.; Sahu, T.; Chowdhury, S.G.; Sahu, P.; Porter, D.A.; Karjalainen, L.P. An effective stacking fault energy viewpoint on the formation of extended defects and their contribution to strain hardening in a Fe–Mn–Si–Al twinning-induced plasticity steel. *Acta Mater.* **2015**, *86*, 69–79. [CrossRef]
173. Gao, S.; Bai, Y.; Zheng, R.; Tian, Y.; Mao, W.; Shibata, A.; Tsuji, N. Mechanism of huge Lüders-type deformation in ultrafine grained austenitic stainless steel. *Scr. Mater.* **2019**, *159*, 28–32. [CrossRef]
174. Song, R.; Ponge, D.; Raabe, D. Mechanical properties of an ultrafine grained C–Mn steel processed by warm deformation and annealing. *Acta Mater.* **2005**, *53*, 4881–4892. [CrossRef]
175. Song, R.; Ponge, D.; Raabe, D.; Speer, J.G.; Matlock, D.K. Overview of processing, microstructure and mechanical properties of ultrafine grained bcc steels. *Mater. Sci. Eng. A* **2006**, *441*, 1–17. [CrossRef]
176. Lee, T.; Park, C.H.; Lee, D.-L.; Lee, C.S. Enhancing tensile properties of ultrafine-grained medium-carbon steel utilizing fine carbides. *Mater. Sci. Eng. A* **2011**, *528*, 6558–6564. [CrossRef]
177. Dini, G.; Najafizadeh, A.; Ueji, R.; Monir-Vaghefi, S.M. Tensile deformation behavior of high manganese austenitic steel: The role of grain size. *Mater. Des.* **2010**, *31*, 3395–3402. [CrossRef]
178. Mao, Q.; Gao, B.; Li, J.; Huang, Z.; Li, Y. Enhanced tensile properties of 316L steel via grain refinement and low-strain rolling. *Mater. Sci. Technol.* **2019**, *35*, 1497–1503. [CrossRef]
179. Jung, Y.-S.; Lee, Y.-K. Effect of pre-deformation on the tensile properties of a metastable austenitic steel. *Scr. Mater.* **2008**, *59*, 47–50. [CrossRef]
180. Höppel, H.W.; Kautz, M.; Xu, C.; Murashkin, M.; Langdon, T.G.; Valiev, R.Z.; Mughrabi, H. An overview: Fatigue behaviour of ultrafine-grained metals and alloys. *Int. J. Fatigue* **2006**, *28*, 1001–1010. [CrossRef]
181. Ueno, H.; Kakihata, K.; Kaneko, Y.; Hashimoto, S.; Vinogradov, A. Enhanced fatigue properties of nanostructured austenitic SUS 316L stainless steel. *Acta Mater.* **2011**, *59*, 7060–7069. [CrossRef]
182. Liu, J.; Deng, X.T.; Huang, L.; Wang, Z.D. High-cycle fatigue behavior of 18Cr-8Ni austenitic stainless steels with grains ranging from nano/ultrafine-size to coarse. *Mater. Sci. Eng. A* **2018**, *733*, 128–136. [CrossRef]

183. Järvenpää, A.; Karjalainen, L.P.; Jaskari, M. Effect of grain size on fatigue behavior of Type 301LN stainless steel. *Int. J. Fatigue* **2014**, *65*. [CrossRef]
184. Man, J.; Chlupová, A.; Kuběna, I.; Kruml, T.; Man, O.; Polak, J. LCF Behaviour of 301LN steel: Coarse-grained vs. UFG-bimodal structure. In Proceedings of the LCF8, the Eighth International Conference on Low Cycle Fatigue, Dresden, Germany, 27–29 June 2017; pp. 27–29.
185. Uusitalo, J.; Karjalainen, L.P.; Retraint, D.; Palosaari, M. Fatigue properties of steels with ultrasonic attrition treated surface layers. *Mater. Sci. Forum* **2009**, *604–605*, 239–248. [CrossRef]
186. Biermann, H.; Glage, A.; Droste, M. Influence of temperature on fatigue-induced martensitic phase transformation in a metastable CrMnNi-steel. *Metall. Mater. Trans. A* **2016**, *47*, 84–94. [CrossRef]
187. Glage, A.; Weidner, A.; Biermann, H. Effect of austenite stability on the low cycle fatigue behavior and microstructure of high alloyed metastable austenitic cast TRIP steels. *Procedia Eng.* **2010**, *2*, 2085–2094. [CrossRef]
188. Glage, A.; Weidner, A.; Biermann, H. Cyclic deformation behaviour of three austenitic cast CrMnNi TRIP/TWIP steels with various Ni content. *Steel Res. Int.* **2011**, *82*, 1040–1047. [CrossRef]
189. Hennessy, D.; Steckel, G.; Altstetter, C. Phase transformation of stainless steel during fatigue. *Metall. Trans. A* **1976**, *7*, 415–424. [CrossRef]
190. Mateo, A.; Fargas, G.; Zapata, A. Martensitic transformation during fatigue testing of an AISI 301LN stainless steel. *IOP Conf. Ser. Mater. Sci. Eng.* **2012**, *31*, 12010. [CrossRef]
191. Topic, M.; Tait, R.B.; Allen, C. The fatigue behaviour of metastable (AISI-304) austenitic stainless steel wires. *Int. J. Fatigue* **2007**, *29*, 656–665. [CrossRef]
192. Smaga, M.; Walther, F.; Eifler, D. Deformation-induced martensitic transformation in metastable austenitic steels. *Mater. Sci. Eng. A* **2008**, *483–484*, 394–397. [CrossRef]
193. Droste, M.; Järvenpää, A.; Jaskari, M.; Motylenko, M.; Weidner, A.; Karjalainen, L.P.; Biermann, H. The role of grain size in the cyclic deformation behavior of laser reversion annealed high-alloy TRIP steel. *Fatigue Fract. Eng. Mater. Struct.* **2020**. in preparation.
194. Hamada, A.S.; Karjalainen, L.P.; Surya, P.K.C.V.; Misra, R.D.K. Fatigue behavior of ultrafine-grained and coarse-grained Cr–Ni austenitic stainless steels. *Mater. Sci. Eng. A* **2011**, *528*, 3890–3896. [CrossRef]
195. Poulon, A.; Brochet, S.; Glez, J.-C.; Mithieux, J.-D.; Vogt, J.-B. Influence of texture and grain size on martensitic transformations occurring during low-cycle fatigue of a fine-grained austenitic stainless steel. *Adv. Eng. Mater.* **2010**, *12*, 1041–1046. [CrossRef]
196. Man, J.; Järvenpää, A.; Jaskari, M.; Kuběna, I.; Fintová, S.; Chlupová, A.; Karjalainen, L.P.; Polák, J. Cyclic deformation behaviour and stability of grain-refined 301LN austenitic stainless structure. *MATEC Web Conf.* **2018**, *165*. [CrossRef]
197. Hamada, A.S.; Järvenpää, A.; Ahmed, E.; Sahu, P.; Farahat, A.I.Z. Enhancement in grain-structure and mechanical properties of laser reversion treated metastable austenitic stainless steel. *Mater. Des.* **2016**, *94*, 345–352. [CrossRef]
198. Järvenpää, A.; Jaskari, M.; Mäntyjärvi, K.; Karjalainen, P. Comparison of the formability of austenitic reversion-treated and temper-rolled 17Cr-7Ni steels. In Proceedings of the ESAFORM19 Conference, Vitoria-Gasteiz, Austria, 8–10 May 2019. [CrossRef]
199. Järvenpää, A.; Jaskari, M.; Keskitalo, M.; Mäntyjärvi, K.; Karjalainen, P. Microstructure and mechanical properties of laser-welded high-strength AISI 301LN steel in reversion-treated and temper-rolled conditions. *Procedia Manuf.* **2019**, *36*, 216–223. [CrossRef]
200. Karjalainen, P.; Oikarinen, T.; Somani, M.; Kyröläinen, A. Softening of temper-rolled austenitic stainless steels in welding. In Proceedings of the Fifteenth International Conference & Exhibition on the Joining of Materials, Helsingor, Denmark, 3–6 May 2009.
201. Cvetkovski, S.; Karjalainen, L.P.; Kujanpää, V.; Ahmad, A. Estimation of heat input in TIG and laser welding of stainless steel sheet. In Proceedings of the IIW International Conference on Advances in Welding and Allied Technologies, Singapore, Singapore, 16–17 July 2009; pp. 323–328.
202. Cvetkovski, S.; Karjalainen, L.P.; Kisko, A.; Lantto, S. Characteristic microstructures in simulated HAZ of temper rolled austenitic stainless steel EN 1.4318. In Proceedings of the 7th European Stainless Steel Conference—Science and Market, Como, Italy, 21–23 September 2011.
203. Lv, Y.; Luo, H.; Tang, J.; Guo, J.; Pi, J.; Ye, K. Corrosion properties of phase reversion induced nano/ultrafine grained AISI 304 metastable austenite stainless steel. *Mater. Res. Bull.* **2018**, *107*, 421–429. [CrossRef]

204. Han, J.; Wang, Z.; Jiang, L. Properties of a 304 Austenitic stainless steel hot strip by TMCP. *J. Iron Steel Res. Int.* **2007**, *14*, 282–287. [CrossRef]
205. Hamada, A.S.; Karjalainen, L.P.; Somani, M.C. Electrochemical corrosion behaviour of a novel submicron-grained austenitic stainless steel in an acidic NaCl solution. *Mater. Sci. Eng. A* **2006**, *431*, 211–217. [CrossRef]
206. Misra, R.D.K.; Girase, B.; Venkata Surya, P.K.C.; Somani, M.C.; Karjalainen, L.P. Cellular mechanisms of enhanced osteoblasts functions via phase-reversion induced nano/submicron-grained structure in a low-Ni austenitic stainless steel. *Adv. Eng. Mater.* **2011**, *13*, B483–B492. [CrossRef]
207. Venkatsurya, P.K.C.; Thein-Han, W.W.; Misra, R.D.K.; Somani, M.C.; Karjalainen, L.P. Advancing nanograined/ultrafine-grained structures for metal implant technology: Interplay between grooving of nano/ultrafine grains and cellular response. *Mater. Sci. Eng. C* **2010**, *30*, 1050–1059. [CrossRef]
208. Nune, C.; Misra, R.D.K. Impact of grain structure of austenitic stainless steel on osteoblasts differention and mineralisation. *Mater. Technol.* **2015**, *30*, 76–85. [CrossRef]
209. Misra, R.D.K.; Nune, C.; Pesacreta, T.C.; Somani, M.C.; Karjalainen, L.P. Understanding the impact of grain structure in austenitic stainless steel from a nanograined regime to a coarse-grained regime on osteoblast functions using a novel metal deformation–annealing sequence. *Acta Biomater.* **2013**, *9*, 6245–6258. [CrossRef] [PubMed]
210. Nune, C.; Misra, R.D.K.; Somani, M.C.; Karjalainen, L.P. Dependence of cellular activity at protein adsorbed biointerfaces with nano- to microscale dimensionality. *J. Biomed. Mater. Res. Part A* **2014**, *102*, 1663–1676. [CrossRef] [PubMed]
211. Gong, N.; Hu, C.; Hu, B.; An, B.; Misra, R.D.K. On the mechanical behavior of austenitic stainless steel with nano/ultrafine grains and comparison with micrometer austenitic grains counterpart and their biological functions. *J. Mech. Behav. Biomed. Mater.* **2020**, *101*, 103433. [CrossRef] [PubMed]
212. Tufan, Y.; Demir, E.C.; Efe, M.; Ercan, B. Efficient fabrication of ultrafine-grained 316L stainless steel surfaces for orthopaedic applications. *Mater. Sci. Technol.* **2019**, *35*, 1891–1897. [CrossRef]
213. Luo, F.; Tang, Z.; Xiao, S.; Xiang, Y. Study on properties of copper-containing austenitic antibacterial stainless steel. *Mater. Technol.* **2019**, *34*, 525–533. [CrossRef]
214. Hong, I.T.; Koo, C.H. Antibacterial properties, corrosion resistance and mechanical properties of Cu-modified SUS 304 stainless steel. *Mater. Sci. Eng. A* **2005**, *393*, 213–322. [CrossRef]
215. Lü, Y.; Hutchinson, B.; Molodov, D.A.; Gottstein, G. Effect of deformation and annealing on the formation and reversion of ε-martensite in an Fe–Mn–C alloy. *Acta Mater.* **2010**, *58*, 3079–3090. [CrossRef]
216. Berrenberg, F.; Haase, C.; Barrales-Mora, L.A.; Molodov, D.A. Enhancement of the strength-ductility combination of twinning-induced/transformation-induced plasticity steels by reversion annealing. *Mater. Sci. Eng. A* **2017**, *681*, 56–64. [CrossRef]
217. Escobar, D.P.; Dafé, S.S.F.; Santos, D.B. Martensite reversion and texture formation in 17Mn-0.06C TRIP/TWIP steel after hot cold rolling and annealing. *J. Mater. Res. Technol.* **2015**, *4*, 162–170. [CrossRef]
218. Dastur, P.; Zarei-Hanzaki, A.; Pishbin, M.H.; Moallemi, M.; Abedi, H.R. Transformation and twinning induced plasticity in an advanced high Mn austenitic steel processed by martensite reversion treatment. *Mater. Sci. Eng. A* **2017**, *696*, 511–519. [CrossRef]
219. Dastur, P.; Zarei-Hanzaki, A.; Rahimi, R.; Moallemi, M.; Klemm, V.; De Cooman, B.C.; Mola, J. Martensite reversion duality behavior in a cold-rolled high Mn transformation-induced plasticity steel. *J. Metall. Mater. Trans. A* **2019**, *50A*, 4550–4560. [CrossRef]
220. Souza Filho, I.R.; Kwiatkowski Da Silva, A.; Sandim, M.J.R.; Ponge, D.; Gault, B.; Sandim, H.R.Z.; Raabe, D. Martensite to austenite reversion in a high-Mn steel: Partitioning- dependent two-stage kinetics revealed by atom probe tomography, in-situ magnetic measurements and simulation. *Acta Mater.* **2019**, *166*, 178–191. [CrossRef]
221. Mohammad-Ebrahimi, M.H.; Zarei-Hanzaki, A.; Abedi, H.R.; Vakili, S.M.; Soundararajan, C.K. The enhanced static recrystallization kinetics of a non-equiatomic high entropy alloy through the reverse transformation of strain induced martensite. *J. Alloys Compd.* **2019**, *806*, 1550–1563. [CrossRef]

© 2020 by the authors. Licensee MDPI, Basel, Switzerland. This article is an open access article distributed under the terms and conditions of the Creative Commons Attribution (CC BY) license (http://creativecommons.org/licenses/by/4.0/).

Review

Repair and Reinforcement of Historic Timber Structures with Stainless Steel—A Review

Marco Corradi [1,*], Adelaja Israel Osofero [2] and Antonio Borri [1]

[1] Department of Engineering, University of Perugia, 06125 Perugia, Italy; antonio.borri@unipg.it
[2] School of Engineering, University of Aberdeen, Aberdeen AB24 3UE, UK; aiosofero@abdn.ac.uk
* Correspondence: marco.corradi@unipg.it; Tel.: +39-(0)-744-492-2908

Received: 20 December 2018; Accepted: 17 January 2019; Published: 21 January 2019

Abstract: Recent trends in the use of stainless steel profiles for repair and reinforcement of historic timber structures, after degradation due to biotic and non-biotic attacks, are discussed in this paper. These structural challenges can vary from inadequate load carrying capacity to complexities involved with choice of repair materials and techniques. Given the recurring requirements of conservation authorities in terms of reversibility of interventions and compatibility between historic and new materials, an increase in the use of non-invasive reinforcement materials and reversible techniques was observed. Subsequently, engineers and researchers have increasingly employed stainless steel alloys in retrofitting historic timber structures. This paper therefore presents the state of the art in the use of stainless steel profiles in retrofitting timber structural elements within historic structures. It includes a review of the development of the retrofitting methods and existing experimental studies on the mechanical behavior of timber structures reinforced with stainless steel. Finally, it presents a number of case studies and draws conclusions on current trends and practices based on reported studies.

Keywords: historic timber structures; stainless steel alloys; connection; reinforcement; repair

---

1. Introduction

The use of stainless steel components, particularly in new constructions, has experienced a huge increase in recent times. Stainless steel is now widely used for both non-structural applications and as ribbed reinforcement bars for concrete, members (beams, columns, ties, etc.) in trusses, bridges and buildings. Similar to structural carbon steel, stainless steel elements can be hot rolled and fabricated or cold rolled formed. However, for many structural and civil engineers, stainless steel can be a confusing material: the term stainless steel refers to a large number of diverse alloys such as the ferritic, the austenitic and duplex stainless steels. Stainless steel often has substantially varying physical and mechanical properties [1–3].

It is well known that the main reason for the use of stainless steel in construction is its resistance to corrosion. However, this property can also vary from one stainless steel alloy to another. Furthermore, there are many other reasons for its application in construction industry: its intrinsic isotropy, the post-elastic plastic behavior, lightness, etc. Most of these properties are in common with carbon steel but are not with other structural materials such as composites.

The use of stainless steel in construction started during the early 20th century with the use of austenitic steel alloys. These alloys were firstly used for reinforcement and stabilization work [4]. In recent times however, there has been an encouraging trend in stainless steel usage, particularly in the construction industry. Approximately 14% of total annual consumption of stainless steel is now used in new constructions [5].

The use of stainless steel alloys for reinforcement of architectural heritage structures has further positive characteristics, making its use even more interesting than in new constructions. In this paper, the state of the art of their use for repair and reinforcement of historic timber constructions will be addressed. In addition to the cited specific strength, corrosion resistance and specific strength, stainless steel possesses other desirable characteristics, for example, its high compatibility with timber makes it suitable for reinforcing timber structures.

The conservation of architectural heritage structures is an important topic in many European countries, with many of them having a government department at national, state or local level, dedicated to this. This should be expected, considering the social, cultural and economic value of these heritage structures. Due to the nature and age of these structures (mostly pre-1920 masonry structures), repair and restoration are often required with the aim of preserving as much of the original structure as possible, i.e., minimal intervention [6,7]. It is also usually required that compatible new materials and reversible retrofitting methods are used. Conservation authorities usually want retrofitting to be done in such a way that reinforcements can be removed in the future, if necessary, without damaging the original timber structure.

The requirements discussed above, in addition to varying loading conditions the structures are often subjected to, call for innovative materials and techniques in retrofitting timber structures. However, it has been challenging to find effective solutions that will also meet the requirements for compatibility and reversibility. Engineers have to make crucial compromises in the meanwhile with ongoing research efforts focusing on the development of new materials and methods for repair and reinforcement. A good example of these efforts is the recent calls for the strategic investments in research and skills, supported by the European Commission (Horizon H2020 call) and national government funding.

One of the most common practices in retrofitting historic timber structures is the use of FRPs (Fiber Reinforced Polymers). FRPs are usually composed of thin fibers of carbon and glass and are, more often than not, epoxy-glued to the deficient timber structure. Several studies have demonstrated that it is possible to reinforce or repair timber structures using composite sheets, bars or strips [8–14]. However, major setbacks in the use of FRPs include their durability, poor compatibility with parent materials, degradation of fiber and poor reversibility [15,16]. These setbacks, which have led some conservation authorities to prohibit or, at least, limit the use of organic adhesives and composite materials on timber structures, is one of the reasons for increased adaptation of stainless steel in retrofitting historic timber structures.

However, there has been limited work done on stainless steel as reinforcement of pre-existing timber structures. This could be due to a number of factors: (1) High cost of stainless steel alloys, (2) Use of competitor materials (FRP) and public perception of FRPs as having excellent mechanical properties, (3) Limited availability, in the construction market, of stainless steel structural profiles.

While composite materials are typically epoxy-glued to timber structures, stainless steel is usually applied by mechanical connectors (screws, bolts, fasteners, etc.). This ensures higher long-term effectiveness of the stainless steel-timber connection compared to FRPs [17]. Other advantages of using stainless steel are reversibility, high durability, compatibility with timber. This is in line with the requirements of the ICOMOS International Wood Committee [18]. Stainless steels are also reversible and high durable. Mechanical connections are removable, and their application causes limited damage to the historic timber structure. While it is well known that high reductions of composites mechanical properties occur in the long run, stainless steel is characterized by negligible mechanical degradation. These characteristics are essentials to satisfying the requirements of conservation authorities and mechanical characteristics when stainless steel is used for repair and reinforcement of heritage timber structures. Stainless steel alloys also possess some aesthetic characteristics, such as an attractive appearance, and minor safety precaution measures are required during application compared to modern solvent-free epoxies and FRPs.

The strength analysis of the connections should always be considered when screwed or bolted connections are used for mechanically attached reinforcement. The resistance of these connections to temperature is of critical importance for timber structures. However, given the relatively low mechanical properties of timber compared to that of the steel screws/bolts, failure usually occurs in the timber material. Tightening parameters and the effects of excess torque in bolts can produce significant damage to historic timber structures, especially when they are made of softwood. Furthermore, defects (knots, shake defects, splits, high-values of grain deviation, etc.) around the area of reinforcement application on the timber structures could compromise the effectiveness of the reinforcement [19].

This study therefore aims to critically review the use of stainless steel for repair and reinforcement of civil timber historic structures. Previously adopted repair and retrofitting techniques are presented and limitations of these techniques are discussed. Finally, suggestions for possible future approaches both in terms of new stainless steel materials and repair techniques are also presented.

2. Stainless Typologies for Structural Applications

The major chemical constituent of Stainless steels is Fe (Iron) alloys with some Cr (Chromium) addition. Chromium acts as the main alloying element (typically between 10–20%) and results in high corrosion resistance due to surface oxidation and protection. Stainless steel corrosion resistance is about 200 times that of normal carbon steel.

Other chemical elements, such as Ni (Nichel), Mo (Molybdenum) and Ti (Titanium) are usually added for special purposes. These added chemical elements have significant effects on the microstructure evolution, mechanical behavior and the corrosion resistance of the resulting stainless steel. Corrosion resistance of stainless steel can be further improved by reducing C (Carbon) content and increasing the content of other elements of the alloy. Generally, stainless steels can be can be categorized according to their chemical compositions as shown in Table 1.

Table 1. Main alloy element compositions according to EN10088 [20,21].

Stainless Steel Grade	1.4016	1.4301	1.4462
Type	Ferritic	Austenitic	Duplex
Molybdenum (%)	-	-	2.5
Nickel (%)	-	8	4.5
Chromium (%)	17	17	21

Austenitic steel is the most common type of stainless steel and is mainly used for food processing equipment, utensils for kitchens and medical equipment. This easily weldable, non-magnetic and not heat-treatable material [22–24] can be divided into three families: Cr-Ni (300 series), Mn-Cr-Ni-N (200 series) and specialty alloys. This family of material is non-magnetic and not heat-treatable. Ferritic steels which usually contain low Ni content, 12–17% Cr and a very low amount of C (<0.1%) could also contain other alloying elements (e.g., Mo, Al, Ti). Although Ferritic steels are known for their good ductility and formability, their behavior at high temperatures is relatively poor when compared to austenitic materials and they are also not heat treatable. For some stainless steel grades (409 and 405), ferritic stainless steels are usually cheaper than many other stainless steels [25–28].

Another type of stainless steels are the martensitic alloys characterized by 11–17% Cr, < 0.4% Ni and relatively high C content (1.2%). Martensitic stainless steels are hardenable and their formability and weldability characteristics are affected by their carbon content. This alloy often requires preheating and post-welding heat treatment to achieve desired properties. They have wide applications in knives, cutting tools, dental and surgical equipment.

Finally, duplex alloys are mainly adopted in chemical plants and piping applications and are characterized by 22–25% Cr and 5% Ni and some Mo and N addition. They have high yield strength and stress corrosion resistance in chloride when compared with austenitic stainless steels. The last type of stainless steel is from precipitation hardening. This type of stainless steel contains Cr-Ni stainless and Al, Cu and Ti as alloying elements. These alloying elements allow the material to harden in a solution and show both austenitic or martensitic microstructure in an aged condition.

Stainless steel prices vary according to their alloy type. Recent prices (Nov. 2018) are as follows: (1) Ferritic steels-1600 €/ton (EN 1.4016); (2) Austenitic steels-2350 €/ton (EN 1.4301); (3) Duplex steels-6500 €/ton (EN 1.4462). Prices will rise (>9000 €/ton) when higher Ni contents (>50%) are required in aggressive environment.

Corrosion resistance of different stainless steel grades is compared by using the Pitting Resistance Equivalent Number (PRE) calculated from Equation (1).

$$PRE = Cr + 3.3 \times Mo \tag{1}$$

where Cr and Mo in Equation (1) are the percentages of Chromium and Molybdenum, respectively.

Stainless steels applications require optimum mechanical properties and corrosion resistance (PRE), both of which strongly depend on the steel microstructure (Figures 1 and 2). The corrosion resistance of stainless steels increases with an increase in the PRE value. For example, for a stainless steel to be considered sea water resistant, it is suggested to have RPE that is greater than 32. Corrosion resistance is also critical for outdoor/unprotected applications. For example, when stainless steel profiles are used to reinforce deficient external masonry structures.

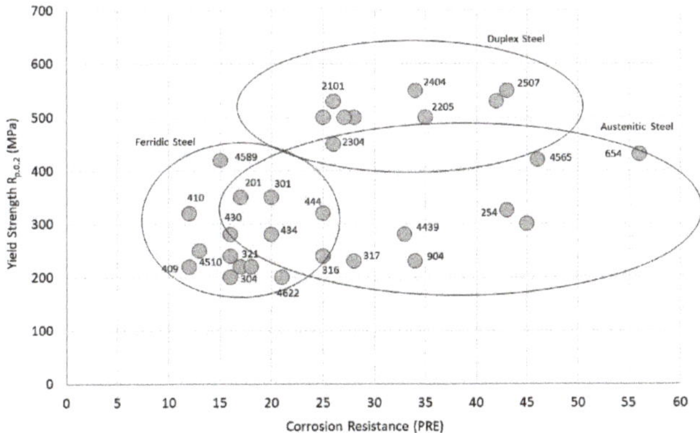

Figure 1. Corrosion resistance (PRE) vs. stainless yield strength (indicative) for different stainless steel grades. Reproduced from [29], with permission from Elsevier, 2018.

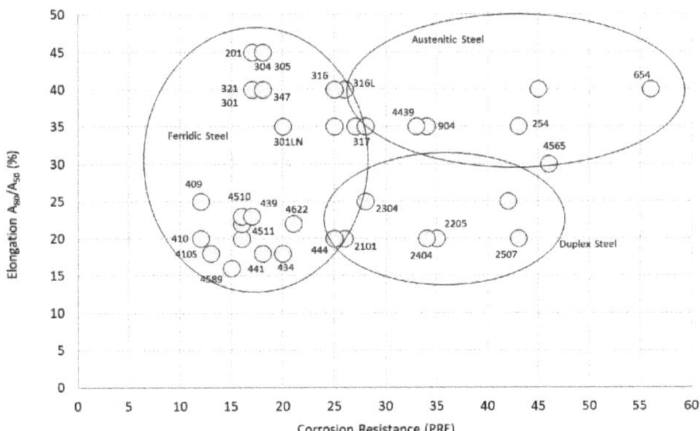

Figure 2. Corrosion resistance (PRE) vs. elongation characteristics (indicative). Reproduced from [29], with permission from Elsevier, 2018.

Yield stress, strain at yield and corrosion resistance all strongly depend on the type of stainless steel: low with ferritic steels and typically high with duplex steels. For example, with a reference value of 275 MPa yield strength (typical of the standard JR275 grade [30]), an elongation at failure of 40% is expected for ferritic stainless steels while about 60% extension is estimated in the case of austenitic steels. Ductility requirement for stainless steel applications varies, i.e., moderate ductility and high modulus are generally required for interventions related to confinement of timber columns or compressed members, while higher ductility is required for bending reinforcement of beams. To avoid stress concentrations, ductility is also a desirable requirement when stainless steels are used to repair rotting wood and for reinforcement interventions [31].

Cost of stainless steel also varies with alloy types as the price depends mainly on the cost of alloying elements rather than on the process cost: Ferritic is the cheapest and duplex steels the most expensive. Irrespective of the alloying material, all stainless steels have characteristic high elongation values which makes it easy to form them and thus commonly used in construction.

3. Typical Timber Historic Structures

Timber has been used for construction for centuries [32,33]. Buildings (both private and public) and infrastructures (bridges, aqueducts, towers, etc.) were usually partially or entirely made of timber structural elements due to their economic benefits (low cost, large supply in nature, ease to transport, work and use, high durability, simple maintenance) (Figures 3–5). The main (technical) reason for its widespread historic use in construction is associated with its fibrous nature: this provide timber material with an excellent tensile and flexural strengths, making this material suitable for roof and floor structures (lintels, ties etc.).

Figure 3. Historic joist floor.

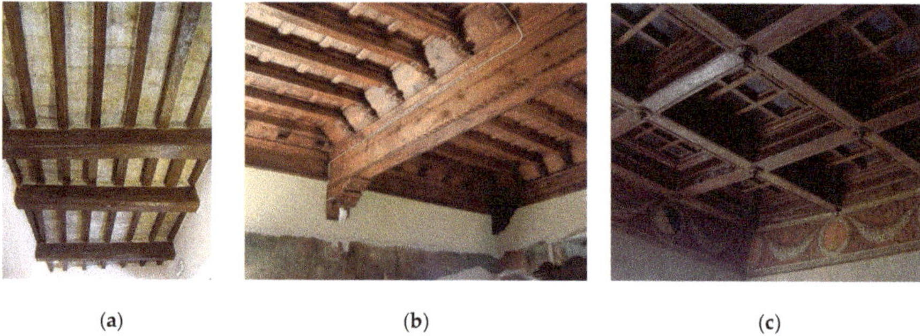

(a) (b) (c)

Figure 4. Different types of horizontal diaphragms (floors): (**a**) common timber-beam floor (solid beams and rafters, tiles); (**b**) decorated and carved timber timber-beam floor; (**c**) bi-directional timber-beam floor.

(**a**)

(**b**)

Figure 5. Different types of roof timber historic structures: (**a**) timber frame for a shed (skillion) roof; (**b**) traditional timber frame for a gable roof.

Timber also exhibits high compressive strength. However, compressive stresses in historic constructions were typically resisted by masonry structural elements. In southern Europe and the Mediterranean basin, buildings often consist of vertical wall elements or pillars, assembled with lime mortar or (rarely) drystone construction, with a frequent use of timber beams for floors and roof structures [34,35]. Although timber is not a heavy material, it wasn't easy or, given its availability, even necessary to transport it in the past from forests. Timber structures were typically made of wood species local to the immediate area.

Rotting timber is the main cause for repair (local) interventions on historic timber structures. Figure 6a,b show two critical points where repair is typically needed: the end (timber-to-masonry connection) of timber beam-floors and the joint between principal rafter and tie-beam for the traditional king post roof truss. As damp and moisture facilitate the attach of biotic agents (*fungi*, insects, *larvae*, etc.), unprotected areas are at high risk.

(**a**) (**b**)

(**c**)

Figure 6. Examples of damage in timber structures: (**a**) and (**b**) Rotting wood at the beam-ends, and at the joists in timber roof truss; (**c**) Deformed timber tie in a frame of a roof structure because of over-loading.

Creep and shrinkage of timber may cause damages (mechanical), leading to excessive deflection, reduction of structural stiffness and strength. Time-dependent deformation under a certain applied dead-load (creep) generally occurs at high temperature, but in timber it can also happen at room conditions. Figure 6c shows a deformed beam due to over-loading. Such deformation is typically produced by bending loads. Differential shrinkage along the three principal directions in timber (radial, tangential and longitudinal) (Figure 7) can cause high reductions of the second moment of inertia of a beam section and also facilitate the attack of biotic agents (insects, fungi, etc.). For mechanical-based damages, interventions with repair or reinforcement are usually required. For the traditional king post roof truss, the relevant terminology and used symbols are reported in Figure 8 [36].

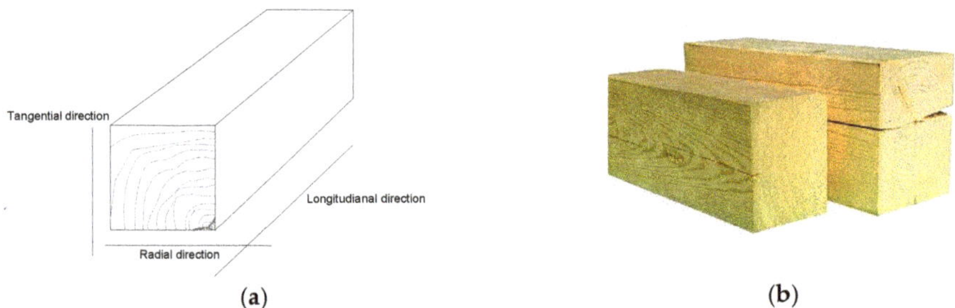

Figure 7. (**a**) Timber is an orthotropic material: the three principal directions (tangential, radial and longitudinal) depend on the grain orientation; (**b**) differential shrinkage can produce significant damage. Shrinkage is typically very high along the tangential direction.

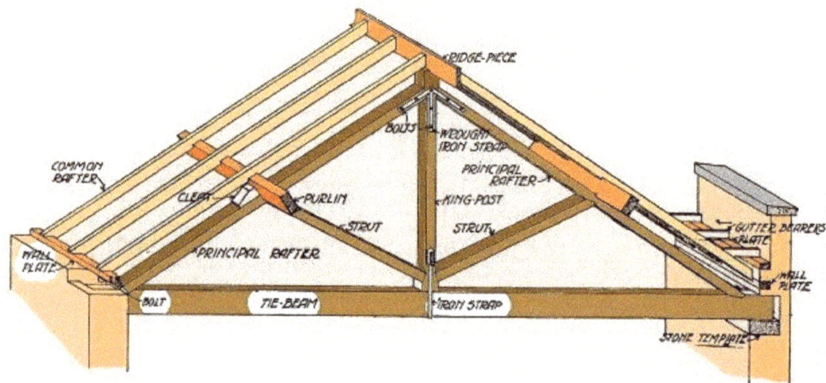

Figure 8. Symbols and terminology used for structural members of the traditional king post roof truss [36].

4. Repair Methods

Damaged timber elements are usually repaired by local interventions [37,38]. Degradation near masonry supports or near the timber-to-timber joints are typical reasons for repair. The use of metal fasteners or profiles is not new, it is also found in repair interventions dated back to the 19th century. Steel and iron have frequently been used in the repair/reinforcement of timber members which may have decayed due to rot and insect infestation or fractured/deformed because of over-loading. Because of the risk of condensation on metals, iron and steel reinforcement should always be visible and open to inspection [39]. This problem could be overcome when stainless steel is used. Although the use of stainless steel often follows the traditional steel repair methods, new technical solutions and methods have been proposed in this area. Table 2 shows the repair methods described in this section, with some information about usability and appropriateness for different repair interventions. It should be remarked that the application of stainless steel elements is often combined with re-construction/repair of pre-existing timber-to-timber lock-joints. The correct use of lock-joints is extremely important in timber connections, following the multi-century-old carpentry tradition of joining timber elements.

Table 2. Repair methods using stainless steel elements.

Method	Appropriate for Historic Structures	Appropriate for Repair of Beam Ends	Appropriate for Repair/Reinforcement of Truss Joints
Fasteners	Yes	No	Yes
Press-bended sheets	Yes	Yes	No
H-, T-, L-, I-shaped Profiles	Yes	Yes	Yes
Rods and prostheses	Yes/No [1]	Yes	Yes
Nail-plates	No	No	Yes

[1] This often depends on the used approach of a local conservation body.

4.1. Stainless Fasteners

The use of metal fasteners to reinforce local timber beams and trusses is a traditional technique (Figure 9). This was very common in the 19th and 20th centuries and it is still applied today [40,41]. The method simply consists in the application of one or more metal fasteners, made of metal strips, to prevent the slippage between two timber elements and increase the resisting section in critical areas. It is typically applied at the beam-to-template connection points and in the timber-to-timber joints of the traditional king post roof trusses (Figure 10).

Figure 9. *Salone dei Cinquecento, Palazzo Vecchio*, Florence, Italy: Detail of timber roof structure and iron fasteners (the iron reinforcement is part of the original structure from 17th century).

(**a**) (**b**) (**c**) (**d**)

Figure 10. The traditional use of metal fasteners in critical sections of the timber beam floor and the king post roof truss: (**a**) timber template-beam (beam-floor); (**b**) joint king post-tie beam (truss); (**c**) joint strut-king post and principal rafter-king post (truss); (**d**) joint principal rafter-tie beam (truss).

A major cause of failures in metal fasteners in timber structures is electrochemical corrosion (rust), caused by the galvanic action. Considering that timber seasoning is not an irreversible process, timber members can absorb moisture. It is well known that air circulation is essential for adequate maintenance of timber structures, but areas under the metal strips typically have higher humidity than the rest of the timber structure and this has deleterious consequences for both the timber and metal materials. Furthermore, the direct contact of damp timber and the fasteners facilitate biotic attacks in the timber and corrosion in the metal. The use of stainless steel strips can partially solve the problem of corrosion. Stainless steel should be carefully selected: high-resilience, high-modulus and high PRE values are sought after characteristics for these applications. Stainless fasteners for repair interventions of historic structures are typically applied on-site by experienced mason. To prevent stainless steel cracking during on-site applications due to high bend angles (for example during wrapping of timber beams with rectangular cross sections), the stainless steel should have high ductile behavior and high resilience strength.

4.2. Re-construction of Beam Ends

4.2.1. Use of Stainless Steel Sections

Figure 3 shows a typical joist floor. This one-way beam-floor is common in the UK, where it is still in use for new construction. To repair the joist ends from rotting at the point where the joists rest in brick or stone walls, pairs of stainless or galvanized steel plates, pre-drilled, are typically used with standard coach screws, which are secured at the end of degraded timber joists (Figure 11). The plates rest in or on the wall and allow for the rotted end to be cut off. To give additional stiffness, for use in rows of continuous end repairs, with no sound joist ends between. A bolted connection through the plates into the parent joist can be adopted [42–44].

(a)

(b)

Figure 11. Re-construction of joist ends with stainless steel L-shaped profiles (press-bended stainless steel sheets) [43].

To reinforce the decayed timber beam ends, stainless steel flitch plates can be inserted along the beam length (Figure 12). This method involves cutting a slot into the timber and making the beam a composite of stainless steel and timber. The plate may also take a T-shape either the right way up (top of the beam) or upside down (underside of the beam). The steel flitch plate is best used at the underside of the beam and fixed upside down, so the wide part of the T is positioned to carry the highest tensile stresses. However, the functionality of the composite beam depends on the type of connection between steel and timber members. When reinforcement is carried out at the ends of simply-supported beams, connection should be able to transfer only the shearing forces (zero bending moment at the supports). In this situation, bolts in the stainless steel web are sufficient. From a practical viewpoint, these are difficult repairs to carry out on site, requiring overhead cutting of the slot by multiple drillings or chain-morticer/chain-saw cutting.

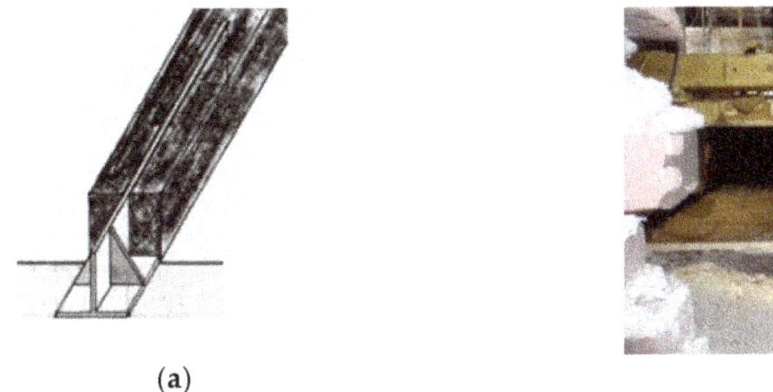

(a) (b)

Figure 12. A stainless steel flitch plate inserted into a timber beam and bolted side-to-side: detail of the steel profile used at the beam ends [44,45].

4.2.2. Use of Stainless Steel Rods

For repairing old solid timber beams, lintels or rafters, it is also common to use timber prostheses to connect with the damaged beam using smooth or ribbed stainless steel rods, connectors or bars. Holes are drilled, or a groove cut, in the undamaged pre-existing timber to accommodate the rod reinforcement. The prosthesis is usually made of a new, laminated structural timber (Figure 13), made in kiln dried wood or epoxy paste (using construction shutters). The rods are typically made of fiberglass, carbon FRP or stainless steel (Figures 14 and 15). Pouring or injection resins are used to fix the rods into the prostheses and the pre-existing beam [45–48].

(a) (b)

Figure 13. Re-construction of the beam ends: (**a**) stainless steel reinforced wooden prosthesis; (**b**) detail of the stainless steel reinforced epoxy prosthesis [48].

Figure 14. Re-construction of joist or beam ends with stainless steel rods [43].

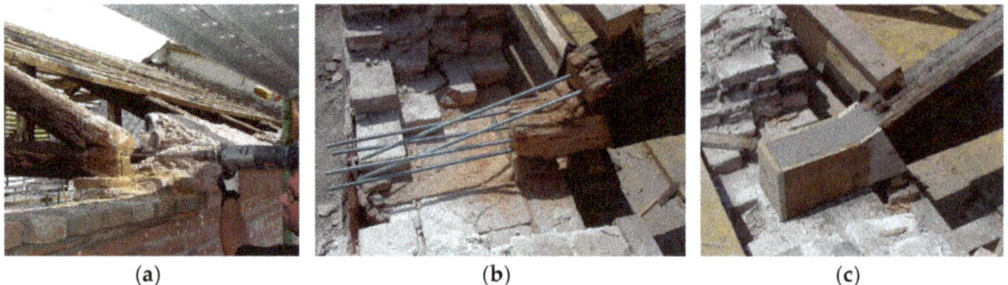

Figure 15. Re-construction of tie or rafter ends with stainless steel rods [45]: (a) drilling the holes for rod installation; (b) insertion of the stainless steel rods; (c) resin casting.

4.2.3. Use of Nail Plates

Nail plates, made of stainless steel, can be used in various applications, mainly for new low-value timber constructions. Nail plates are typically used for repairing or joining timber elements or to increase the flexural stiffness of timber beams and joists. However, these plates are often used to connect wood elements in the same plain. Low carpentry skills are needed in order to construct a timber structure using nail-plated connections.

Although the use of nail plates can be considered a versatile and cost-effective method to improve the connection between wood elements (Figure 16), their use for historic or listed timber structures is rare and not often authorized by conservation bodies. This is mainly due to the fact that a nail-plate connecting system can cause damage to the timber structure, and facilitate biotic attack by saprophytic insects and fungi. For historic structures, it is always preferable to repair/restore the pre-existing historic connecting devices (post-and-beam, tie-and-rafter and other construction techniques). Furthermore, nail-plate connections are rarely able to transfer to maximum allowable internal force from one element to another [49,50].

Figure 16. Examples of connections made with nail-plates [51,52].

5. Reinforcement Methods

Historic timber structures have to transfer external loads to the walls, dealing with the corresponding internal axial load, shearing force and bending moment. The stresses and deformations in the timber structures should not exceed the strength and deformation limits given by modern standards and building codes. Furthermore, the resisting sections of the structure and the mechanical properties of the timber has to be reduced due to mechanical and biological damage. In many situations, except for dismantling and reconstruction of the timber structure, the only option available for structural engineers is to apply a reinforcement. In this section, some traditional and innovative retrofitting solutions will be described and discussed. It should be recommended here that the on-site application of a bending reinforcement should always be preceded by a total or partial removal of the external loads, with the aim of facilitating the stress transfer to the reinforcement during future loading. When this is difficult to carry out, bending moment could be reduced by applying opposite external forces, using hydraulic loading cylinders.

5.1. Reinforcement by Converting a Beam Element into a Trussed Girder

The conversion of a beam-floor into a trussed system is a traditional method, introduced in the 19th century to reinforce deficient timber-beam floors [53–58]. Figure 17a shows a retrofitting intervention carried out in the 1920s on a timber-beam floor in Italy. A more recent application is shown in Figure 17b,c. Nowadays, it is possible to use stainless steel bars, thus highly reducing corrosion problems. A critical aspect of this intervention is the creep deformation of both the timber and the stainless steel materials. Mechanical devices (turnbuckles) are often needed for the stainless ties and low-creep, high-modulus stainless steel alloys should be preferred.

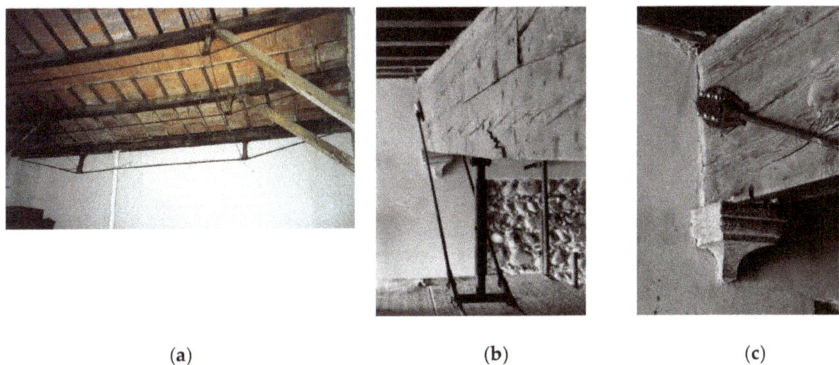

(**a**) (**b**) (**c**)

Figure 17. Examples of bending reinforcement by conversion of a timber-beam floor into a trussed girder by adding steel struts and tension rods underneath ([53] for (**b**) and (**c**)).

Figure 18 shows different retrofitting interventions, designed by professor L. Jurina [54]. The use of stainless steel wire ropes may represent an interesting solution when the reinforcement is exposed and unprotected. The installation of the ropes is relatively easy, the intervention is reversible, and the damage to the timber structure is very limited. However, particular attention should be given to stress concentration problems at the rope-timber joints and it is recommended to take periodic measurements of the creep deformation of the stainless steel ropes. The joint between the wire rope (tie) and the timber rafter is made using two steel plates (Figure 18b). These are inserted into a glulam element connected to the diagonal timber rafter with a bolted connection. The wall-to-wire connection can be effectively realized using anchor bolts (Figure 18c).

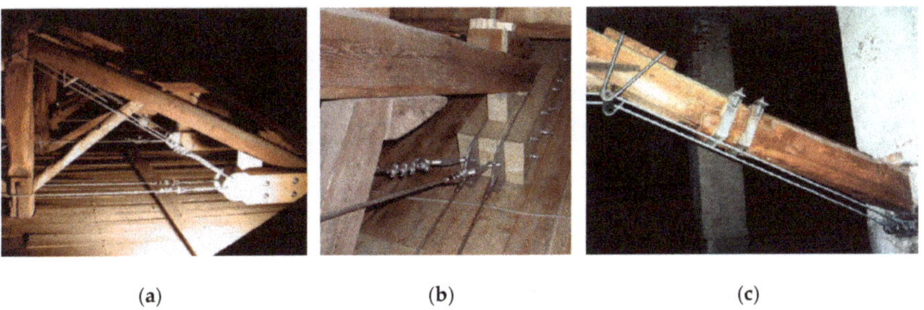

(**a**) (**b**) (**c**)

Figure 18. Reinforcement interventions using stainless steel wire ropes (design, Prof. L. Jurina): (**a**) and (**b**) Residence Masciadri in Arcene, Italy; (**c**) Residence Masciadri in Arcene, Italy [55].

5.2. Reinforcement by Addition of Stainless Steel H-shaped Profiles

When conservation bodies or local authorities do not authorize retrofitting interventions, on listed buildings, at beam intrados (under the timber beams), for example, for the presence of decorations, it is possible to effectively reinforce timber-beam floors by operating on the beam extrados (over the beams) [58] (Figures 19 and 20). In fact, to increase both the bending capacity and stiffness characteristics of existing wood beams, the use of stainless steel profiles placed in the compression side is an interesting solution. Stainless steel elements do not function as a substitute for the timber beams but rather effect an increase in their capacity and flexural stiffness through the creation of a mixed timber–stainless steel structure [59]. The application of stainless steel profiles can be done using epoxy resins, fasteners and/or metal screws or bolts. The use of mechanically attached connectors (screws or bolts) is usually preferred, given the high stress level, under loading, at interface between the timber beam and stainless steel beam. Stainless steel profiles can be also notched to better accommodate the floor joists, without highly affecting the effectiveness of the steel reinforcement. Furthermore, it should be noted that the reinforcement can be completely invisible behind the timber beams, it does not cause significant damage to the timber beams, and it can be easily removed, if needed, providing the method with reversibility characteristics. The reinforcement with H-type elements is characterized by this property and by high machinability of stainless steel elements to be carried out on-site. The reinforcement of timber beams using stainless steel beams is capable of resulting in a significant increase in strength, stiffness, and ductility and the application of the reinforcement is extremely fast and effective.

Figure 19. Reinforcement of a timber beam using a stainless steel H-shaped beam applied on the compressed area.

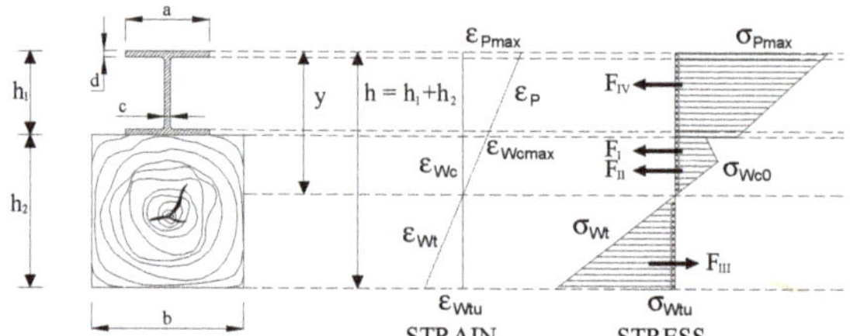

Figure 20. Reinforced timber beam cross section: stress and strain distribution, after timber yielding in compression. Reproduced from [58], with permission from Elsevier, 2007.

5.3. Reinforcement with Side Tension Rods or Plates

Stainless steel rods or plates can also be used as side tension reinforcement. This method simply consists of the application of a reinforcement at the bottom or either side of a timber beam under bending. This can be externally attached or bonded-in. When conversion into a trussed girder is not possible due to space constraints, a possible variation is to use side rods or plates. This method should be used with more caution as the smaller height of the timber-steel cross section produces higher stresses in both materials and problems related to stress concentration are likely to occur near the bolted joints.

5.3.1. Reinforcement with Externally Attached Steel Plates or Cords

Reinforcement of deficient timber beams using epoxy-bonded or screwed stainless steel plates or profiles is not a new technique. Several methods have been developed since the 1960s, which are particularly focused on reinforcing timber beams with steel elements, applied to the beam tension side and bonded by epoxy adhesives [60,61] (Figure 21). More recently, mechanically attached (bolted) stainless steel plates have been proposed with the aim of facilitating the removal of the reinforcement, when needed (the so-called requirement for "reversibility") [62,63]. These reinforcement methods use wide steel plates, steel cords [64–66], I- or L-shaped profiles which cover a large part of the lateral or intrados (bottom) surfaces of the timber beam (Figure 22): this represents a limitation for the application of this reinforcement method, especially when the beams are painted, decorated or carved. Because wood is hygroscopic, its cellulose molecules attract water and, when stainless steel plates are used for reinforcement, areas under the plates could remain more humid and this could facilitate biotic attacks to the timber material.

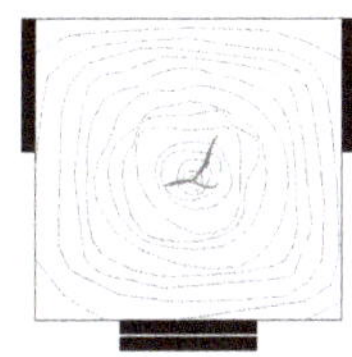

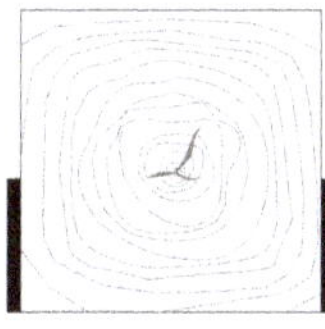

Figure 21. Examples of flexural reinforcement using externally-bonded or screwed stainless steel plates, strips or profiles.

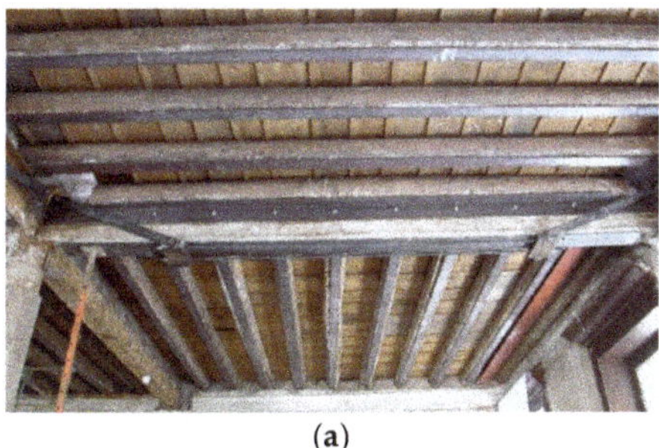
(a)

(b)

Figure 22. Steel plates and profiles as flexural reinforcement of timber beams [64].

When bolted joints are used to connect the plates to the timber, hardwood timber is more appropriate and suitable compared to softwood given higher parallel- and perpendicular-to-grain compressive strengths.

Furthermore, to avoid the risk of delamination of the stainless steel plate under bending loads, the timber surface should be sufficiently flat and smooth and the quantity of epoxy resin (bond-line thickness) should be as small as possible (0.5–3 mm). Delamination phenomena could also be induced by a defect in the timber material (usually a knot located on the tension side) and by the swelling and shrinkage, from cyclic moisture content variations in timber. The long term behavior of the epoxy bonding agent and its phase transformation could highly affect this reinforcement method: for these reasons it is always recommended to use mechanical devices (metal screws or bolts) to connect the plates with the timber beam. The main advantage of this reinforcement method is its rapidity of application. The timber beams are usually not notched or damaged during the reinforcement application.

The recent use of screwed connections is a revival of an old method in use long before strong adhesives became available in the construction market. This method includes the application of a stainless reinforcement using high-strength screws or bolts. This method has several advantages: it eliminates the use of organic adhesive (typically, an epoxy resin) and thus meets the requirement of reversibility of the reinforcement and compatibility with the wood material. For listed buildings, these are often essential conditions for authorization of reinforcement interventions by a statutory conservation body. However, one of the main limitations of the use of mechanical devices is the stress concentration that can occur near the screws/bolts. Because stress transfer from timber to stainless steel reinforcement is guaranteed by these devices, stress peaks can cause local failures and slippage phenomena between the two materials. Such phenomena can compromise the action of the reinforcement and should be studied and avoided. In order to prevent this, it is recommended to avoid the use of connecting mechanical devices for reinforcement of softwood beams.

Figure 23a shows the application of stainless steel cords. The application of the cords at the beam bottom is a relatively simple operation: an epoxy paste is needed to glue and protect the fibers on the beam surface (Figure 23b). The negligible bending stiffness of the stainless steel fibers also allows for pre-stressing them: a mechanical device, consisting of a metal cylinder, positioned at one beam end, can be used for this purpose. Fibers are wound around the metal cylinder (Figure 23c) using a clamp and fastened to the other end, until the design target-value of the clamping couple is reached. At the end of curing time of the epoxy paste (typically 48–72 h), the mechanical pre-stressing device can be removed. The final result is a timber beam strengthened, at the tension side, with pre-tensioned stainless steel cords.

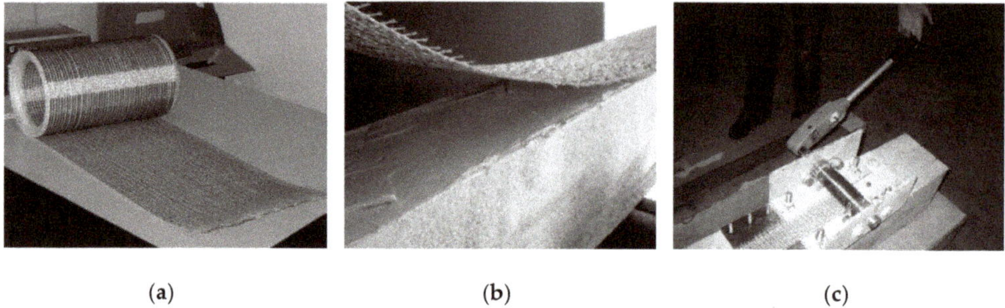

(**a**) (**b**) (**c**)

Figure 23. Use of steel cords: (**a**) steel-fiber coil, (**b**) application of the cords using an epoxy paste, (**c**) pre-stressing device. Reproduced from [66], with permission from Elsevier, 2011.

Considering that the state of the timber material near the bonded surface is of critical importance for the stress transfer between timber and stainless steel, the wood decay should always be assessed by means of on-site penetration tests or similar non-destructive techniques.

5.3.2. Reinforcement with Bonded-In Steel Rods or Strips

The introduction of FRPs in Civil Engineering in the 1980s facilitated the diffusion of bonded-in reinforcement methods. Glass or Carbon FRP rods were widely used to reinforce, on the tension side, deficient timber beams [67,68] (Figure 24). These composite elements are associated with high strength and stiffness to weight ratios. However, among the drawbacks of FRP rods is the linear elastic stress–strain relationship (brittle behavior) and the weak long-run response with high reductions of their tensile strength and stiffness. In such context, the replacement of FRP rods with stainless steel ones could be considered as a viable alternative solution. The reinforcement method consists of the application of stainless steel rods or strips in internal grooves, connected to the timber beam using structural adhesives (typically epoxy resins) [69–72]. A large variety of reinforcement configurations can be used, depending on many factors: presence of decorative ceilings, carved beams, fire protection requirements. The fire requirement for non-exposed surfaces is often the main reason for the exclusion of the use of externally bonded stainless steel plates.

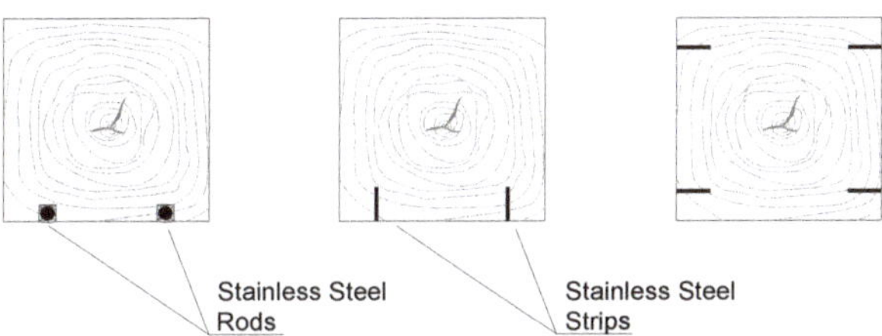

Figure 24. Examples of flexural reinforcement using bonded-in stainless steel rods or strips.

It is worth noting that, because timber and resins differ in their reactions to variations in humidity and loading, high shear stresses and local failures could result from differential shrinkage and swelling at the adhesive bond.

Bonded-in stainless steel reinforcements has some drawbacks (Figure 25): timber beams are usually damaged due to the use of a groove cutter for the installation of the rods/strips, and the use of adhesives is not recommended, given their unsatisfactory long-term behavior. However, several positive characteristics can also be highlighted: because the reinforcement is confined in the groove and it is completely embedded in the resin, the risk of the delamination of the steel reinforcement is very low. Reinforcement delamination is typically induced both by bending loads (and resulting shear stresses at interface) and by the swelling and shrinkage of the wood. Furthermore, unlike the externally-applied reinforcements, glued to the beam's sides or to the bottom surface, the choice of bonded-in methods allows the epoxy adhesive layer to be partially protected in case of fire by applying a longitudinal wooden board, which covers the groove.

(a) **(b)**

Figure 25. Bonded-in steel bar reinforcement: (**a**) detail of the reinforcement before resin application; (**b**) filling of the grooves with epoxy paste.

When metal strips are used, it is recommended to apply them vertically. Figure 26 shows an on-site experimental investigation [71]: the 15th-century timber beams of an historic building in Brescia, Italy. The width of the grooves was 16 mm and the depth varied from 60 mm at the mid-span to 0 mm at the groove ends in order to minimize wood removal. Strips were 50 mm × 4 mm × 4200 mm in dimensions. They were applied vertically using both an epoxy resin and diagonal high strength steel nails.

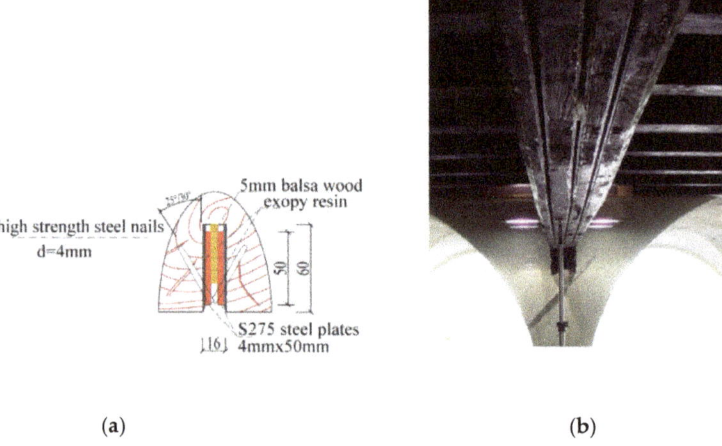

(**a**) (**b**)

Figure 26. Bonded-in steel strip reinforcement: (**a**) detail of the groove; (**b**) layout of the reinforced beam. Reproduced from [71], with permission from Elsevier, 2016.

6. Reinforcement of Wooden Floors

The rehabilitation of historic buildings both for new use and occupancy often requires the reinforcement of the wooden floors. Reinforcement could be needed to increase the flexural capacity and stiffness (vertical static loading) or to improve the structural response against in-plane seismic loads (horizontal dynamic loading). In seismic prone areas, horizontal diagrams in buildings have the critical function to transfer the seismic loads from the walls perpendicular-to-the-seismic- direction to the parallel-to-the-seismic-direction walls (Figure 27). To achieve this, an increment of the lateral in-plane stiffness of wooden floors is often necessary.

(**a**) (**b**)

Figure 27. In seismic prone areas, wooden floors may prevent the out-of-plane mechanism of external walls (overturning). To facilitate this function, the wooden floors should be stiff enough (in-plane stiffness) and effectively connected to the walls to transfer the seismic load to the walls parallel to the seismic direction.

To increase both the flexural and in-plane stiffnesses, an effective strengthening method consists in the use of stainless steel screws to connect a new Reinforced Concrete (RC) slab, typically applied over the floor, or a double-layer wooden floor boards to the existing timber joists or beams [73,74]. By preventing relative sliding between the new slab/boards and the underlying pre-existing timber structure, it is possible to highly increase the second moment of inertia of the resisting cross section [75–80].

In respect of the traditional reinforcement of one-way wooden floors with a steel mesh reinforced concrete (RC) slab (Figure 28), it is important to document research efforts on the use of stainless steel studs. These are typically used to connect the new RC slab with the pre-existing underlying timber structure [81–83]. The advantages of this method are numerous: the bending stiffness and capacity of the floor are highly enhanced (given the increment of the second moment of inertia of the cross section); the in-plane response of the floor is also improved as the steel-mesh reinforcement is able to provide the needed tensile strength to the RC slab. If the floor or the RC slab is properly connected to the load-bearing walls, the seismic response of the building can be significantly enhanced.

(a) (b) (c)

Figure 28. Traditional reinforcement of one-way wooden floor. A steel mesh reinforced concrete slab is applied over the wooden floors (there are no connections between the slab and the timber beams or rafters): (**a**) schematic arrangement; (**b**) RC slab application; (**c**) use of autoclaved aerated concrete; to reduce the RC slab weight (dead loads).

The damage produced to the timber beams/rafter due to the application of the stainless steel studs is negligible. However, it should be remarked that the application of RC slab can also increase the magnitude of the dead load (Figure 28b). A mitigation measure of this problem is the use of autoclaved aerated concrete (Figure 28c), but this should be applied with caution given the reduced mechanical properties of this type of concrete. However, the use of autoclaved aerated concrete can produce savings in terms of energy consumption, given its highly thermally insulating properties. Different types of connectors are available in the construction market. Examples are shown in Figure 29.

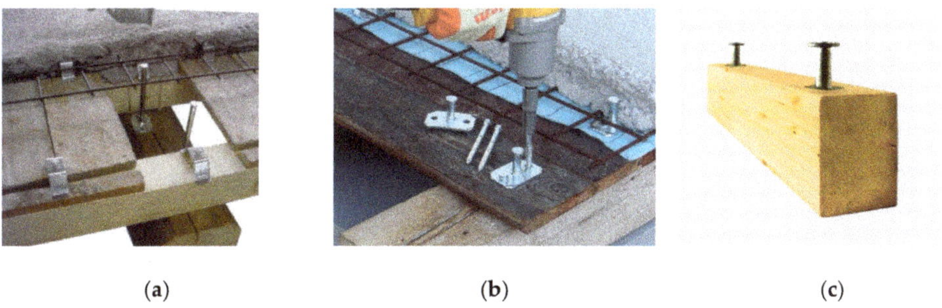

(a) (b) (c)

Figure 29. Examples of metal and shear stainless steel studs [82,83].

The use of low-cost boards is an interesting solution, given also its positive characteristics in increasing the thermal insulation of the building and thus reducing its energy usage (Figure 30). However, while this method is usually effective for static vertical loads, its effectiveness is limited in seismic prone areas, as the boards are typically made of plywood or are particle-made boards. Their in-plane stiffness and strength is usually low and the boards are usually subjected to swelling due to moisture absorption. The use of large solid-wood boards, effectively connected to the underneath joists, can solve the problem, but this usually lead to an increase in the cost of the intervention.

(a) (b)

Figure 30. Examples of use of wooden boards.

Figure 31 shows a schematic layout of the reinforcement of wooden floor by adding a new layer of wood boards. These can be effectively connected to the underneath wooden floor boards and timber beams by using inclined high-strength stainless screws. A variation of the reinforcement method in Figure 31 consists of the application of diagonal stainless steel strips [80] (Figure 32), screwed to the pre-existing underlying wood boards and beams/joists.

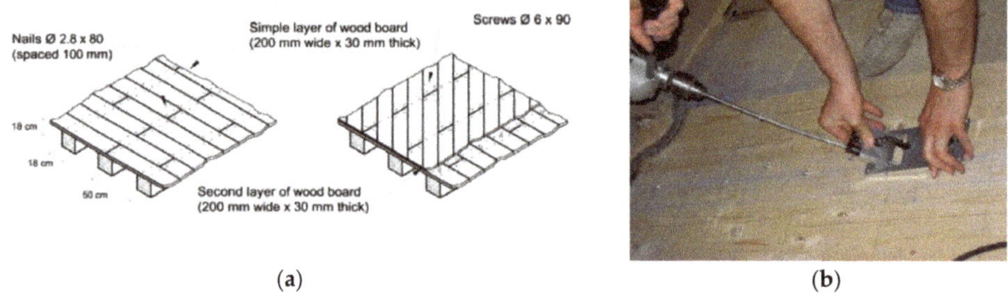

(a) (b)

Figure 31. Reinforcement of timber floor: two layers of wood boards can be added over a pre-existing timber beam floor. The use of stainless inclined steel screws to connect the boards to the beams may highly increase both flexural and in-plane stiffnesses. Reproduced from [80], with permission from Elsevier, 2015.

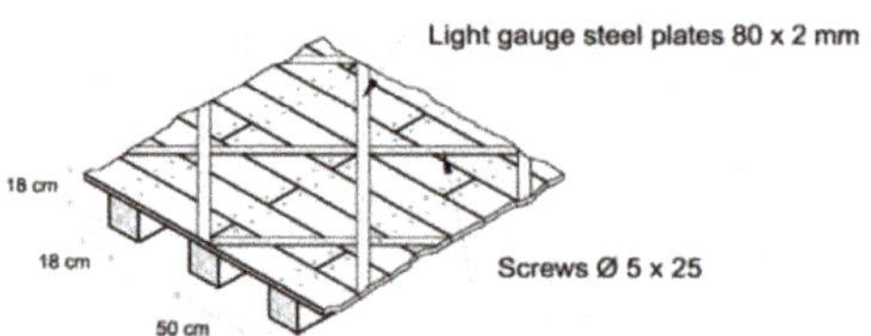

Figure 32. A possible alternative solution to the use of a double layer of timber boards—application of stainless steel strips over the new or pre-existing wooden floor boards. Reproduced from [80], with permission from Elsevier, 2015.

7. Future Developments and Possibilities

Stainless steel does not refer to a single material but to a family of corrosion resistant steels and its use for repair and retrofit of timber structures may vary with different solutions and possibilities. The main obstacle to its widespread diffusion is not its high cost (from 2 to 20 times higher compared to carbon steel), but limitations in engineers' and architects' knowledge of its mechanical properties and characteristics. In many situations, this material is lumped with standard carbon steel with the additional property of higher corrosion resistance. This is somewhat simplistic. Its yield strength can vary from 180 to >580 MPa. Worldwide demand for stainless steel is increasing at a rate of about 5% per annum and annual consumption is now well over 20 million tonnes and still rising. New uses are being continuously found for its attractive appearance, corrosion resistance, low maintenance and high strength. Furthermore, stainless steel has no need for painting or other protective coatings. Reinforcements of timber structures are often left unprotected and exposed. In this situation the use of stainless steel fits well with the need for corrosion resistance and low maintenance. The more specific requirement for reversibility of the repair/reinforcement intervention emphasizes the importance of screwed and mechanically attached applications of reinforcement: in this area, the use of stainless steel has several advantages over bonded composite reinforcements. Finally, there are two additional positive characteristics of stainless steel when compared to composite materials: its negligible mechanical degradation with time (ageing) and its isotropic behavior. Unlike composite materials, stainless steel has a satisfactory response when loaded in different directions, and also when bended or loaded over the yield strength.

Another important factor to consider in civil applications is the ductile behavior. Ductility tends to be defined by the % elongation during a standard tensile test. The elongation for austenitic stainless steels is quite high (Figure 2) and this is considered positive for timber structures exposed to earthquake actions. Furthermore, cryogenic (low temperature) resistance is also high and the use of stainless steel for reinforcement and repair of outdoor timber structures is quite straightforward: at cryogenic temperatures the tensile strengths of austenitic stainless steels are substantially higher than at ambient temperatures. All these factors mean stainless steel can be economically viable for repair and reinforcement of timber structures once service life and life-cycle costs are considered.

8. Conclusions

Metal methods of strengthening and repairing timber structures have been in use for about 200 years. Surveys, analyses and recordings of these methods have been undertaken. Information on the ways in which metals-timber joints behave is also presented. A major cause of failures in metal reinforcements of timber structures is electrochemical corrosion (rust), caused by the galvanic action, facilitated by the absorption of the air humidity from timber members.

Today, the use of stainless steel profiles, strips, screws, rods represent an interesting solution to the problem of corrosion of metal reinforcements. Stainless steel can provide an efficient and durable method of reinforcement and for making connections in timber structures. Furthermore, for pre-existing, old and historic timber structures, recent seismic codes allow internal stresses to be higher than the elastic limit and typically require that the timber members exhibit a sufficient post elastic behavior, in terms of deformation and dissipation capacities. In such context, the use of stainless steel could be considered as an interesting solution, given the intrinsic plasticity of the stainless steel material. Brittle materials (i.e., glass or carbon fibres) should be avoided, as much as possible, in areas of high stress concentration or dissipating zones.

This paper summarized the actual use of stainless steel in repair and reinforcement of deficient timber structures. It has been demonstrated that the use of stainless steel is nowadays common in such applications. However, structural engineers often have a superficial technical knowledge of stainless steel. The different stainless typologies for structural applications, the pitting resistance, the post-elastic behavior and mechanical properties remain sometimes unclear to them. This information could be even more important when numerical models are implemented to capture the structural response of stainless steel repaired or reinforced timber structures.

In addition to more traditional solutions (use of stainless steel fasteners, conversion of a timber beam element into a trussed girder), more recent solutions have been discussed in this paper. Most of these solutions are not new and were introduced in the 1980s and 1990s. However, it should be remarked that a repair or reinforcement intervention on historic structures should not cause a substantial modification of the original structural conception and adhere with the principle of "minimum intervention and maximum retention of materials" and low invasiveness.

Author Contributions: Conceptualization, M.C. and A.B.; writing—original draft preparation, A.I.O. and M.C.; writing—review and editing; supervision, M.C. and A.I.O.

Funding: This research received no external funding.

Acknowledgments: The authors wish to thank Marco Frigo of Outokumpu.

Conflicts of Interest: The authors declare no conflict of interest.

References

1. Gedge, G. Structural uses of stainless steel—Buildings and civil engineering. *J. Constr. Steel Res.* **2008**, *64*, 1194–1198. [CrossRef]
2. Gedge, G. Duplex steels for durable bridge construction. In Proceedings of the International Association for Bridge and Structural Engineering (IABSE) Conference, Weimar, Germany, 16–21 September 2007.
3. Baddoo, N.R.; Burgan, B. *Structural Design of Stainless Steel*; Steel Construction Institute: London, UK, 2001; SCI P291.
4. McGurn, J.F. Stainless steel reinforcing bars in concrete. In Proceedings of the Intern. Conference of Corrosion and Rehabilitation of Reinforced Concrete Structures FHWA, Orlando, FL, USA, 7–11 December 1998.
5. Baddoo, N.R. Stainless steel in construction: A review of research, applications, challenges and opportunities. *J. Constr. Steel Res.* **2008**, *64*, 1199–1206. [CrossRef]
6. Venice Charter. International charter for the conservation and restoration of monuments and sites. In Proceedings of the 2nd International Congress of Architects and Technicians of Historical Monuments, Venice, Italy, 25–31 May 1964.
7. Krakow Charter. Principles for conservation and restoration of built heritage. In Proceedings of the International Conference on Conservation, Krakow, Poland, 23–24 November 1998.
8. Pleviris, N.; Triantafillou, T.C. FRP-reinforced wood as structural material. *J. Mater. Civ. Eng.* **1992**, *4*, 300–317. [CrossRef]
9. Triantafillou, T.C. Shear reinforcement of wood using FRP materials. *J. Mater. Civ. Eng.* **1997**, *9*, 65–69. [CrossRef]
10. Borri, A.; Corradi, M.; Grazini, A. FRP reinforcement of wood elements under bending loads. In Proceedings of the 10th International Conference on Structural Faults + Repair, London, UK, 1–3 July 2003.
11. Gentile, C.; Svecova, D.; Rizkalla, S.H. Timber beams strengthened with GFRP bars: Development and applications. *J. Compos. Constr.* **2002**, *6*, 11–20. [CrossRef]
12. Borri, A.; Corradi, M.; Grazini, A. A method for flexural reinforcement of old wood beams with CFRP materials. *Compos. Part B Eng.* **2005**, *39*, 143–153. [CrossRef]
13. Miceli, F.; Scialpi, V.; La Tegola, A. Flexural reinforcement of glulam timber beams and joints with Carbon Fiber-Reinforced Polymer rods. *J. Compos. Constr.* **2005**, *9*, 337–347. [CrossRef]
14. Gilfillan, J.; Gilbert, S.; Patrick, G. The use of FRP composites in enhancing the structural behavior of timber beams. *J. Reinf. Plast. Compos.* **2003**, *22*, 1373–1388. [CrossRef]
15. Chu, W.; Wu, L.; Karbhari, V.M. Durability evaluation of moderate temperature cured E-glass/vinylester systems. *Compos. Struct.* **2004**, *66*, 367–376. [CrossRef]
16. Karbhari, V.M.; Chin, J.W.; Hunston, D.; Benmokrane, B.; Juska, T.; Morgan, R.; Lesko, J.J.; Sorathia, U.; Reynaud, D. Durability gap analysis for Fiber-Reinforced Polymer composites in civil infrastructure. *J. Compos. Constr. ASCE* **2003**, *7*, 238–247. [CrossRef]
17. Krempl, E. An experimental study of room-temperature rate sensitivity, creep and relaxation of AISI 304 stainless steel. *J. Mech. Phys. Solids* **1979**, *27*, 363–375. [CrossRef]
18. ICOMOS International Wood Committee. *Principles for the Preservation of Historic Timber Structures*; ICOMOS General Assembly: Guadalajara, Mexico, 1999.
19. Camanho, P.; Hallett, S. *Composite Joints and Connections: Principles, Modelling and Testing*; Woodhead Publishing: Cambridge, UK, 2011.
20. *EN10088 Part 1 Stainless Steels Part 1—List of Stainless Steels*; BSI: London, UK, 1995.

21. *EN10088 Part 2 Stainless Steels. Technical Delivery Conditions for Sheet/Plate and Strip of Corrosion Resisting Steels for General Purposes*; BSI: London, UK, 2005.
22. Sanstrom, R.; Bergqvist, H. Temperature dependence of tensile properties and strengthening of nitorgen alloyed austenitic stainless steels. *Scand. J. Metall.* **1977**, *6*, 156–169.
23. Di Schino, A.; Longobardo, M.; Turconi, G.L.; Porcu, G.; Scoppio, L. Metallurgical design and development of C125 grade for mild sour service application. In Proceedings of the NACE—International Corrosion Conference, San Diego, CA, USA, 12–16 March 2006; pp. 061251–0612514.
24. Di Schino, A.; Kenny, J.M.; Salvatori, I.; Abbruzzese, G. Modelling primary recrystallization and grain growth in a low nickel austenitic stainless steel. *J. Mater. Sci.* **2001**, *36*, 593–601. [CrossRef]
25. International Molybdenum Association. *Practical Guidelines for the Fabrication of Duplex Stainless Steels*; IMOA: London, UK, 2014.
26. Outokumpu. Available online: http://www.outokumpu.com/SiteCollectionDocuments/Outokumpu-Product-Range-Wallchart.pdf (accessed on 14 November 2018).
27. *European Standard EN ISO 15156-1:2015, NACE MR0175 Petroleum and Natural Gas Industries—Materials for Use in H2S-Containing Environments in Oil and Gas Production—Part 1: General Principles for Selection of Cracking-Resistant Materials*; ISO: Geneva, Switzerland, 2015.
28. National Association of Corrosion Engineers (NACE). *TM-01-69:1995 (R2000)—Laboratory Corrosion Testing of Metals*; NACE: Houston, TX, USA, 2000.
29. Corradi, M.; Di Schino, A.; Borri, A.; Rufini, R. A review of the use of stainless steel for masonry repair and reinforcement. *Constr. Build. Mater.* **2018**, *181*, 335–346. [CrossRef]
30. General Technical Delivery Conditions. *European Standard EN 10025-1:2004 Hot Rolled Products of Structural Steels*; CEN European Committee for Standardisation: Brussels, Belgium, 2004.
31. Di Caprio, G. *Gli Acciai Inossidabili*; Hoepli: Milan, Italy, 2003.
32. Panshin, A.J.; deZeeuw, C. *Textbook of Wood Technology*, 4th ed.; McGraw–Hill: New York, NY, USA, 1980.
33. Parisi, M.A.; Piazza, M. Restoration and strengthening of timber structures: Principles, criteria, and examples. *Pract. Period. Struct. Des. Constr.* **2007**, *12*, 177–185. [CrossRef]
34. Barbaro, D. *I Dieci Libri Dell'architettura di M. Vitruvio*; Il Polifilo: Milan, Italy, 1997.
35. Tampone, G. *Conservation of Historic Wooden Structures*; Hoepli: Florence, Italy, 2005.
36. Unalansusam. Available online: http://unalansusam.com/roof-struts/kptruss-digital-art-gallery-roof-struts/ (accessed on 23 November 2018).
37. Tampone, G.; Mannucci, N.; Macchioni, N. *Strutture di Legno. Cultura Conservazione Restauro*; De Lettera: Rome, Italy, 2002.
38. Mariani, M. *Consolidamento Delle Strutture Lignee con L'acciaio*; DEI: Rome, Italy, 2004.
39. Charles, F.W.B. Dismantling, Repairing and Rebuilding as a Means of Conservation. In Proceedings of the ICOMOS UK: Timber Engineering Conference, London, UK, 8 April 1992.
40. Bulleit, W.M. Reinforcement of wood materials: A review. *Wood Fiber Sci.* **2007**, *16*, 391–397.
41. Giordano, G. *Tecnica Delle Costruzioni in Legno*; Hoepli: Milan, Italy, 1999.
42. Franke, S.; Franke, B.; Harte, A.M. Failure modes and reinforcement techniques for timber beams–State of the art. *Constr. Build. Mater.* **2015**, *97*, 2–13. [CrossRef]
43. Buildingconservation. Available online: http://www.buildingconservation.com/articles/structural-timber-repairs/structural-timber-repairs.htm (accessed on 21 October 2018).
44. Joist-repair. Available online: https://www.joist-repair.co.uk/joistendrepair_diydoc.htm (accessed on 3 December 2018).
45. Fiorentecnica. Available online: http://www.fiorentecnica.it/paginaintx02.html (accessed on 25 November 2018).
46. Alessandri, C.; Mallardo, V.; Pizzo, B.; Ruocco, E. The roof of the Church of the Nativity in Bethlehem: Structural problems and intervention techniques. *J. Cult. Herit.* **2012**, *13*, 70–81. [CrossRef]
47. Pizzo, B.; Gavioli, M.; Lauriola, M.P. Evaluation of a design approach to the on-site structural repair of decayed old timber end beams. *Eng. Struct.* **2013**, *48*, 611–622. [CrossRef]
48. Resine-epossidiche. Available online: https://resine-epossidiche.it/edilizia-conservativa/ (accessed on 25 November 2018).

49. Blass, H.J.; Schmid, M.; Litze, H.; Wagner, B. Nail plate reinforced joints with dowel-type fasteners. In Proceedings of the 6th World Conference on Timber Engineering, British Columbia, Canada, 31 July–3 August 2000.
50. Karadelis, J.N.; Brown, P. Punched metal plate timber fasteners under fatigue loading. *Constr. Build. Mater.* **2000**, *14*, 99–108. [CrossRef]
51. Fordevr. Available online: https://www.fordevr.com/nail-plate-construction.html (accessed on 10 January 2019).
52. Beauty. Available online: http://beauty.fzl99.com/gang-nail-truss-plates/ (accessed on 10 January 2019).
53. Studiofornale. Available online: http://www.studiofornale.it/intervento-di-consolidamento-trave-in-legno-lesionata-con-.contraffisso-metallico/ (accessed on 15 December 2018).
54. Larsen, K.E.; Marstein, N. Conservation of historic timber structures: An ecological approach. In *Butterworth-Heinemann Series in Conservation and Museology*; Butterworth Heinemann: Oxford, UK, 2000.
55. Jurina, L. L'uso Dell'acciaio nel Consolidamento delle Capriate e dei Solai in Legno. 2004. Available online: http://www.jurina.it/10/2012/02/2011_L%E2%80%99uso-dell%E2%80%99acciaio-nel-consolidamento-delle-capriate-e-dei-solai-in-legno.pdf (accessed on 21 January 2019).
56. New South Wales State Heritage Office. *The Maintenance Series, Timber Repairs, Information Sheet 5.2*; New South Wales State Heritage Office: Sidney, Australia, 1998.
57. Song, J.K.; Kim, S.Y.; Yang, K.H.; Han, S.A. Flexural behavior of post-tensioned glued-laminated timber with wire rope. In Proceedings of the International Conference on Sustainable Building Asia, Seoul, Korea, 27–29 June 2007.
58. Corradi, M.; Borri, A. Fir and chestnut timber beams reinforced with GFRP pultruded Elements. *J. Compos. Part B* **2007**, *38*, 172–181. [CrossRef]
59. Borgin, K.B.; Loedolff, G.F.; Saunders, G.R. Laminated wood beams reinforced with steel strips. *J. Struct. Div.* **1968**, *94*, 1681–1706.
60. Alam, P.; Ansell, M.P.; Smedley, D. Mechanical repair of timber beams fractured in flexure using bonded-in reinforcements. *Compos. Part B Eng.* **2009**, *40*, 95–106. [CrossRef]
61. Jasieńko, J.; Nowak, T.P. Solid timber beams strengthened with steel plates—Experimental studies. *Constr. Build. Mater.* **2014**, *63*, 81–88. [CrossRef]
62. Corradi, M.; Borri, A.; Castori, G.; Speranzini, E. Fully reversible reinforcement of softwood beams with unbonded composite plates. *Compos. Struct.* **2016**, *149*, 54–68. [CrossRef]
63. Corradi, M.; Vo, T.P.; Poologanathan, K.; Osofero, A.I. Flexural behaviour of hardwood and softwood beams with mechanically connected GFRP plates. *Compos. Struct.* **2018**, *206*, 610–620. [CrossRef]
64. Resinproget. Available online: http://www.resinproget.it/bverona.html (accessed on 21 January 2019).
65. Ghanbari Ghazijahani, T.; Jiao, H.; Holloway, D. Composite timber beams strengthened by steel and CFRP. *J. Compos. Constr.* **2017**, *21*, 04016059. [CrossRef]
66. Borri, A.; Corradi, M. Strengthening of timber beams with high strength steel cords. *Compos. Part B Eng.* **2011**, *42*, 1480–1491. [CrossRef]
67. Raftery, G.M.; Kelly, F. Basalt FRP rods for reinforcement and repair of timber. *Compos. Part B Eng.* **2015**, *70*, 9–19. [CrossRef]
68. Bainbridge, R.; Mettem, C.; Harvey, K.; Ansell, M. Bonded-in rod connections for timber structures-development of design methods and test observations. *Int. J. Adhes. Adhes.* **2002**, *22*, 47–59. [CrossRef]
69. Wu, W.; Gou, H. Analyzing of the timber beams of ancient buildings strengthened with NSM stainless steel bars. *Int. J. Sci.* **2015**, *2*, 77–87.
70. Lantos, G. The flexural behavior of steel reinforced laminated timber beams. *Wood Sci.* **1970**, *2*, 136–143.
71. Metelli, G.; Preti, M.; Giuriani, E. On the delamination phenomenon in the repair of timber beams with steel plates. *Constr. Build. Mater.* **2016**, *102*, 1018–1028. [CrossRef]
72. Bulleit, W.M.; Sandberg, L.B.; Woods, G.J. Steel-reinforced glued laminated timber. *J. Struct. Eng.* **1989**, *115*, 433–444. [CrossRef]
73. Branco, J.M.; Tomasi, R. Analysis and strengthening of timber floors and roofs. In *Structural Rehabilitation of Old Buildings*; Costa, A., Miranda Guedes, J., Varum, H., Eds.; Springer: Berlin, Germany, 2013; pp. 235–258.
74. Turrini, G.; Piazza, M. Una tecnica di recupero dei solai in legno. *Recuperare* **1983**, *5*, 396–407.

75. Modena, C.; Valluzzi, M.R.; Garbin, E.; da Porto, F. A strengthening technique for timber floors using traditional materials. In Proceedings of the 4th structural analysis of historical constructions, Padua, Italy, 10–13 November 2004.
76. Piazza, M.; Tomasi, R.; Baldessari, C.; Acler, E. Behaviour of refurbished timber floors characterized by different in-plane stiffness. In Proceedings of the 6th International Conference on Structural Analysis of Historic Construction (SAHC), Bath, UK, 2–4 July 2008.
77. Valluzzi, M.; Garbin, E.; Benetta, M.D.; Modena, C. Experimental assessment and modelling of in-plane behavior of timber floors. In Proceedings of the 6th International Conference on Structural Analysis of Historic Construction (SAHC), Bath, UK, 2–4 July 2008.
78. Angeli, A.; Piazza, M.; Riggio, M.; Tomasi, R. Refurbishment of traditional timber floors by means of wood-wood composite structures assembled with inclined screw connectors. In Proceedings of the 11th world conference on timber engineering—WCTE 2010, Riva del Garda, Italy, 20–24 June 2010.
79. Tomasi, R.; Baldessari, C.; Piazza, M. The refurbishment of existing timber floors: Characterization of the in-plane behaviour. In Proceedings of the 1st International Conference on Protection of Historical Constructions (PROHITECH), Rome, Italy, 21–24 June 2009.
80. Parisi, M.A.; Piazza, M. Seismic strengthening and seismic improvement of timber structures. *Constr. Build. Mater.* **2015**, *97*, 55–66. [CrossRef]
81. Gelfi, P.; Giuriani, E.; Marini, A. Stud shear connection design for composite concrete slab and wood beams. *J. Struct. Eng.* **2002**, *128*, 1544–1550. [CrossRef]
82. Tecnaria. Available online: https://www.tecnaria.com/ (accessed on 13 December 2018).
83. Albertani. Available online: https://www.albertani.com/sistemi-costruttivi/#h.s.b (accessed on 20 December 2018).

© 2019 by the authors. Licensee MDPI, Basel, Switzerland. This article is an open access article distributed under the terms and conditions of the Creative Commons Attribution (CC BY) license (http://creativecommons.org/licenses/by/4.0/).

Review

Duplex and Superduplex Stainless Steels: Microstructure and Property Evolution by Surface Modification Processes

Alisiya Biserova Tahchieva [1,*], Núria Llorca-Isern [1] and José-María Cabrera [2]

1 PROCOMAME, Departament de Ciència de Materials i Química Física, Facultat de Química, Universitat de Barcelona, 08028 Barcelona, Spain; nullorca@ub.edu

2 PROCOMAME, Departament de Ciència dels Materials i Enginyeria Metal·lúrgica, EEBE, Universitat Politècnica de Catalunya 2, 08019 Barcelona, Spain; jose.maria.cabrera@upc.edu

* Correspondence: abiserova.tahchieva@ub.edu; Tel.: +34-934-039-621

Received: 7 February 2019; Accepted: 13 March 2019; Published: 19 March 2019

Abstract: The paper presents an overview of diffusion surface treatments, especially nitriding processes, applied to duplex and superduplex stainless steels in the last five years. Research has been done mainly to investigate different nitriding processes in order to optimize parameters for the most appropriate procedure. The scope has been to improve mechanical and wear resistance without prejudice to the corrosion properties of the duplex and superduplex stainless steels. Our investigation also aimed to understand the effect of the nitriding layer on the precipitation of secondary phases after any heating step.

Keywords: duplex stainless steels; superduplex stainless steels; nitriding; secondary phases; surface treatments

---

1. Duplex and Super Duplex Stainless Steels

Duplex stainless steels (DSSs) are stainless steels classified according to their composition and thermomechanical treatment. DSSs are characterized by their microstructure, which contains both delta ferrite and austenite phases with approximately equal volume fractions, and an unusual combination of mechanical strength, toughness, and corrosion resistance [1]. Because of their content of Cr, Mo, and N, these steels are particularly resistant to pitting corrosion and can be differentiated in terms of pitting resistance equivalent number (PREN), which is often defined as:

$$PREN = \%Cr + 3.3\ (\%Mo) + 16\ (\%N). \quad (1)$$

Typically, DSSs contain 17–30 wt% Cr and 3–13 wt% Ni [2], and according to Equation (1), the addition of chromium, molybdenum, and nitrogen raises the pitting corrosion resistance. PREN is useful for comparing and ranking the grades of duplex stainless steels, but it can rarely be useful as a prediction for a suitable application. Duplex stainless steels with a PREN higher than 40 are extremely resistant to pitting corrosion and are designated as super DSS (SDSS) [3].

Due to the promising corrosion properties, a high mechanical strength combined with a good toughness, SDSS are widely used in chemical tankers, in the petrochemical industry as well as in marine and nuclear power industries [4–6]. However, studies for new alloys with even higher strength than that of the DSSs and SDSSs have increased in the past few years with new demanding applications, especially in the oil and gas industry [7]. For these challenges, new high alloyed duplex stainless steels have been developed and are characterized by a PREN close to 50, termed as hyper duplex stainless steels (HDSS) [8,9].

With respect to applications, one of the main restrictions of duplex stainless steels has been the exposure to elevated temperatures. The unstable ferrite phase at higher temperatures leads to its decomposition and degradation of both mechanical and corrosion properties [3]. In the temperature range 250–500 °C, decomposition of ferrite and embrittlement may occur after long-term exposure, inducing a decrease in toughness. At higher temperatures, various types of precipitates may form in the range of 600–900 °C, leading to a reduction in corrosion resistance and/or reduction of toughness as well [5]. A variety of secondary phases, mostly undesirable, may form during isothermal aging or inappropriate heat treatments, essentially from the instability mentioned before [5]. The following phases have been observed and have been under investigation for a long time: σ phase [10–12], Cr_2N, CrN [13,14], secondary austenite, χ phase [14–16], R phase, π phase, M_7C_3, $M_{23}C_6$, and τ phase. A lot of literature can be found regarding their formation as well as conditions and properties influencing mostly the mechanical and corrosion resistance of duplex and super duplex stainless steels [5,17–20].

Despite the excellent mechanical strength and corrosion resistance of DSS and SDSS, they have limited applications in industries where high wear resistance together with corrosion resistance are required. Thus, another restriction can be found in the tribological applications. Fortunately, in the past years, exponential development in surface engineering has been done, and consequently, an important improvement in surface hardness and tribological properties is possible, even in unusual alloys.

2. Surface Modification Technologies

Diffusion surface treatments including nitriding, carburizing, or nitrocarburizing can be performed under a controlled gas atmosphere as well as under plasma conditions [21–23]. Nitriding can be applied to unalloyed steels and irons to produce a corrosion and wear-resistant nitride layer. In alloyed steels containing nitride-forming elements (Cr, Mn, Mo, V, W, Al, Ti) a deeper diffusion layer develops. Nitriding techniques are suitable for the enhancement of wear resistance, improving hardness up to 1400 HV. However, due to the high content of chromium, nitride layers tend to be formed as a result of the strong bonding energy between chromium and nitrogen. Thus, chromium nitride precipitation occurs, which decreases the corrosion resistance. The precipitation of the other phases already mentioned is also critical for the corrosion and/or mechanical properties. Nevertheless, a moderate formation of the sigma phase is tolerable and can imply an improvement in the wear resistance. Among the possible solutions to this problem, the development of an appropriate coating is critical. The goal is to produce layers of high hardness on the steel to improve tribological performance without degrading the corrosion resistance. Such coating can be achieved by accurate nitriding. Nitriding techniques can differ among others in the nitriding media and process parameters such as temperature, time, composition of the nitriding atmosphere, and pressure [24–26].

Nitriding processes of austenitic stainless steels (fcc phase) are the most studied for the enhancement of hardness and wear properties. Thermochemical surface engineering has been investigated as an effective path by alloying the surface with interstitial elements, such as C and N, at elevated temperatures via diffusion. The nitrogen inlet produces a supersaturated austenite phase, which enhances the surface tribology and corrosion resistances. [27,28]. This phase has been the object of extensive investigation. In the 1980s and 1990s, research demonstrated that the surface hardness was improved without loss of corrosion resistance by low temperature nitriding or carburizing [29–33], mainly because of the lack of Cr nitrides or carbides precipitated at temperatures below 500 °C [23]. Ichii et al. [34] studied the evolution of phase transformation that takes place during the process. The XRD pattern of low temperature plasma nitride austenitic stainless steel contained five extra peaks shifted to the lower diffraction angles relative to the corresponding γ peaks. As they were not listed in the ASTM (American Society for Testing and Materials) index, Ichii et al. denoted the term S-phase to describe the new phase. The term "expanded austenite" (γ_N) associated to this S-phase was used for the first time by Leyland et al. in 1993 [35]. S-phase, or expanded austenite, has received wide scientific interest and is considered as one of the most important issues in surface engineering research. During the last decade, several techniques to form S-phase in austenitic stainless steels have been developed. It has been also demonstrated that this phase can be generated in Ni-Cr, Co-Cr, and Co-Cr-Mo alloys [36–39].

The expanded austenite phase (γ_N) evolved with the understanding of its crystallographic structure. It was described as a duplex layer consisting of a microstructure of fcc γ' [(Fe, Cr, N)$_4$ N] [23] and austenite or a compound layer with a structure of M_4N [M = (Fe, Cr, Ni, ...] [34]. It distinguishes from γ by the high N-content, its concentration in solid solution reaches up to 25–38% in γ_N in comparison to the equilibrium solubility of <0.65 at % in austenitic AISI316 [23,29,40,41]. γ_N is also characterized by lattice rotation processes in the expanded fcc structure, which cause anisotropic distortions in the normal surface direction [42,43]. A very interesting microstructural analysis concerning N content and the properties associated to S-phase was carried out by Dahm and Dearnley [44].

On the other hand, the nitrogen diffusion to ferritic α stainless steel (bcc crystal structure) leads to the formation of iron nitrides (ε-$Fe_{2\text{-}3}N$, γ'-Fe_4N) and nitrides such as CrN [45,46]. The nitrogen solubility in a ferrite structure increases with chromium content, but decreases its diffusivity and is mostly found along the grain boundaries instead of being homogenously distributed interstitially [47,48]. Hence, the main differences observed between austenite and ferrite structures regarding nitriding process are the result of the thermodynamic conditions, since both phases present different nitrogen diffusion rates.

Duplex stainless steel and super duplex stainless steels contain both austenite and ferrite phases. Eventually, expanded phases in austenite grains and nitride precipitates in ferrite grain boundaries are expected [49]. However, this cannot be a general rule because it depends on the processing conditions and the varied compositions in different alloys [28].

In the last five years, some scientific publications have been done with the objective of understanding the morphology, structure, and composition of nitride layers in nitrided duplex stainless steels. The corrosion response by varying process parameters, or trying to combine different methods of nitriding and/or nitrocarburizing, has been mainly examined. The objective was to study the nature of the nitrogen-rich layer. One of the most important variables during the process is the temperature. Thus, low and high nitriding temperature processes can be differentiated among the works. For instance, several authors have demonstrated the formation of expanded austenite in DSS by nitriding at low temperature, assuring good corrosion resistance and improvement in hardness. This is due to the fact that chromium nitride formation is avoided at lower temperatures. Paijan et al. [26] revealed this fact by analyzing DSS samples after nitriding in a tubular furnace in the temperature range of 400 to 500 °C, holding for 6 h. Several researchers [26,50,51] found that above 500 °C chromium nitride precipitates are formed in the nitrogen-enriched layer, significantly reducing the corrosion resistance. It was also observed by Blawert et al. [52] and Larisch et al. [46] that after low temperature plasma nitriding of duplex stainless steel, transformation of ferrite into the expanded austenite phase after nitrogen incorporation took place. Bielawski et al. [24] and Chiu et al. [53] reported the generation of expanded ferrite (α_N) and expanded austenite (γ_N) in both ferrite and austenite phases, respectively. Pinedo et al. [49] also explained the low-temperature plasma nitriding of duplex stainless steels carried out at 400 °C, which led to the formation of expanded austenite on the austenite stringers and expanded ferrite on the ferrite stringers of the microstructure. Moreover, some deformation bands were detected inside the expanded ferrite, which were assigned to compressive residual stresses, without chromium or iron nitride precipitation. On the contrary, in a more recent paper by Tschiptschin et al. [54] it was found that iron nitride (ε-Fe_3N) needles with a particular orientation relationship precipitate coherently inside the expanded ferrite regions of the nitride layer, increasing the hardness considerably. Similar ε-Fe_3N needles were precipitated from nitrogen-saturated ferrite grains in a lean duplex stainless steel, as determined by Li et al. [55]. Both reports revealed by TEM that the S-phase could not be formed on lean duplex stainless steel by low-temperature plasma nitriding. Tschiptschin et al. [54] also revealed a duplex thermochemical diffusion treatment consisting of two steps, one gas nitriding conducted at high temperature and a final plasma nitriding at low temperature (HTGN + LTPN). UNS

S31803 duplex stainless steel was the alloy selected by these researchers, and they found an almost 4 μm-thick nitrogen-rich expanded austenite layer 1227 ± 78 HV00.1 formed on top of a high nitrogen austenitic layer of 330 HV hardness. This experiment ensured a more homogenous microstructure and gentler hardness gradients, with suitable properties for wear-bearing applications. Similar duplex thermochemical treatments were previously experimented by Mesa et al. in order to improve the cavitation resistance of the same duplex stainless steel [56].

Plasma nitrocarburizing influences the corrosion behavior in a similar way as plasma nitriding. The formation of a larger amount of CrN in the alloy leads to a poor corrosion resistance at temperatures between 450 and 500 °C. It was found by Alphonsa et al. [50] that treatments at 350 and 400 °C are beneficial, but 450 and 500 °C treatments are better suited to duplex stainless steels for applications requiring a combination of both high hardness (1010 HV0.01) and corrosion resistance, for which a better corrosion resistance maintaining high hardness values compared to the bare duplex stainless steel is observed.

Concerning the morphology of the layer formed after nitriding, some authors reported that the thickness of the layer is a function of the temperature. It was observed that the thickness over the two phases (austenite and ferrite) is different in the case of plasma nitride specimens, whereas such a difference was not found in the plasma nitrocarburized samples. However, it was also described that treatments above 500 °C lead to similar thicknesses over both phases [57].

The reason for the difference in nitrogen diffusion into the austenite and ferrite phases was discussed by Pinedo et al. [49] and Bobadilla and Tschiptschin [57]. Diffusion of nitrogen not only occurs in the volume of the substrate but also through the grain boundaries. The lateral flow of nitrogen tends to move from ferrite to the austenite in order to reach equilibrium.

Some authors studied the plasma variables (such as frequency, pulse width, and voltage) with the surface modification in order to understand nitriding plasma-based ion implantation [28]. Li et al. [55] suggested an optimized plasma nitriding process as the best treatment for lean duplex stainless steel for only 10 h at 450 °C and at 420 °C for dry-wear and corrosion-wear applications, respectively.

Most of the reported research mainly investigated the surface layer structure, morphology, hardness, corrosion, and mechanical resistance as well as tribological and tribocorrosion properties as a function of the type of nitriding processes mostly under low temperature procedures. Hence, our research aimed to study the plasma nitriding process applied to duplex stainless steel (S32205) and super duplex stainless steel (S32750) at high temperature, 520 °C, during a short nitriding time (10 h).

All samples were previously annealed to 1080 °C for 30 min and immediately quenched with water in order to homogenize the structure before performing the nitriding process. Characterization of the microstructure of the formed layers and their potential to precipitate secondary phases when a heat treatment is carried out on nitride steels is reported. Table 1 summarizes the chemical composition of both DSS and SDSS.

Table 1. Chemical composition of duplex and super duplex stainless steels.

Sample Identification	**Chemical Composition wt %**									
UNS	C	Si	Mn	P	S	Cr	Ni	Mo	N	Cu
S32205	0.015	0.4	1.5	0.018	0.001	22.49	5.77	3.21	0.184	0.18
S32750	0.018	0.26	0.84	0.019	0.001	25.08	6.880	3.82	0.294	0.17

Layer morphology of treated samples under the same conditions is shown in Figure 1. The layer structure appears with the outer zone enriched in nitrogen (Figure 1a,c,e). A line scan of some diffusion elements confirmed the nitride layer formation and the excellent continuous interphase within the substrate.

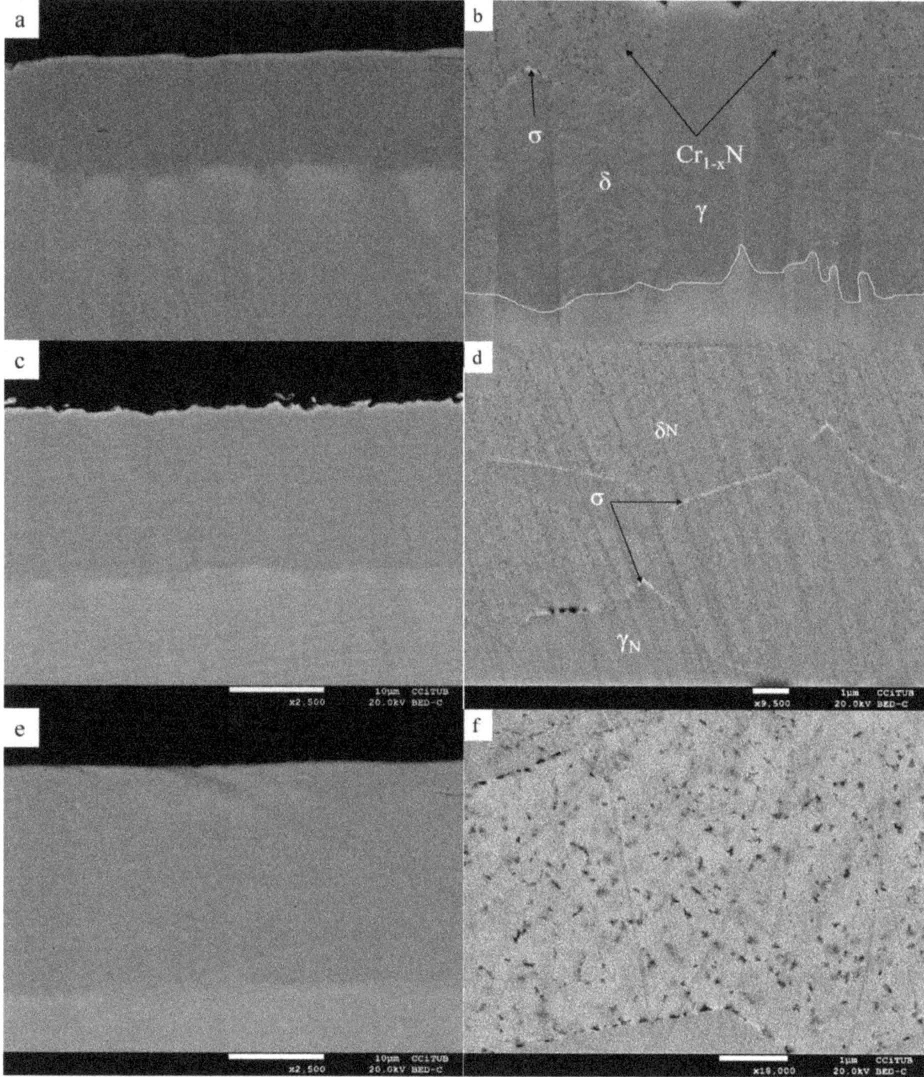

Figure 1. Backscattered electron (BSE) micrographs of super duplex stainless steel (SDSS) and duplex stainless steel (DSS) specimens after 10 h plasma nitriding at 520 °C. (**a**) Cross-section nitride layer of the SDSS tube. (**b**) Details of the nitride layer indicating the austenite (γ) and ferrite (δ) phases, as well as the sigma phase and chromium nitrides. (**c**) Cross-section of the nitride layer of the SDSS plate. (**d**) Higher magnification of the nitride layer in (**c**) showing expanded austenite γ_N and expanded ferrite δ_N phases, as well as sigma phase in the boundaries. (**e**) Longitudinal section of the nitride layer of the DSS plate, and (**f**) higher magnification of the nitride layer (expanded ferrite δ_N) on the DSS plate showing the distribution of chromium nitrides inside.

Backscattered electron (BSE) micrographs at higher magnification (Figure 1b,d,f) show the microstructure of samples extracted from the SDSS tube, SDSS plate, and DSS plate, respectively. Presence of a brighter phase at the grain boundaries of the expanded austenite and expanded ferrite phases is detected. This phase is associated with the sigma precipitate characteristic of duplex stainless steels after thermal treatments in the range of 600 to 850 °C. This phase was identified through the energy-dispersive X-ray spectroscopy (EDS) microanalyses showing a higher content of Mo than that of the ferrite phase. Precipitates of chromium nitrides (black dots) are also present and can be seen in Figure 1b as expected after nitriding at a temperature of 520 °C. These chromium nitride precipitates are apparently distributed only in the expanded austenite. Figure 1c,d shows the microstructure of SDSS samples extracted from the SDSS plate. Figure 1d shows the sigma precipitation linking the expanded austenite phase within the expanded ferrite phase. Chromium nitrides are also present inside both phases. The composition of the different phases was determined using a JEOL (Peabody, MA, USA) J-7100 field emission scanning electron microscope (FE-SEM) with an energy-dispersive X-ray spectroscopy system (EDS) INCA PentaFETx3 detector. In addition, a JEOL JXA-8230 microprobe (with five WDS spectrometers) allowed us to obtain a higher chemical composition accuracy. Different chemical compositions (such as differences in Mo, Cr) led to identify each phase.

A certain discordance was found between the literature and the present results, as the nitride layer showed an irregular thickness at the same temperature. Furthermore, the nitriding-substrate interface on austenite and ferrite phases is not regular (Figure 1b). It was predicted that for duplex stainless steel, a homogenous interphase and an almost equal thickness of nitride layer on austenite and ferrite phases of the substrate should be found above 500 °C [42]; however, SDSS does not follow this trend.

On the other hand, a Mo-enriched phase was not detected in the samples extracted from the DSS plate. This was probably due to the chemical composition of the alloy, as DSS had almost 0.5%wt less Mo than SDSS. Instead, DSS nitride layer accommodated more black spots, identified as nitrides. These nitrides are mainly found in the outer part of the nitrided steel layer and help raise the hardness of the nitrided steels (Figure 1f).

Hardness of the nitride layer was measured in order to check the improvement of the resistance strength obtained at a higher temperature of the plasma nitriding process. Minimum and maximum values of $HV_{0.3}$ and medium thickness with a standard deviation of the nitride layers are summarized in Table 2.

Table 2. Sample identification, $HV_{0.3}$ hardness (maximum and minimum range values), and medium thickness of the nitride layers.

Material	$HV_{0.3}$	Thickness Layer
Super Duplex S32750 tube	1263 ÷ 1393	13.7 ± 0.0006 μm
Super Duplex S32750 plate	1379 ÷ 1402	16.5 ± 0.0004 μm
Duplex S32205 plate	1263 ÷ 1329	23.5 ± 0.0003 μm

After the plasma nitriding process, thermal treatment at 830 °C for up to 3 min was carried out for the superduplex stainless steels. This treatment was selected in order to compare the precipitation of secondary phases before and after nitriding and the influence of a large concentration of nitrogen in that precipitation [58,59]. As observed in the images from backscattered electron microscopy (BSE) of SDSS plate cross-section (Figure 2a,b), precipitates of chromium nitride are identified inside the expanded ferrite phase. The sigma phase can also be seen in the ferrite–austenite grain boundaries, and inside the ferrite as well. The SDSS tube sample also shows the sigma phase precipitates. Less detectable is the presence of chromium nitrides, as can be observed in Figure 2c,d.

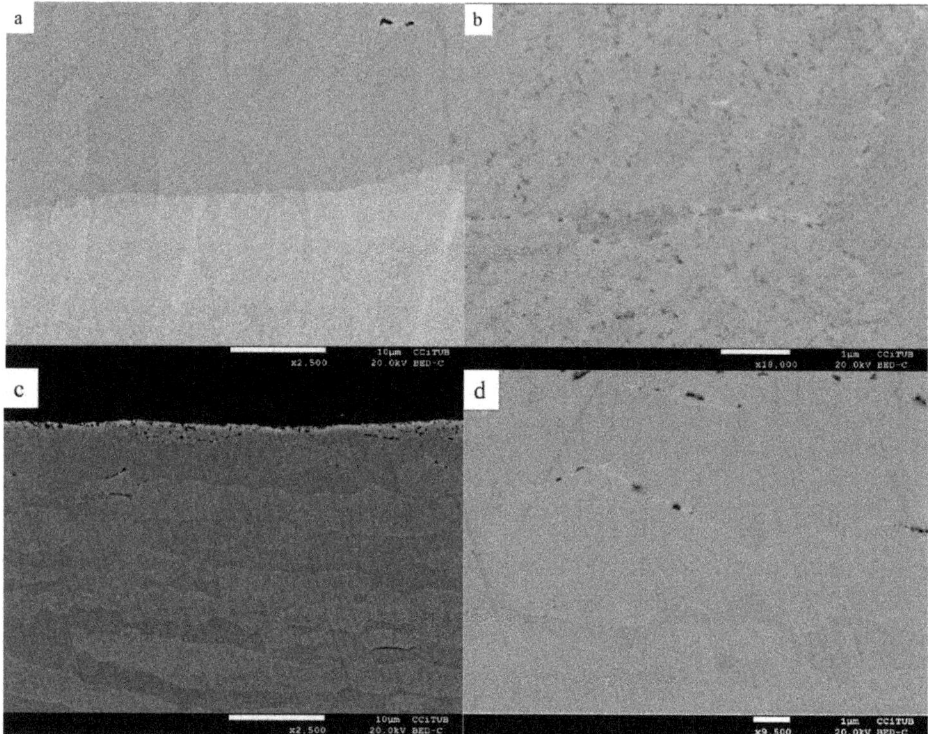

Figure 2. BSE micrographs of the SDSS plate and the SDSS tube samples after 10 h plasma nitriding at 520 °C and post heat treatment at 830 °C for up to 3 min. (**a**) Cross-section of the SDSS plate sample. (**b**) Higher magnification of (**a**). (**c**) Longitudinal section of the SDSS tube after 10 h, and (**d**) higher magnification of (**c**).

In light of the results, it can be observed that the thickness of the nitride layer formed at the same processing conditions was mainly influenced by the composition of the stainless steel. For instance, the DSS sample showed almost double the thickness of the SDSS samples. Nevertheless, both SDSS samples eventually presented higher hardness values because of the higher amount of chromium nitrides, which enhanced the hardness of the outer surface. Due to the high amount of nitrogen and chromium in SDSS compared with that of DSS, chromium nitrides were consequently formed in higher amounts. Another fact observed was that the microstructure resulting from the forming process was somehow influencing the diffusion through the steel. In this sense, rolled samples were previously annealed at 1080 °C for 30 min in order to homogenize them before performing the nitriding process, hence, only the formed microstructure remained before nitriding. The plate layer thickness was around 20% thicker than the tube layer thickness with the same processing conditions.

Secondary phases, mostly the sigma phase, were already formed within the nitriding process before any heat treatment was applied to the SDSS samples. Such behavior could not only be due to the temperature, but it can be regarded as high nitrogen ferrite and austenite help to induce its formation, even at temperatures as low as 520 °C. The nucleation and growth kinetics could be influenced by the previous forming process history of the specimen related to the texture and anisotropy developed in each case, as in the SDSS plate and the SDSS tube. Moreover, chromium nitrides act as a favorable site for the nucleation of the σ-phase.

However, heating the sample after nitriding sigma phase formation was clearly enhanced; the latter occurred mostly in the SDSS tube, as it was detected not only at the phase boundaries

(Figure 1b,d), but also in the austenite phase (Figure 2b). Surprisingly, the SDSS plate prepared longitudinally (Figure 2c,d) did not show the sigma phase inside the austenite phase. This again confirms the importance of the microstructure facing nitrogen diffusion: the diffusion path involving grain boundaries (favored by cross-sections facing the outer surface) will enhance nitrogen diffusion and, consequently, the induced formation of secondary phases. On the contrary, a longitudinal microstructure will limit the amount of nitrogen able to diffuse through austenite or ferrite phases.

3. Conclusion

Surface property enhancement of duplex stainless steels has been observed and supports the idea of an increasing demand on nitriding in order to enhance wear resistance. Plasma nitriding at 520 °C on duplex and superduplex stainless steels (UNS S32205 and S32750, respectively) has been carried out, and the characterization of the nitride layer formed has been studied. This layer is dense and well-formed and follows the microstructure of the steel. One of the main conclusions found is that layer thickness is dependent on the composition of the steel, hence, duplex and superduplex stainless steels show different thicknesses (~20%). Furthermore, nitrogen diffuses more intensively in ferrite, enhancing the formation of secondary phase precipitation (mainly sigma-phase) and its distribution along grain boundaries. Another interesting conclusion is that the oriented microstructure of ferrite and austenite from previous forming processes would influence the diffusion path of nitrogen. In this sense, plate samples showed a thicker nitriding layer than tube ones.

Author Contributions: All authors contributed to the conceptualization of the research, A.B.T., N.L.-I., and J.-M.C.; methodology of the review was created by A.B.T.; formal analyses were done by A.B.T., N.L.-I., and J.-M.C.; investigation was done by A.B.T.; resources were provided by all three authors, A.B.T., N.L.-I., and J.-M.C.; writing—original draft preparation was contributed by A.B.T.; writing—review and editing were done by A.B.T., N.L.-I., and J.-M.C.; supervision was carried out by all the authors, A.B.T., N.L.-I., and J.-M.C.

Funding: This research received no external funding.

Acknowledgments: The authors would like to thank S.A. Metalografica for the plasma nitriding process.

Conflicts of Interest: The authors declare no conflict of interest.

References

1. Alvarez-Armas, I.; Degallaix-Moreuil, S. *Duplex Stainless Steels*; John Wiley and Sons: Hoboken, NJ, USA, 2009.
2. Lo, K.H.; Kwok, C.T.; Chan, W.K.; Kuan, H.C.; Lai, K.K.; Wang, K.Y. Duplex Stainless Steels. In *Encyclopedia of Iron, Steel, and Their Alloys*; CRC Press: Boca Raton, FL, USA, 2016; pp. 1150–1160.
3. Nilsson, J.O.; Wilson, A. Influence of isothermal phase transformations on toughness and pitting corrosion of super duplex stainless steel SAF 2507. *Mater. Sci. Technol.* **1993**, *9*, 545–554. [CrossRef]
4. Gunn, R. *Duplex Stainless Steels: Microstructure, Properties and Applications*; Woodhead Publishing: Cambridge, UK, 1997.
5. Nilsson, J.O. Super duplex stainless steels. *Mater. Sci. Technol.* **1992**, *8*, 685–700. [CrossRef]
6. Solomon, H.D.; Devine, T.M., Jr. *Duplex Stainless Steels—A Tale of Two Phases*; American Society for Metals: Metals Park, OH, USA, 1982; pp. 693–656.
7. Kangas, P.; Chai, G. Use of advanced austenitic and duplex stainless steels for applications in Oil & Gas and Process industry. *Century Stainl. Steels* **2013**, *794*, 645–669.
8. Chail, G.; Kangas, P. Super and hyper duplex stainless steels: Structures, properties and applications. *Procedia Struct. Integr.* **2016**, *2*, 1755–1762. [CrossRef]
9. Salvio, F.; da Silva, B.R.S. On the Role of HISC on Super and Hyper Duplex Stainless Steel Tubes. In Proceedings of the Offshore Technology Conference, Rio de Janeiro, Brazil, 29–31 October 2013.
10. Wan, J.; Ruan, H.; Wang, J.; Shi, S. The Kinetic diagram of sigma phase and its precipitation hardening effect on 15Cr-2Ni duplex stainless steel. *Mater. Sci. Eng. A* **2017**, *711*, 571–578. [CrossRef]
11. Atamert, S.; King, J.E. Sigma-phase formation and its prevention in duplex stainless steels. *J. Mater. Sci. Lett.* **1993**, *12*, 1144–1147. [CrossRef]

12. Chen, T.H.; Weng, K.L.; Yang, J.R. The effect of high-temperature exposure on the microstructural stability and toughness property in a 2205 duplex stainless steel. *Mater. Sci. Eng. A* **2002**, *338*, 259–270. [CrossRef]
13. Magnabosco, R.; Santos, D.C.d. Intermetallic Phases Formation During Short Aging between 850 °C and 950 °C of a Superduplex Stainless Steel. *J. Mater. Res. Technol.* **2012**, *26*. [CrossRef]
14. He, Y.-L.; Zhu, N.-Q.; Lu, X.-G.; Li, L. Experimental and computational study on microstructural evolution in 2205 duplex stainless steel during high temperature aging. *Mater. Sci. Eng. A* **2010**, *528*, 721–729. [CrossRef]
15. Escriba, D.M.; Materna-Morris, E.; Plaut, R.L.; Padilha, A.F. Chi-phase precipitation in a duplex stainless steel. *Mater. Charact.* **2009**, *60*, 1214–1219. [CrossRef]
16. Pohl, M.; Storz, O.; Glogowski, T. Effect of intermetallic precipitations on the properties of duplex stainless steel. *Mater. Charact.* **2006**, *58*, 65–71. [CrossRef]
17. Sieurin, H.; Sandström, R. Sigma phase precipitation in duplex stainless steel 2205. *Mater. Sci. Eng. A* **2007**, *444*, 271–276. [CrossRef]
18. Nilsson, J.O.; Liu, P. Aging at 400–600 °C of submerged arc welds of 22Cr–3Mo–8Ni duplex stainless steel and its effect on toughness and microstructure. *Mater. Sci. Technol.* **1991**, *7*, 853–862. [CrossRef]
19. Kim, Y.J.; Chumbley, L.S.; Gleeson, B. Determination of isothermal transformation diagrams for sigma-phase formation in cast duplex stainless steels CD3MN and CD3MWCuN. *Metall. Mater. Trans. A* **2004**, *35*, 3377–3386. [CrossRef]
20. Nilsson, J.O.; Karlsson, L.; Andersson, J.O. Secondary austenite for mation and its relation to pitting corrosion in duplex stainless steel weld metal. *Mater. Sci. Technol.* **1995**, *11*, 276–283. [CrossRef]
21. Kuwahara, H.; Matsuoka, H.; Takada, J.; Kikuchi, S.; Tomii, Y.; Tamura, I. Plasma nitriding of Fe-18Cr-9Ni in the range of 723?823 K. *Oxid. Met.* **1991**, *36*, 143–156. [CrossRef]
22. Flis, J.; Mańkowski, J.; Roliński, E. Corrosion Behaviour of Stainless Steels Aner Plasma and Ammonia Nitriding. *Surf. Eng.* **1989**, *5*, 151–157. [CrossRef]
23. Zhang, Z.L.; Bell, T. Structure and Corrosion resistance of plasma nitrided Stainless Steel. *Surf. Eng.* **1985**, *1*, 131–136. [CrossRef]
24. Bielawski, J.; Baranowska, J. Formation of nitrided layers on duplex steel—influence of multiphase substrate. *Surf. Eng.* **2010**, *26*, 299–304. [CrossRef]
25. Jasinski, J.J.; Fraczek, T.; Kurpaska, L.; Lubas, M.; Sitarz, M. Investigation of nitrogen transport in active screen plasma nitriding processes—Uphill diffusion effect. *J. Mol. Struct.* **2018**, *1164*, 37–44. [CrossRef]
26. Paijan, L.H.; Berhan, M.N.; Adenan, M.S.; Yusof, N.F.M.; Haruman, E. Structural Development of Expanded Austenite on Duplex Stainless Steel by Low Temperature Thermochemical Nitriding Process. *Adv. Mater. Res.* **2012**, *576*, 260–263. [CrossRef]
27. Dong, H. S-phase surface engineering of Fe–Cr, Co–Cr and Ni–Cr alloys. *Int. Mater. Rev.* **2010**, *55*, 65–98. [CrossRef]
28. De Oliveira, W.R.; Kurelo, B.C.E.S.; Ditzel, D.G.; Serbena, F.C.; Foerster, C.E.; de Souza, G.B. On the S-phase formation and the balanced plasma nitriding of austenitic-ferritic super duplex stainless steel. *Appl. Surf. Sci.* **2018**, *434*, 1161–1174. [CrossRef]
29. Sun, Y.; Li, X.; Bell, T. Structural characteristics of low temperature plasma carburised austenitic stainless steel. *Mater. Sci. Technol.* **1999**, *15*, 1171–1178. [CrossRef]
30. Mingolo, N.; Tschiptschin, A.P.; Pinedo, C.E. On the formation of expanded austenite during plasma nitriding of an AISI 316L austenitic stainless steel. *Surf. Coat. Technol.* **2006**, *201*, 4215–4218. [CrossRef]
31. Borgioli, F.; Fossati, A.; Galvanetto, E.; Bacci, T. Glow-discharge nitriding of AISI 316L austenitic stainless steel: Influence of treatment temperature. *Surf. Coat. Technol.* **2005**, *200*, 2474–2480. [CrossRef]
32. Fewell, M.P.; Mitchell, D.R.G.; Priest, J.M.; Short, K.T.; Collins, G.A. The nature of expanded austenite. *Surf. Coat. Technol.* **2000**, *131*, 300–306. [CrossRef]
33. Blawert, C.; Kalvelage, H.; Mordike, B.L.; Collins, G.A.; Short, K.T.; Jiraskova, Y.; Schneeweiss, O. Nitrogen and carbon expanded austenite. *Surf. Coat. Technol.* **2001**, *136*, 181–187. [CrossRef]
34. Ichii, T.; Fujimura, K.; Takase, K. Structure of the ion-nitrided layer of 18-8 stainless steel. *Tech. Rep. Kansai Univ.* **1986**, *27*, 134–144.
35. Leylanda, A.; Lewis, D.B.; Stevenson, P.R.; Matthewsa, A. Low temperature plasma diffusion treatment of stainless steels for improved wear resistance. *Surf. Coat. Technol.* **1993**, *62*, 608–617. [CrossRef]
36. Makishi, T.; Nakata, K. Surface Hardening of Nickel Alloys by Means of Plasma Nitriding. *Metall. Mater. Trans. A Phys. Metall. Mater. Sci.* **2004**, *35*, 227–238. [CrossRef]

37. Williamson, D.L.; Davis, J.A.; Wilbur, P.J. Effect of austenitic stainless steel composition on low-energy, high-flux, nitrogen ion beam processing. *Surf. Coat. Technol.* **1998**, *104*, 178–184. [CrossRef]
38. Li, X.Y.; Habibi, N.; Bell, T.; Dong, H. Microstructural characterisation of a plasma carburised low carbon Co–Cr alloy. *Surf. Eng.* **2007**, *23*, 45–51. [CrossRef]
39. Liu, R.; Li, X.; Hu, X.; Dong, H. Surface & Coatings Technology Surface modi fi cation of a medical grade Co–Cr–Mo alloy by low-temperature plasma surface alloying with nitrogen and carbon. *Surf. Coat. Technol.* **2013**, *232*, 906–911.
40. Williamson, D.L.; Wei, R.; Wilbur, P.J. Metastable phase formation and enhanced diffusion in f. c. c. alloys under high dose, high flux nitrogen implantation at high and low ion energies. *Surf. Coat. Technol.* **1994**, *65*, 15–23. [CrossRef]
41. Foerster, C.E.; Souza, J.F.P.; Silva, C.A.; Ueda, M.; Kuromoto, N.K.; Serbena, F.C.; Silva, S.L.R.; Lepienski, C.M. Effect of cathodic hydrogenation on the mechanical properties of AISI 304 stainless steel nitrided by ion implantation, glow discharge and plasma immersion ion implantation. *Nucl. Instrum. Methods Phys. Res. Sect. B Beam Interact. Mater. Atoms* **2007**, *257*, 727–731. [CrossRef]
42. Manova, D.; Mändl, S.; Neumann, H.; Rauschenbach, B. Formation of metastable diffusion layers in Cr-containing iron, cobalt and nickel alloys after nitrogen insertion. *Surf. Coat. Technol.* **2017**, *312*, 81–90. [CrossRef]
43. Stinville, J.C.; Villechaise, P.; Templier, C.; Rivière, J.P.; Drouet, M. Lattice rotation induced by plasma nitriding in a 316L polycrystalline stainless steel. *Acta Mater.* **2010**, *58*, 2814–2821. [CrossRef]
44. Dearnley, P.; Dahm, K.L.; Dearnley, P.A. On the Nature, Properties and Wear Response of S-phase (nitrogen-alloyed Stainless Steel) Coatings on Aisi 316l. *J. Mater. Des. Appl.* **2000**, *21*, 181–198.
45. Blawert, C.; Mordike, B.L.; Rensch, U.; Schreiber, G.; Oettel, H. Nitriding Response of Chromium Containing Ferritic Steels on Plasma Immersion Ion Implantation at Elevated Temperature. *Surf. Eng.* **2002**, *18*, 249–254. [CrossRef]
46. Larisch, B.; Brusky, U.; Spies, H.-J. Plasma nitriding of stainless steels at low temperatures. *Surf. Coat. Technol.* **1999**, *116–119*, 205–211. [CrossRef]
47. Gontijo, L.C.; Machado, R.; Casteletti, L.C.; Kuri, S.E.; Nascente, P.A.P. X-ray diffraction characterisation of expanded austenite and ferrite in plasma nitrided stainless steels. *Surf. Eng.* **2010**, *26*, 265–270. [CrossRef]
48. Panicaud, B.; Chemkhi, M.; Roos, A.; Retraint, D. Theoretical modelling of iron nitriding coupled with a nanocrystallisation treatment. Application to numerical predictions for ferritic stainless steels. *Appl. Surf. Sci.* **2012**, *258*, 6611–6620. [CrossRef]
49. Pinedo, C.E.; Varela, L.B.; Tschiptschin, A.P. Low-temperature plasma nitriding of AISI F51 duplex stainless steel. *Surf. Coat. Technol.* **2013**, *232*, 839–843. [CrossRef]
50. Alphonsa, J.; Raja, V.S.; Mukherjee, S. Study of plasma nitriding and nitrocarburizing for higher corrosion resistance and hardness of 2205 duplex stainless steel. *Corros. Sci.* **2015**, *100*, 121–132. [CrossRef]
51. Maleque, M.A.; Lailatul, P.H.; Fathaen, A.A.; Norinsan, K.; Haider, J. Nitride alloy layer formation of duplex stainless steel using nitriding process. *IOP Conf. Ser. Mater. Sci. Eng.* **2018**, *290*, 012015. [CrossRef]
52. Blawert, C.; Weisheit, A.; Mordike, B.L.; Knoop, R.M. Plasma immersion ion implantation of stainless steel: Austenitic stainless steel in comparison to austenitic-ferritic stainless steel. *Surf. Coat. Technol.* **1996**, *85*, 15–27. [CrossRef]
53. Chiu, L.H.; Su, Y.Y.; Chen, F.S.; Chang, H. Microstructure and Properties of Active Screen Plasma Nitrided Duplex Stainless Steel. *Mater. Manuf. Process.* **2010**, *25*, 316–323. [CrossRef]
54. Tschiptschin, A.P.; Varela, L.B.; Pinedo, C.E.; Li, X.Y.; Dong, H. Development and microstructure characterization of single and duplex nitriding of UNS S31803 duplex stainless steel. *Surf. Coat. Technol.* **2017**, *327*, 83–92. [CrossRef]
55. Li, X.; Dou, W.; Tian, L. Combating the Tribo-Corrosion of LDX2404 Lean Duplex Stainless Steel by Low Temperature. *Lubricants* **2018**, *6*, 93. [CrossRef]
56. Mesa, D.H.; Pinedo, C.E.; Tschiptschin, A.P. Improvement of the cavitation erosion resistance of UNS S31803 stainless steel by duplex treatment. *Surf. Coat. Technol.* **2010**, *205*, 1552–1556. [CrossRef]
57. Bobadilla, M.; Tschiptschin, A. On the Nitrogen Diffusion in a Duplex Stainless Steel. *Mater. Res.* **2015**, *18*, 390–394. [CrossRef]

58. Llorca-Isern, N.; López-Luque, H.; López-Jiménez, I.; Biezma, M.V. Identification of sigma and chi phases in duplex stainless steels. *Mater. Charact.* **2016**, *112*, 20–29. [CrossRef]
59. Llorca-Isern, N.; Biserova-Tahchieva, A.; Lopez-Jimenez, I.; Calliari, I.; Cabrera, J.M.; Roca, A. Influence of severe plastic deformation in phase transformation of superduplex stainless steels. *J. Mater. Sci.* **2019**, *54*, 2648–2657. [CrossRef]

© 2019 by the authors. Licensee MDPI, Basel, Switzerland. This article is an open access article distributed under the terms and conditions of the Creative Commons Attribution (CC BY) license (http://creativecommons.org/licenses/by/4.0/).

Article

Mechanical Spectroscopy Investigation of Point Defect-Driven Phenomena in a Cr Martensitic Steel

Alessandra Fava, Roberto Montanari * and Alessandra Varone

Department of Industrial Engineering, University of Rome "Tor Vergata", 00133 Rome, Italy; alessandra.fava@uniroma2.it (A.F.); alessandra.varone@uniroma2.it (A.V.)

* Correspondence: roberto.montanari@uniroma2.it; Tel.: +39-06-7259-7182

Received: 3 October 2018; Accepted: 18 October 2018; Published: 24 October 2018

Abstract: The paper presents and discusses results of mechanical spectroscopy (MS) tests carried out on a Cr martensitic steel. The study regards the following topics: (i) embrittlement induced by Cr segregation; (ii) interaction of hydrogen with C–Cr associates; (iii) nucleation of Cr carbides. The MS technique permitted characterising of the specific role played by point defects in the investigated phenomena: (i) Cr segregation depends on C–Cr associates distribution in as-quenched material, in particular, a slow cooling rate (~150 K/min) from austenitic field involves an unstable distribution, which leads to Cr concentration fluctuations after tempering at 973 K; (ii) hydrogen interacts with C–Cr associates, and the phenomenon hinders hydrogen attack (HA) because hydrogen atoms bound by C–Cr associates are not able to diffuse towards grain boundaries and dislocation where CH_4 bubbles may nucleate, grow, and merge to form the typical HA cracks; (iii) C–Cr associates take part in the nucleation mechanism of Cr_7C_3 carbides, and specifically these carbides form by the aggregation of C–Cr associates with 1 Cr atom.

Keywords: mechanical spectroscopy; Cr martensitic steel; point defects

1. Introduction

The role played by point defects is of the utmost relevance in several phenomena occurring in pure metals and alloys. Point defects (vacancies, interstitials, substitutional atoms, etc.) alter the elastic constants of a solid, strongly affect diffusion and ordering, and are relevant for irradiation, cold work, and recovery; thus, they have received a lot of attention since the 1960s (see, for example refs. [1,2]). Vacancies and self-interstitials play an important role also in metal melting [3–5] and, recently, some works demonstrated that a remarkable increase of self-interstitials takes place in a temperature range of about 10 K before melting, weakens interatomic bonds, and favours the final avalanche of vacancies, leading to the solid–liquid transformation [6–11].

Investigations by means of conventional techniques (hardness tests, resistivity measurements, X-ray diffraction, and dilatometry) give information only on the presence and annihilation of point defects. A much more complete description, including concentration, mobility, and symmetry, is obtained through mechanical spectroscopy (MS). The theory of point defect relaxations and the applications of the technique are described in the classical book of Nowick and Berry [12].

This paper presents and discusses results obtained through MS tests on a Cr martensitic steel. The experimental approach and analysis method can be extended to all the steels of this type, materials of great interest for many industrial applications. In particular, the focus of the study is on the following topics: (i) Cr segregation phenomena; (ii) interaction of hydrogen with C–Cr associates; (iii) nucleation of Cr carbides.

Cr martensitic steels are materials of great interest for structural applications in future nuclear fusion reactors, and a lot of work has been done to design and produce low activation steels with higher resistance to swelling and embrittlement under irradiation. One of the authors has been involved in extensive experimental campaigns for investigating the microstructure of Cr martensitic steels with different Cr contents, and prepared under different conditions of quenching and tempering treatments. The scope was to understand how specific microstructural features affect swelling and brittle behaviour [13–25]. It was demonstrated that states of strain remain in non-irradiated steels after prolonged heat treatments, sufficient to anneal out dislocations and sub-boundaries, and their origin was attributed to the non-homogeneous Cr distribution in the matrix.

Different cooling rates from the austenitic field (1348 K) do not seem to have effects on the microstructure and mechanical properties of the steel, however, successive steps of tempering treatment determine different fracture modes. The present paper shows that such apparently anomalous behaviour depends on elementary structures of point defects (C–Cr associates).

Furthermore, the same structures affect both the hydrogen attack and the precipitation of Cr carbides.

2. Material and Methods

The nominal chemical composition of the investigated steel is reported in Table 1.

Table 1. Chemical composition of the investigated Cr martensitic steel (wt %).

C	Cr	Mo	Ni	Mn	Nb	V	Si	Al	N	P	Fe
0.17	10.50	0.50	0.85	0.60	0.20	0.25	0.32	0.05	0.003	0.005	to balance

MS experiments were carried out using bar-shaped samples (60 mm × 7 mm × 0.5 mm) excited by flexural vibrations and operating in conditions of resonance (frequencies in the range 150–400 Hz). The VRA 1604 apparatus (CANTIL Srl, Bologna, Italy) used in MS experiments has been described in detail in [26]. MS measurements are quite sensitive to experimental conditions, in particular, the heating rate and the strain amplitude, which should be taken strictly as constant. VRA 1604 allows for measurement of dynamic modulus and damping (Q^{-1}) with high precision and accuracy, characteristics of fundamental importance to perform the successive analyses.

The damping factor Q^{-1} was determined from the logarithmic decay δ of flexural vibrations:

$$Q^{-1} = \frac{\delta}{\pi} = \frac{1}{k\pi} \ln \frac{A_n}{A_{n+k}}, \quad (1)$$

A_n and A_{n+k} being the amplitude of the n-th and $n + k$-th vibration, respectively. The dynamic modulus E was calculated from the resonance frequency f:

$$\mathrm{E} = \frac{48\pi^2 L^4 \rho}{m^4 h^2} f^2, \quad (2)$$

with m being a constant depending on the specific sample geometry (for a reed m = 1.875), ρ the material density, and L and h the length and thickness of the sample, respectively. The heating rate was 2 K/min, while the strain amplitude was kept lower than 1×10^{-5}.

Scanning electron microscopy (SEM Hitachi S-2460N, Hitachi, Tokyo, Japan) observations have been carried out on the fracture surfaces of probes broken in Charpy tests and on carbides extracted from the metallic matrix. Extraction was carried out through an electrochemical method (solution of HCl (10%) in methanol, cathode of AISI 316 steel, operative temperature 300 K, tension 1.5 V, and current 0.5 A). The filtering was made using a membrane with pores of 0.1 μm. The extracted carbides were then examined by X-ray diffraction (Philips, Eindhoven, The Netherlands) and SEM observations with EDS microanalysis (Thermo Fisher Scientific, Madison, WI, USA). XRD spectra have been recorded using Mo-Kα radiation, in the 2θ range 5–20° by scanning with 2θ steps of 0.05° and a counting time of 5 s per step.

3. Results and Discussion

3.1. Q^{-1} Spectra Analysis

As shown by the examples in Figure 1, the Q^{-1} vs. temperature spectra of the Cr martensitic steel exhibit a complex scenario with more peaks of different position and intensity [27], depending on the quenching rate from austenitic field. Dynamic modulus E, normalised to the room temperature value E_0, exhibits slope changes in correspondence of the Q^{-1} peaks, indicating that they are relaxation peaks.

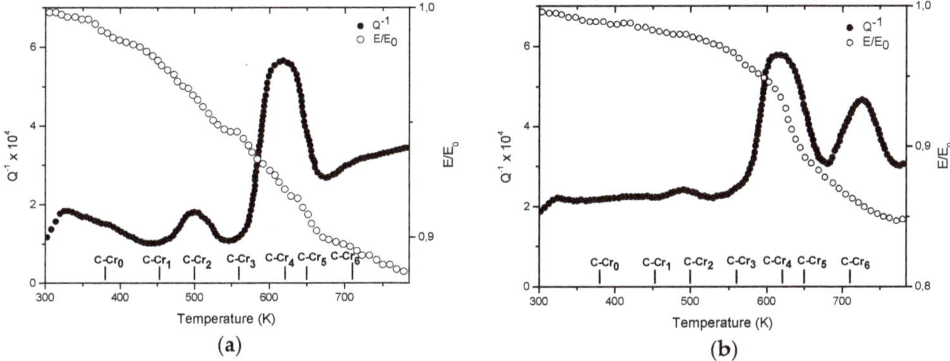

Figure 1. Q^{-1} and *E/Eo* curves of the Cr martensitic steel after quenching from the austenitic field (1348 K) with a cooling rate of 3600 K/min (**a**) and 150 K/min (**b**). The resonance frequency f = 250 Hz. Data of these materials where already published in ref. [27] but the present curves obtained through VRA 1604 are more pricise in particular in the initial part.

To understand the origin of the damping peaks, the interaction of C atoms in interstitial positions, with the atoms (Fe and Cr) forming the bcc lattice, has been considered. As shown in Figure 2, C atoms in solid solution are assumed to occupy the octahedral interstices in the bcc lattice of the steel.

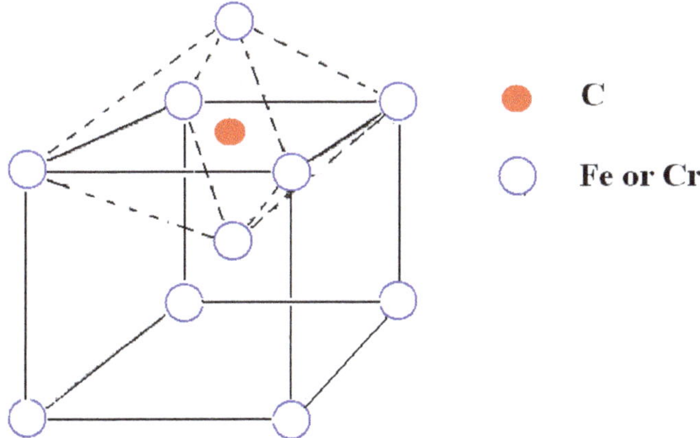

Figure 2. C atoms are allocated in the octahedral interstices of bcc lattice. The corners of the octahedron can be occupied by either Fe or Cr atoms, giving rise to seven possible C–Cr configurations.

In the Fe–C system, all the interstitial positions are energetically equivalent, and any inhomogeneity in C atoms distribution only depends on possible interactions with lattice defects, for instance, dislocations (Cottrell's atmospheres). If other alloying elements are present, they can substitute Fe atoms at the octahedron corners, giving rise to different configurations. In the Fe–Cr–C system, typical of the investigated steel, the corners of each octahedron can be occupied by a number i of Cr atoms, varying from 0 to 6, thus 7 different C–Cr configurations may occur, each of them with a specific binding energy for the C atom.

Some theoretical models describing the effects of substitutional atoms on the Snoek peak have been developed, and a lot of data can be found in the handbook of Blanter et al. [28]. One of the first theories describing the effects of Cr atoms in the bcc Fe lattice was proposed by Tomilin et al. [29]. According to this model, H_0 = 20 kcal/mol is the activation energy when only Fe atoms are present at the corners of the octahedron, and a contribution ΔH = 3.1 kcal/mol has to be added to H_0 for every Cr atom substituting a Fe atom. On this basis, the Q^{-1} curves consist of the overlapping of seven Snoek-type peaks corresponding to the aforesaid configurations. The activation energies and the positions of the peaks calculated for a resonance frequency of 250 Hz, typical of present experiments, are summarised in Table 2.

Table 2. Activation energies and peak positions of the seven $Q^{-1}{}_i$ peaks corresponding to octahedral configurations with a number of Cr atoms i varying from zero to six. (Resonance frequency f = 250 Hz.)

Number of Cr Atoms i	0	1	2	3	4	5	6
Activation energy (kcal/mol)	20.0	23.1	26.2	29.3	32.4	35.5	38.6
Peak position (K)	380	453	500	560	621	650	710

On these grounds, the Q^{-1} curves can be considered as the sum of seven contributions:

$$Q^{-1} = \sum_{i=0}^{6} Q^{-1_i} = \sum_{i=0}^{6} \Delta_i \frac{\omega\tau_i}{1+\omega^2\tau_i{}^2} = \sum_{i=0}^{6} \Delta_i sech\left[\left(\frac{H_0+i\Delta H}{R}\right)\left(\frac{1}{T}-\frac{1}{T_i}\right)\right], \tag{3}$$

where R is the gas constant, $\omega = 2\pi f$, Δ_i the relaxation strength of peak $Q^{-1}{}_i$, proportional through a factor $\lambda \cong 50$ to the concentration C_i of C atoms in octahedral sites with i Cr atoms, T_i the peak temperature, and τ_i its relaxation time, given by

$$\tau_i = \tau_0 \exp -\left(\frac{H_0+i\Delta H}{RT}\right), \tag{4}$$

with τ_0 being about 10^{-15} s.

The Tomilin's model assumes that, at the thermodynamic equilibrium, the number P_i of configurations with i atoms of Cr obeys a binomial distribution:

$$P_i = \frac{6!(1-N)^{6-i}N^i}{i!(6-i)!}, \tag{5}$$

where N is the Cr molar concentration. The concentration of C atoms in interstices with i Cr neighbours results in being

$$C_i = P_i\xi_0 / \left[\exp -\left(\frac{i\Delta H}{RT}\right) + (1-\exp -(i\Delta H/RT))\xi_0\right] \tag{6}$$

where ξ_0 is the probability of occupation of interstices with i = 0.

Figure 3 shows the Q^{-1} curve calculated on the basis of Equation (3) for an alloy with the same Cr and C content of the examined steel, according to the assumptions of Tomilin's model expressed by Equations (5) and (6).

The comparison between the curves in Figure 1a,b and that in Figure 3 (re-drawn from ref. [27]) clearly shows that both the experimental curves differ from that predicted by the model. The experimental curves substantially exhibit peaks centred at the temperatures foreseen by the model and reported in Table 2, however, the intensities of the peaks (relaxation strengths Δ_i) do not correspond. This means that the distributions of C–Cr associates in both cases are not those of thermodynamic equilibrium, and the different distributions depend on the cooling rate from the austenitic field.

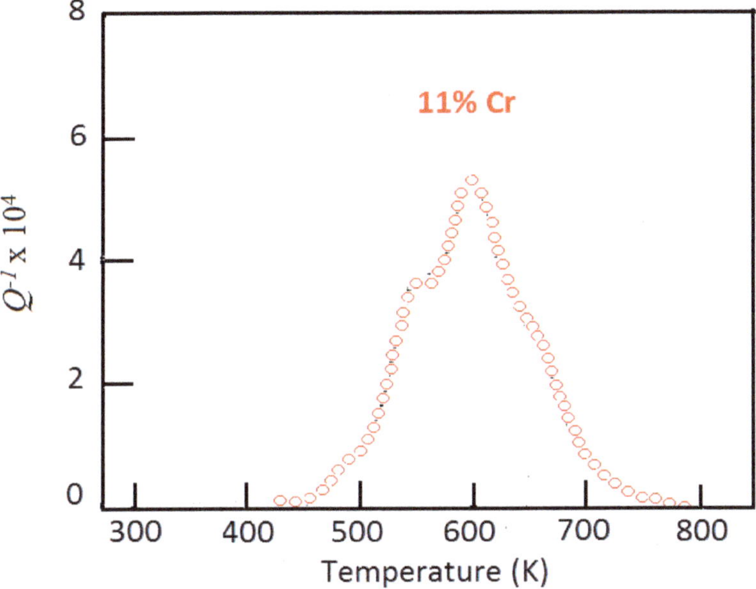

Figure 3. Q^{-1} curve for an alloy with the same Cr and C content of the examined steel, calculated on the basis of Equation (3), according to the assumptions of Tomilin's model. These data were already published in ref. [27], but the present curve obtained through VRA 1604 is more precise in particular in the initial part.

To analyse the Q^{-1} curves of the steel under different conditions of quenching, heat treatment etc., they have been fitted by considering the possible anelastic processes contributing to the damping. In addition to the relaxation peaks due to C–Cr associates, further contributions to the curves come from background and peaks connected to N relaxation processes.

The background is mainly due to the interactions of dislocations with interstitial atoms (C and N) and exhibits an increasing trend with temperature that can be described by an exponential function.

The damping contributions of N are, in fact, the same Q^{-1} peaks observed in Fe–Cr–N alloys by Ritchie and Rawlings [30]. Four peaks have been observed: the first and second peaks, attributed by these investigators to N/Cr–Cr interactions, have activation energies of 16.19 and 16.98 kcal/mol, respectively, and are centred for our resonance frequency (~250 Hz) at 324 K and 339 K. The third one, ascribed to Fe–N interactions, has an activation energy of 18.58 kcal/mol and is centred at 360 K. The fourth peak, due to N/Fe–Cr interactions, has an activation energy of 20.78 kcal/mol, and its temperature is 410 K.

In fact, Q^{-1} spectra are mainly affected by N contributions in the low temperature range, and by C contributions in the high temperature range.

The curve fitting of the experimental data is the sum of all the peaks due to C and N, and the background, therefore, it can be written as

$$Q^{-1} = \sum_{i=0}^{6} \Delta_i \sec h\left[\left(\frac{H_0+i\Delta H}{R}\right)\left(\frac{1}{T}-\frac{1}{T_i}\right)\right] + \sum_{k=1}^{4} \Delta_k \sec h\left[\left(\frac{H_k}{R}\right)\left(\frac{1}{T}-\frac{1}{T_k}\right)\right] + A\exp\left(-\frac{T}{B}\right) + C, \quad (7)$$

where H_k and T_k are the activation energies and peak temperatures of N peaks, while A, B, and C are constants.

Since the background depends on microstructural features, such as dislocations and point defects, it is not the same in samples prepared in a different way, thus, the values of A, B, and C have to be determined for each single spectrum.

Once determined, the background curve, the activation energies (H_i and H_k), and the corresponding peak temperatures (T_i and T_k) of C and N peaks, are introduced into Equation (7), and the relaxation strengths Δ_i and Δ_k are suitably adjusted to find the best fitting curve.

Moreover, it has been assumed that the activation energy of each single relaxation process could be affected by small variations with respect that considered by the Tomilin's model, owing to a possible different environment surrounding the octahedra involved in the process, in particular, the number of Cr atoms present in the nearest atomic shells. Therefore, some shifts with respect to the temperatures T_i, reported in Table 2, have been considered to get the best fitting.

For example, Figure 4 displays the analysis of the Q^{-1} spectra in Figure 1. Table 3 reports the peak positions and relaxation strengths, Δ_i, used to calculate the different contributions which allow the best fit. The temperature of C–Cr_4 and C–Cr_6 peaks are a little shifted with respect to the values predicted by the Tomilin's model; the shifts correspond to variations of activation energies of about 1%.

Table 3. Peak positions (T) and relaxation strengths (Δ) used to fit experimental data in Figure 4.

T, Δ	C Peaks							N Peaks			
	C–Cr_0	C–Cr_1	C–Cr_2	C–Cr_3	C–Cr_4	C–Cr_5	C–Cr_6	N/Cr–Cr	N/Cr–Cr	Fe/N	N/Fe–Cr
Figure 4a											
T (K)	380	453	500	560	617	650	714	324	339	360	410
$\Delta \times 10^4$	0.65	0.4	1.3	0	4.9	0	1.25	0.7	0.85	0.45	0.55
Figure 4b											
T (K)	380	453	495	560	617	650	725	324	339	360	410
$\Delta \times 10^4$	0.2	0.45	0.65	0.35	4	0	2.6	0.4	0.2	0.3	0.5

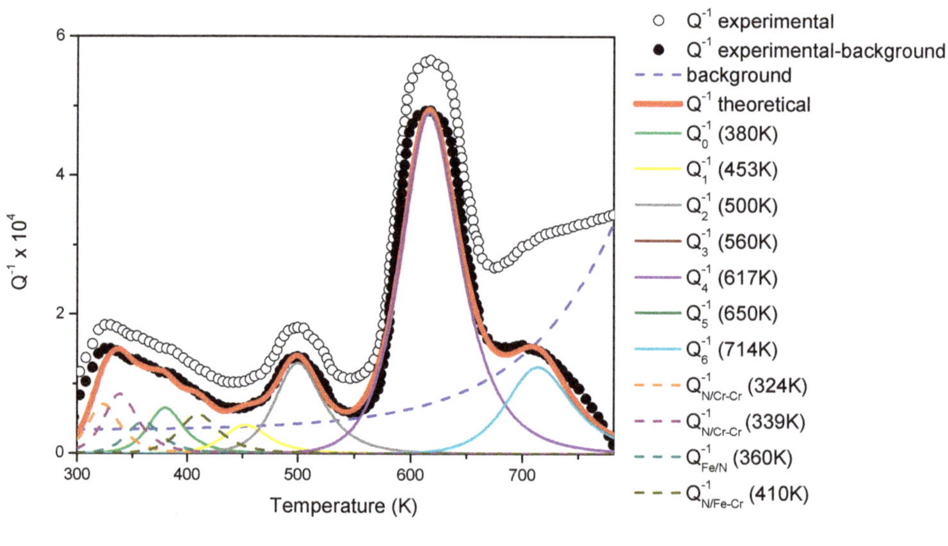

(a)

Figure 4. *Cont.*

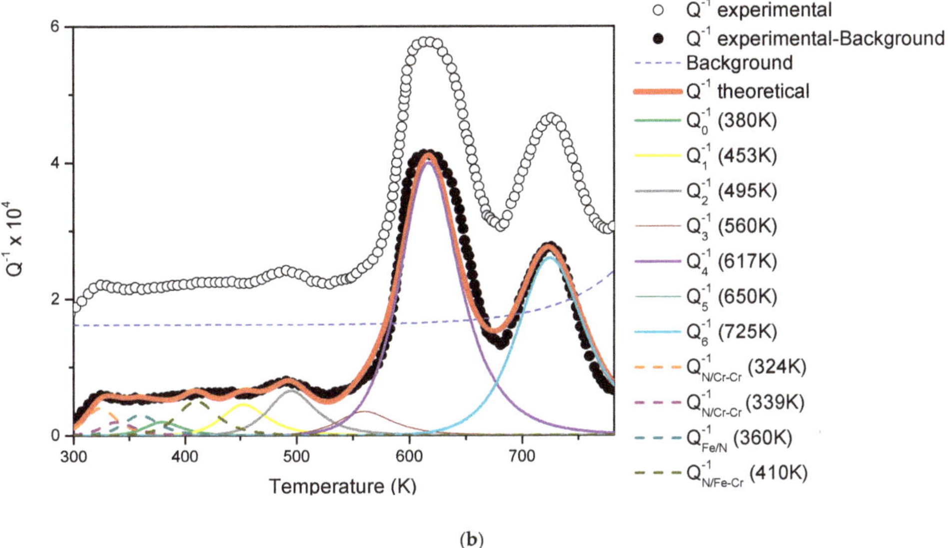

(b)

Figure 4. Fitting of the Q^{-1} curves of the steel quenched from 1348 K with cooling rates of 3600 K/min (**a**) and 150 K/min (**b**). The best fit curve is the sum of the background curve, seven peaks corresponding to C–Cr associates with i values from 0 to 6, and four peaks due to N.

3.2. Cr Segregation Phenomena

Different cooling rates from the austenitic field (1348 K) do not seem to have effects on the microstructure and mechanical properties of the steel, however, successive steps of tempering treatment determine a quite different behaviour, in particular, of the fracture mode. Figure 5, re-drawn from ref. [22], compares the ductile to brittle transition temperature (DBTT) of samples quenched from 1348 K with cooling rates of 150 and 3600 K/min and, then, submitted to heat treatments at 973 K for increasing time. DBTT was determined from Charpy tests performed on standard V-notched probes in the temperature range from 173 K (−100 °C) to + 423 K (+150 °C).

After quenching, DBTT is 285 K (+12 °C) for both materials, however, the curves exhibit a different trend as the time of tempering treatment at 973 K increases. In the case of the samples quenched at 150 K/min, a minimum is observed after 2 hours, followed by a maximum at 6 hours, and by a progressive decrease for longer tempering time. On the contrary, DBTT of the samples quenched at 3600 K/min shows an initial decrease (up to 4 h), then it remains substantially stable.

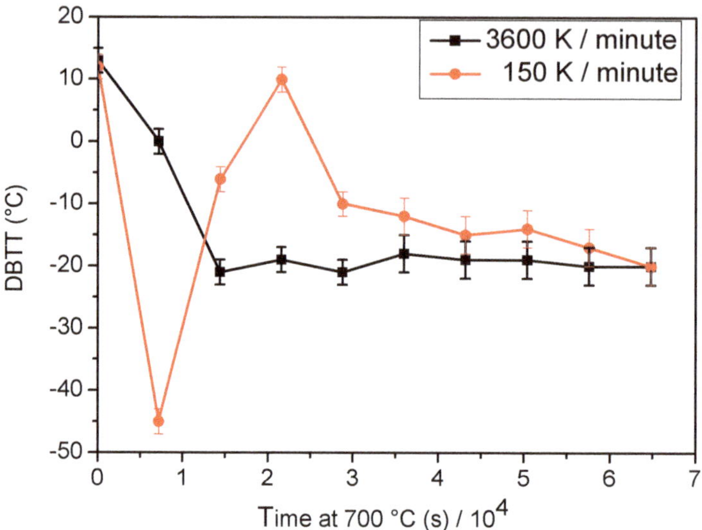

Figure 5. DBTT of the Cr martensitic steel vs. time of heat treatment at 973 K: Comparison of samples quenched from 1348 K with two cooling rates (150 and 3600 K/min).

SEM observations on fracture surfaces of probes broken in ductile field showed no significant differences in samples cooled with different rates. In brittle field, the morphological features for both the materials in as-quenched condition are the same, namely, those typical of a quasi-cleavage fracture, but differences arise and become gradually more pronounced at increasing time of tempering. After 7.2×10^4 s (20 h) at 973 K, the fracture surfaces of samples cooled at 3600 and 150 K/min show different morphologies (Figure 6): the fracture mode is always of quasi-cleavage in (Figure 6a), whereas is mixed (quasi-cleavage plus intercrystalline) in (Figure 6b).

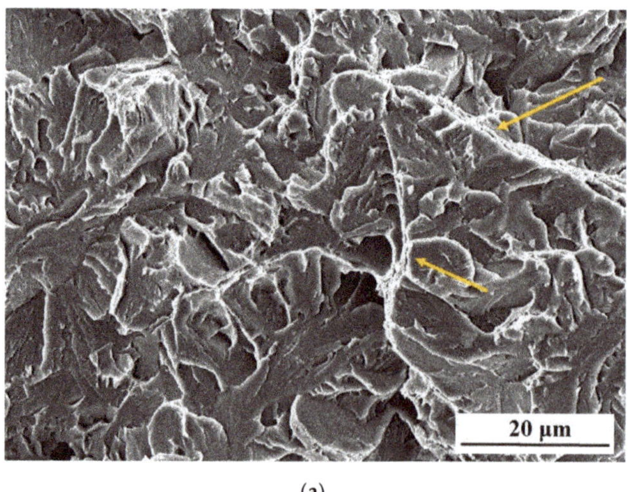

(a)

Figure 6. *Cont.*

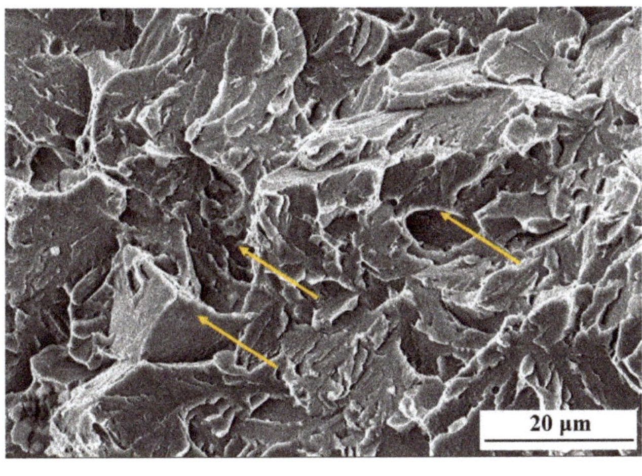

(b)

Figure 6. Fracture surfaces of samples treated 20 h at 973 K, and broken in Charpy tests at 233 K (−40 °C), namely, in the brittle field: quenching rate from austenitic field of 3600 K/min (**a**) and 150 K/min (**b**). Arrows indicate tear ridges in (**a**), and the zones with intercrystalline fracture morphology in (**b**).

The image of Figure 6a shows the typical quasi-cleavage fracture surface: the cracks are multiple and transgranular, with tear ridges in the zones where moving cracks join together. A similar morphology is observed in Figure 6b, however, in this case, zones are also present where the fracture front travelled along the boundaries of primary austenitic grains (typical features of intercrystalline fracture). In fact, the fracture is mixed, occurring in both quasi-cleavage and intercrystalline modes.

The different evolution of the two materials, whose microstructure and mechanical properties after quenching are apparently similar, has been related to the different C–Cr associates distribution shown in Figure 1, namely, large numbers of C–Cr associate with i = 4 and 2 after faster cooling, i = 4 and 6 after slower cooling. Both cases are consistent with an original Cr clustering in austenitic field (before cooling), which exhibits a specific evolution depending on cooling rate. Cr clustering in austenite was examined in detail in a previous work [15]. Monte Carlo simulations demonstrated that Cr clustering occurs through a two-step process owing to the different diffusivities of C and Cr: 1st step—rapid diffusion of C atoms to form C–Cr associates, 2nd step—slower diffusion of Cr leading to the C–Cr associates coalescing, giving rise to Cr-rich clusters.

Figure 7 shows the Q^{-1} curves of samples quenched at 3600 and 150 K/min, and treated at 973 K for 3.6×10^4 s (10 h) and 7.2×10^4 s (20 h). The curves of the steel quenched with the faster cooling rate (Figure 7a,b) do not show relevant variations after the heat treatments, and maintain prevalent peaks referable to C–Cr associates with i = 2 and i = 4, like that of as-quenched steel (see Figure 4a and Table 3).

On the contrary, those of the steel cooled at a rate of 150 K/min (Figure 7c,d) significantly change in two particular branches, at lower and higher temperature (substantially below or above 500 K), and can be distinguished referable to the occurrence of Cr-depleted and Cr-enriched zones, respectively (see Table 4).

From the results, it is clear that the C–Cr associates distribution depends on the cooling rate from austenitic field; the distribution obtained with the lower cooling rate is not stable, and evolves when the material is heat treated at 973 K, giving rise to zones with different Cr content, i.e., Cr segregation. In particular, Cr segregation at grain boundaries is connected to phenomena of embrittlement and DBTT variations (see Figure 5). The distribution of C–Cr associates in the steel, submitted to the higher cooling rate, is substantially stable, thus, segregation does not take place and, consequently, the mechanical behaviour is not affected by intercrystalline ruptures.

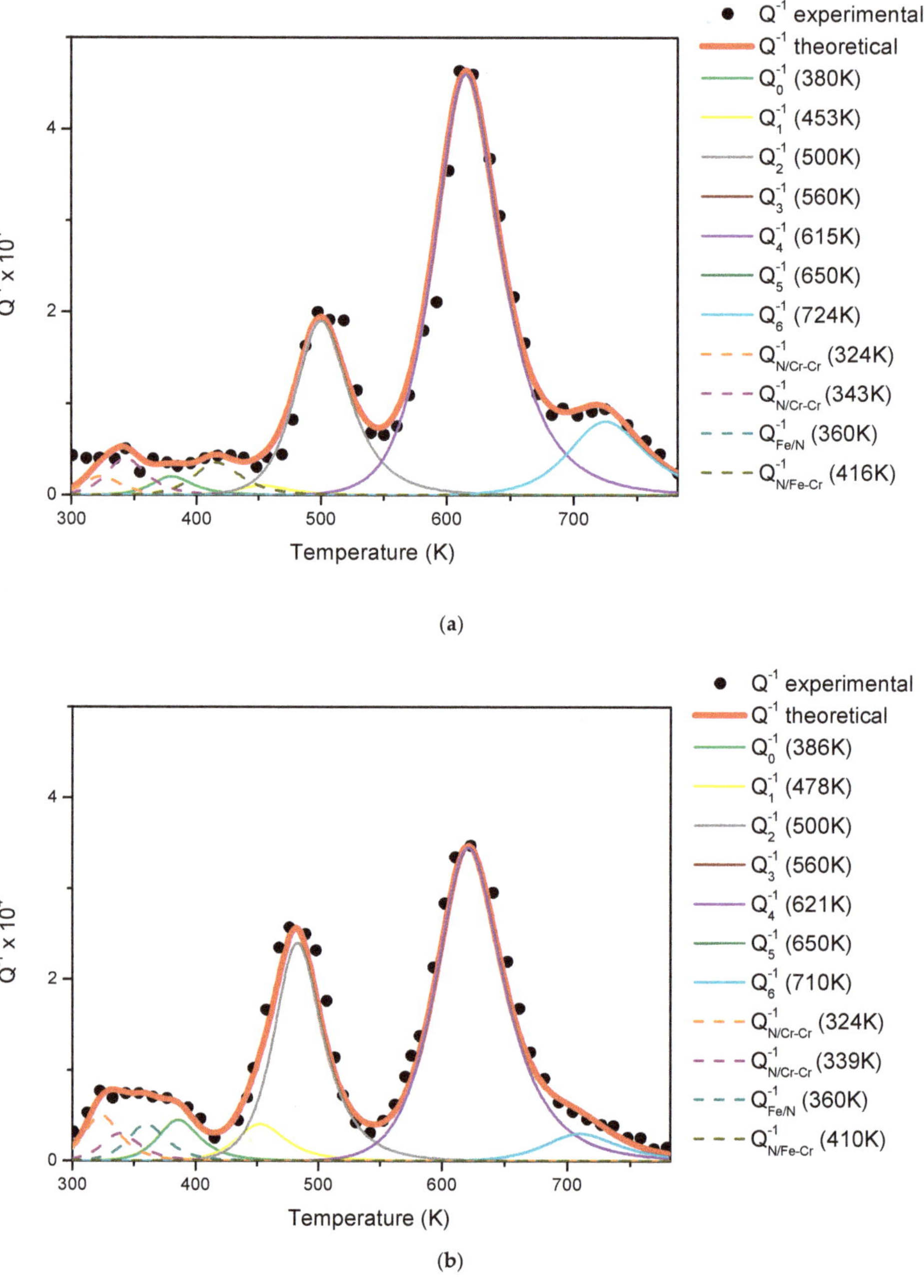

(a)

(b)

Figure 7. *Cont.*

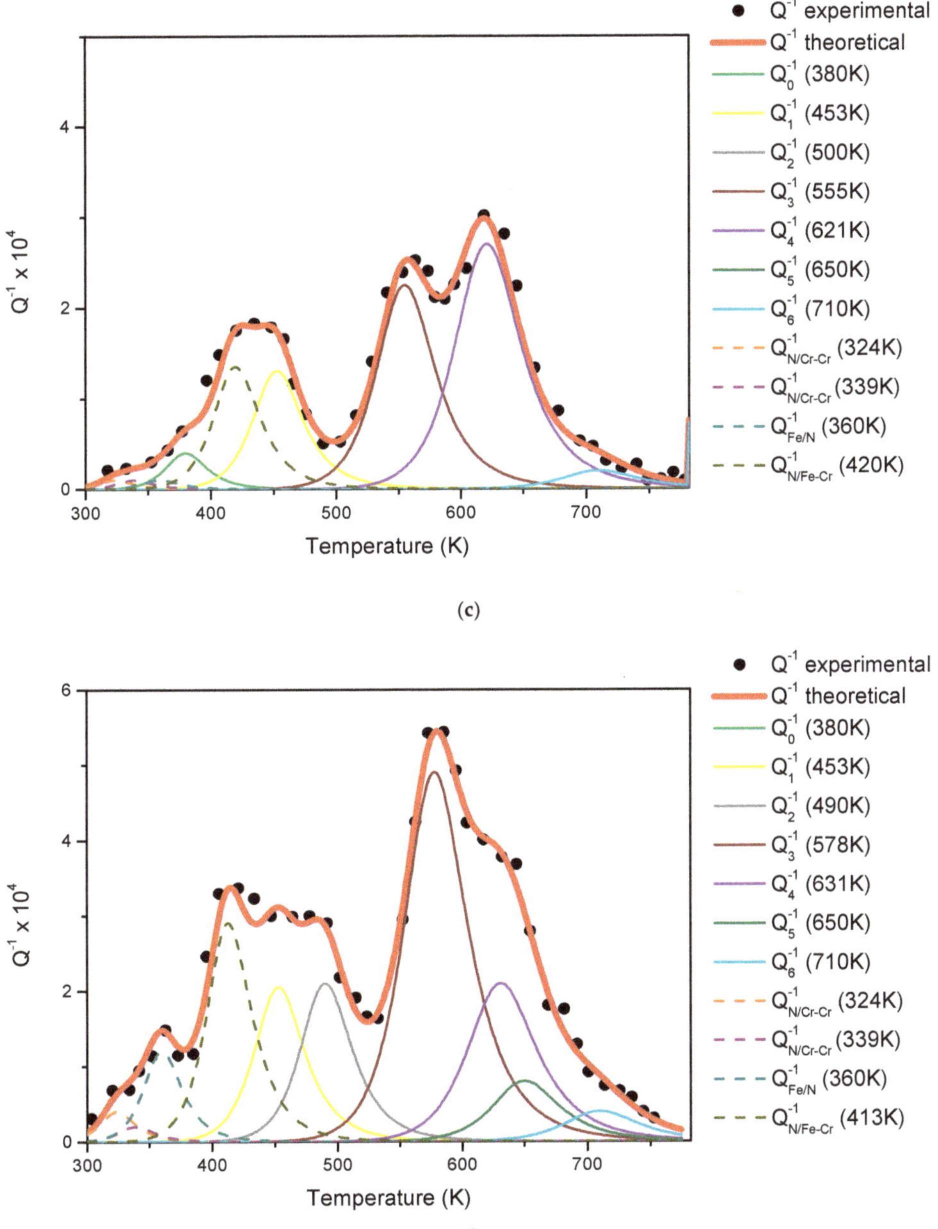

Figure 7. Q^{-1} curves of samples quenched at 3600 (**a**,**b**) and 150 K/min (**c**,**d**), and treated at 973 K for 3.6×10^4 s (**a**,**c**) and 7.2×10^4 s (**b**,**d**). Experimental data are reported after background subtraction.

Table 4. Temperatures T and relaxation strengths Δ of the Q^{-1} peaks used to fit MS experimental data displayed in Figure 7, quenched at 3600 (Figure 7a,b) and 150 K/min (Figure 7c,d), and treated at 973 K for 3.6×10^4 s (Figure 7a,c) and 7.2×10^4 s (Figure 7b,d).

T, Δ	C Peaks							N Peaks			
	C–Cr_0	C–Cr_1	C–Cr_2	C–Cr_3	C–Cr_4	C–Cr_5	C–Cr_6	N/Cr–Cr	N/Cr–Cr	Fe/N	N/Fe–Cr
					Figure 7a						
T (K)	380	453	500	560	615	650	724	324	343	360	416
$\Delta \times 10^4$	0.2	0.1	1.9	0	4.6	0	0.8	0.2	0.4	0	0.35
					Figure 7b						
T (K)	386	453	483	560	621	650	710	324	339	360	410
$\Delta \times 10^4$	0.45	0.4	2.4	0	3.45	0	0.3	0.5	0.3	0.4	0
					Figure 7c						
T (K)	380	453	500	555	621	650	710	324	339	360	420
$\Delta \times 10^4$	0.4	1.3	0	2.25	2.7	0	0.2	0.1	0.1	0.1	1.35
					Figure 7d						
T (K)	380	453	490	578	631	650	710	324	339	360	413
$\Delta \times 10^4$	0	2.05	2.1	4.9	2.1	0.8	0.4	0.4	0.2	1.2	2.9

3.3. Interaction of Hydrogen with C–Cr Associates

Hydrogen in steels represents the cause of severe failures in mechanical components. Hydrogen embrittlement (HE) and hydrogen attack (HA) are different phenomena related to the presence of hydrogen in steels, and both of them have detrimental effects.

HE decreases the ductility, toughness, true stress at fracture, and accelerates crack growth [31–34], leading to catastrophic failure or delayed rupture of structural materials. Hydrogen can be pre-existing in the matrix, or picked up from the environment; in any case, HE effects depend on the concentration, temperature, composition, and microstructure of the material. Turnbull et al. [35] analysed the combined effects of reversible and irreversible trap sites on hydrogen transport in Cr martensitic steels. Typical HE mechanisms are bubble formation, reduction in lattice cohesive forces, reduction in surface energy, and hydride formation.

HA is an irreversible process, occurring only above 473 K, and consists in the reaction of H with C to form methane (CH_4). The phenomenon, described and discussed in a classical paper by Shewmon [36], causes swelling and brittle behaviour in plain carbon steels, whereas it does not occur in Cr steels if the Cr content is higher than 7%. Lacombe et al. [37] demonstrated by means of autoradiographic technique that HE is suppressed because H concentrates around Cr-enriched grain boundaries and Cr carbides. The problem investigated here regards HA: how do C–Cr associates interact with H introduced into the steel at high temperature?

The steel has been submitted to an austenitisation treatment at 1348 K for 7.2×10^3 s (2 h), and cooled to room temperature with a rate of 150 K/min. To examine the effect of H permeation on the material in different conditions, two sets of samples have been examined by MS measurements: as-quenched and annealed for 7.2×10^4 s (20 h) at 973 K. Both sets have been H-permeated by a three-step treatment consisting of (i) heating up to 873 K at a rate of 6.7×10^{-3} K/s, (ii) permeation at 873 K for 8.6×10^5 s in flowing gas with a pressure of 10^5 Pa, (iii) cooling down to room temperature at the same rate used for heating. MS measurements were carried out on both sets of samples before and after H permeation.

Q^{-1} curves of the steel, quenched and H-permeated (QH), quenched and annealed (QA), quenched, and annealed and H-permeated (QAH), are reported in Figure 8. Experimental data are reported after background subtraction.

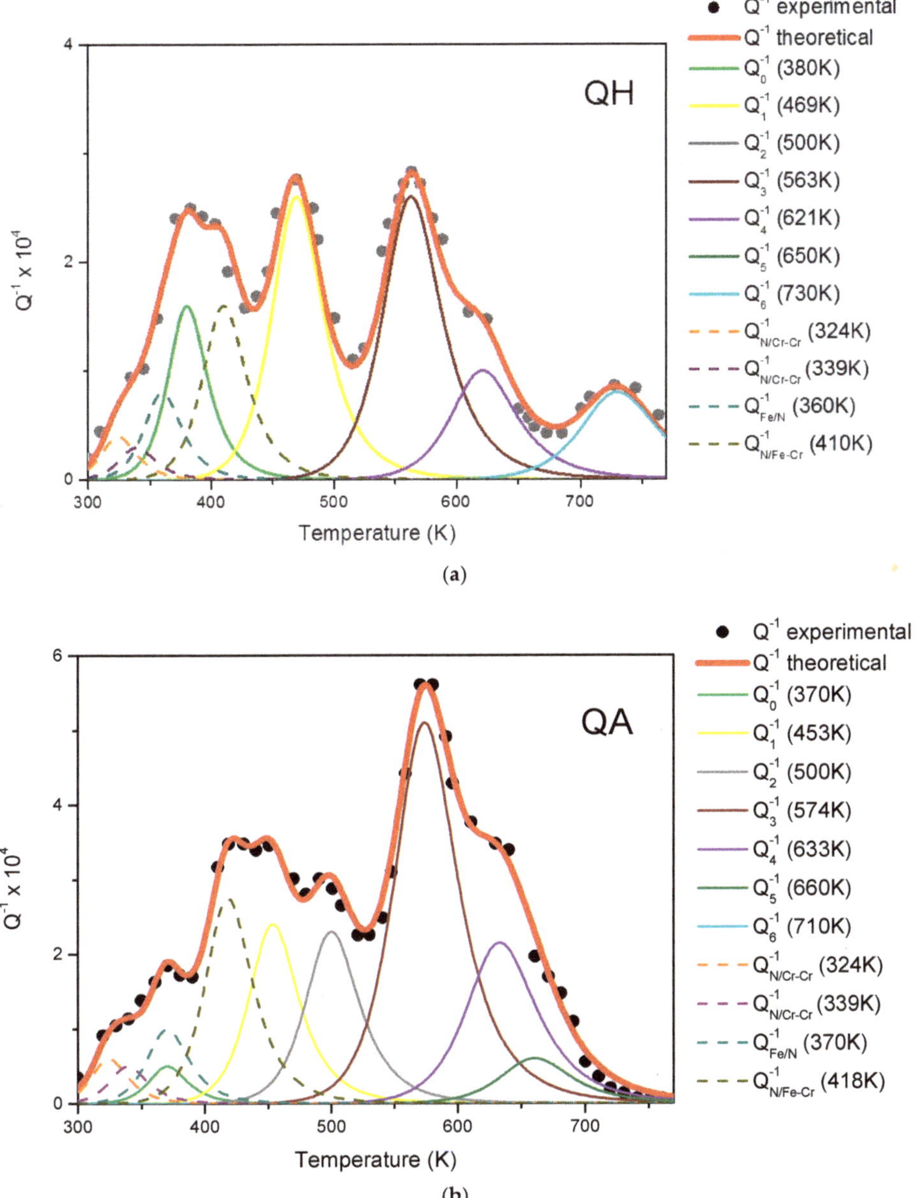

(a)

(b)

Figure 8. *Cont.*

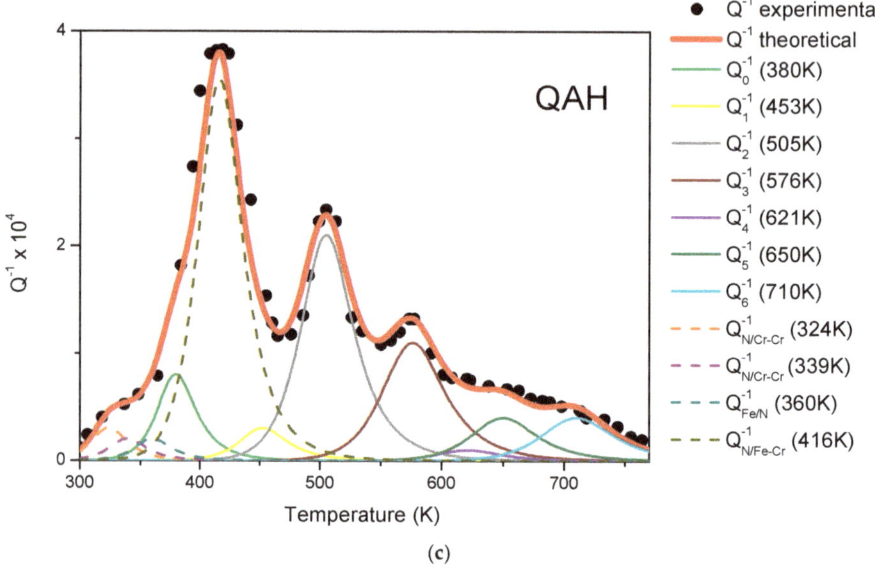

(c)

Figure 8. Q^{-1} spectra of the steel, quenched and H-permeated (**QH**), quenched and annealed (**QA**), quenched, and annealed and H-permeated (**QAH**). Experimental data are reported after background subtraction.

The relaxation strengths of the peaks used to fit the curves in Figure 8 are reported in Table 5. The Q^{-1} curve of the quenched (Q) steel is displayed in Figure 4b, and the corresponding data are reported in Table 3 and, also, in Table 5, to make the comparison easier.

Table 5. Temperatures T and relaxation strengths Δ of the Q^{-1} peaks used to fit MS experimental data displayed in Figure 8, for the samples quenched (Q), quenched and H-permeated (QH), quenched and annealed (QA), and quenched, annealed, and H-permeated (QAH). Data of the quenched (Q) steel (Table 3) are also reported for comparison.

T, Δ	C Peaks							N Peaks			
	C–Cr_0	C–Cr_1	C–Cr_2	C–Cr_3	C–Cr_4	C–Cr_5	C–Cr_6	N/Cr–Cr	N/Cr–Cr	Fe/N	N/Fe–Cr
				Figure Q							
T (K)	380	453	495	560	617	650	725	324	339	360	410
$\Delta \times 10^4$	0.2	0.45	0.65	0.35	4	0	2.6	0.4	0.2	0.3	0.5
				Figure QA							
T (K)	370	453	500	574	633	660	710	324	339	370	418
$\Delta \times 10^4$	0.5	2.4	2.3	5.1	2.15	0.6	0	0.6	0.5	1	2.75
				Figure QH							
T (K)	380	469	500	563	621	650	730	324	339	360	410
$\Delta \times 10^4$	1.6	2.6	0	2.6	1	0	0.8	0.4	0.3	0.8	1.6
				Figure QAH							
T (K)	380	453	505	576	621	650	710	324	339	360	416
$\Delta \times 10^4$	0.8	0.3	2.1	1.1	0.1	0.4	0.4	0.3	0.2	0.2	3.55

As shown in Figure 8 and Table 5, H permeation changes the relaxation strengths Δ of Q^{-1} peaks, and the total amount of C and N atoms giving rise to relaxation processes.

From the comparison of the relaxation strengths of the as-quenched steel before and after H permeation, it is evident that the amount of interstitials (C and N) involved in relaxation processes is higher in H-charged samples. The total amount of interstitials contributing to relaxation processes is proportional to the area under the Q^{-1} curve, and the ratio between the areas of Q and QH samples is 0.97. The increasing amount of C and N atoms contributing to relaxation processes depends on the martensitic structure evolution. As-quenched martensite consists of laths with a high density of dislocations (~10^{11} cm^{-2}); such microstructure changes if the material is treated at the temperature of 873 K involved in H permeation, in particular, laths disappear, and dislocation density strongly decreases [38]. Since laths and dislocations are sinks for interstitials, a large number of C and N atoms become free and, consequently, even if a part of them forms carbides and nitrides, and another part is bound by H atoms, the overall effect of H permeation treatment is to increase the amount of interstitial atoms participating in anelastic processes.

A different behaviour is exhibited by the samples annealed for 7.2 × 10^4 s at 973 K: the ratio between the areas under the Q^{-1} curves of QA and QAH samples is 2.03. In this case, the material has, in origin, a fully recovered structure, and the precipitation of carbides and nitrides has been substantially completed [38], therefore, C and N atoms contributing to relaxation peaks decrease, due to the H blocking effect.

It is noteworthy that the relative intensities of Q^{-1} peaks, due to C–Cr clusters with a higher number of Cr atoms, are strongly reduced after H permeation in both as-quenched (Q) and quenched and annealed steels (QA). In as-quenched steel, the intensities of the ${Q^{-1}}_6$ and ${Q^{-1}}_4$ peaks decrease from 2.6 × 10^{-4} and 4.0 × 10^{-4} (Q) to 0.8 × 10^{-4} and 1.0 × 10^{-4} (QH), respectively, while in quenched and annealed samples, the relaxation strengths of ${Q^{-1}}_4$ and ${Q^{-1}}_3$ peaks pass from 2.15 × 10^{-4} and 5.1 × 10^{-4} (QA) to 0.1 × 10^{-4} and 1.1 × 10^{-4} (QAH).

Each relaxation process consists of the alternative jumping of a single interstitial atom from an octahedral position to another one. If an atom of C is bound to an atom of H, it is no more able to jump because the process would not involve the single C atom, but the C–H pair with a tremendous lattice distortion. Since C atoms strongly bound by H atoms are hindered to participate in relaxation processes, the intensities of the corresponding ${Q^{-1}}_6$ peaks are expected to decrease. Therefore, MS results indicate that a preferred interaction between H atoms and C–Cr associates with a high number of Cr atoms does exist. These associates play a significant role in HA mechanism because, if H atoms are attracted and bound by C–Cr associates, they are not able to diffuse towards grain boundaries, sub-boundaries, and dislocation tangles where CH_4 bubbles could nucleate (first stage of HA), grow, merge and, finally, form fissures and cracks (final stage of HA).

In conclusion, the present experiments demonstrate how C–Cr associates with a high number of Cr atoms contribute, together with Cr carbides and Cr-enriched boundaries, to block H atoms and suppress HA. The result is in agreement with the existence of a critical limit, around 7%, below which the Cr amount is not sufficient to avoid H attack. As a matter of fact, the Q^{-1} peaks corresponding to C–Cr associates with a high number of Cr atoms are absent in the curves of low Cr steels (see ref. [30]).

3.4. Nucleation of Cr Carbides

The aim of the present work is to determine whether the C–Cr associates affect the precipitation of Cr carbides and, in this case, to assess their specific role in the process.

The samples underwent austenitisation treatment (1.8 × 10^3 s at 1348 K), then, were cooled to room temperature at a rate of 150 K/min, and subjected to a multi-step heat treatment consisting of successive steps of 3.6 × 10^3 s at 673, 773, and 873 K. MS measurements have been carried out on the same samples after each heating step, while a set of different samples were used for destructive tests (carbide extraction). The examined temperature range is that of M_7C_3 and $M_{23}C_6$ carbides (M = metal, C = carbon) precipitation.

Carbides were extracted from the metallic matrix through an electrochemical method. For example, Figure 9 shows the carbides extracted after the heating step at 873 K. The SEM image displays some squared particles (side ~7–8 µm), and a lot of small round particles (~0.1 µm); EDS microanalysis showed that the larger particles are rich in Cr, while those of smaller size are rich in Nb.

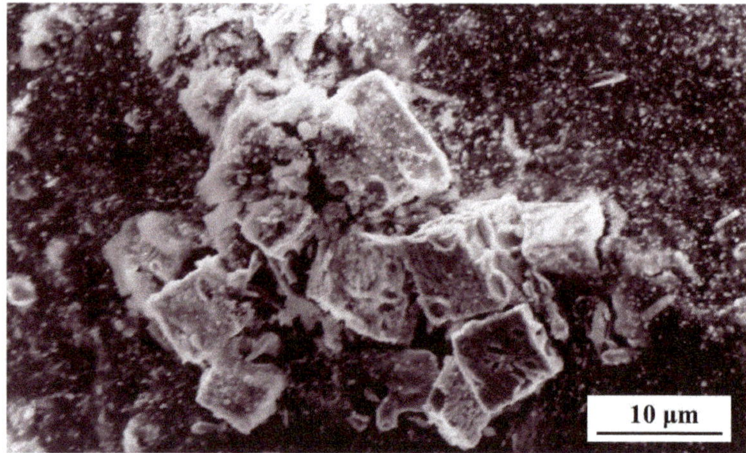

Figure 9. SEM image of the carbides extracted after the heating step at 873 K.

From XRD spectra, the carbides have been identified using the JCPDS-ICDD database [39]:

i. NbC, cubic (a = 0.44698 nm), JCPDS-ICDD file 38–1364,
ii. Nb_2C, orthorhombic (a = 1.0920 nm, b = 0.4974 nm, c = 0.3000 nm), JCPDS-ICDD file 19–858,
iii. Cr_7C_3, hexagonal (a = 1.398 nm, c = 0.452 nm), JCPDS-ICDD file 11–550,
iv. $Cr_{23}C_6$, cubic (a = 1.066 nm), JCPDS-ICDD file 35–783.

In fact, the XRD peaks exhibit small shifts with respect to the angular positions reported by the database, because it refers to monometallic carbides whereas, in the present case (see Table 6), carbides contain also other elements in addition to the main ones (Nb and Cr). Therefore, the crystal structure is the same, but lattice plane distances are a little changed.

Table 6. EDS measurements: Average atomic fractions of metals present in the extracted carbides.

Carbides Type	Cr (atom %)	Fe (atom %)	Nb (atom %)	Mo (atom %)
NbC	3	8	79	10
Nb_2C	1	1	91	7
Cr_7C_3	50	45	3	2
$Cr_{23}C_6$	63	28	4	5

Table 7 summarises the XRD results and reports the most intense reflections of the carbides extracted after each heating step and their relative intensities. Only the XRD reflections of Nb-rich carbides (NbC and Nb_2C) are present after the 673 K heating step. These carbides, already present in austenitic field, are retained after quenching, and are substantially unaffected by the treatments in the examined temperature range. After the step at 773 K, M_7C_3 Cr-rich carbides form, whereas $M_{23}C_6$ carbides appear after the 873 K step. The change in relative intensities of XRD reflections indicate that the precipitation of $M_{23}C_6$ takes place at the expense of M_7C_3 carbides.

Table 7. Angular positions Θ and relative intensities I/I_0 of the XRD reflections of carbides extracted from samples heat treated up to 673, 773, and 873 K.

Treatment Temperature	673 K		773 K		873 K	
Θ (deg)	Carbide type	I/I_0	Carbide type	I/I_0	Carbide type	I/I_0
7.620	M_2C (400)	35	M_2C (400)	28	M_2C (400)	58
7.900	MC (111)	100	MC (111)	100	MC (111)	100
8.610	M_2C (211)	33	M_2C (211)	27	M_2C (211)	60
8.995	M_7C_3 (411)	-	M_7C_3 (411)	20	M_7C_3 (411)	-
9.130	MC (200)	74	MC (200)	77	MC (200)	70
9.380	$M_{23}C_6$ (422)	-	$M_{23}C_6$ (422)	-	$M_{23}C_6$ (422)	32
9.950	$M_{23}C_6$ (511)	-	$M_{23}C_6$ (511)	-	$M_{23}C_6$ (511)	42
12.635	$M_{23}C_6$ (511)	-	$M_{23}C_6$ (511)	-	$M_{23}C_6$ (511)	28
12.650	M_7C_3 (332)	-	M_7C_3 (332)	17	M_7C_3 (332)	-
12.970	MC (220)	40	MC (220)	44	MC (220)	42
14.550	M_2C (231)	28	M_2C (231)	11	M_2C (231)	19
15.460	M_7C_3 (641)	-	M_7C_3 (641)	17	M_7C_3 (641)	5
16.045	M_7C_3 (423)	-	M_7C_3 (423)	13	M_7C_3 (423)	4

Figure 10a–c compares the Q^{-1} spectra after successive heating steps; each spectrum has been fitted according to the aforesaid analysis procedure, and the different relaxation strengths are displayed. Also, in this case, experimental data are reported after background subtraction. In general, it can be observed that the spectra undergo relevant changes depending on the heat treatment temperature with variations of the peak relaxation strengths (see Table 8).

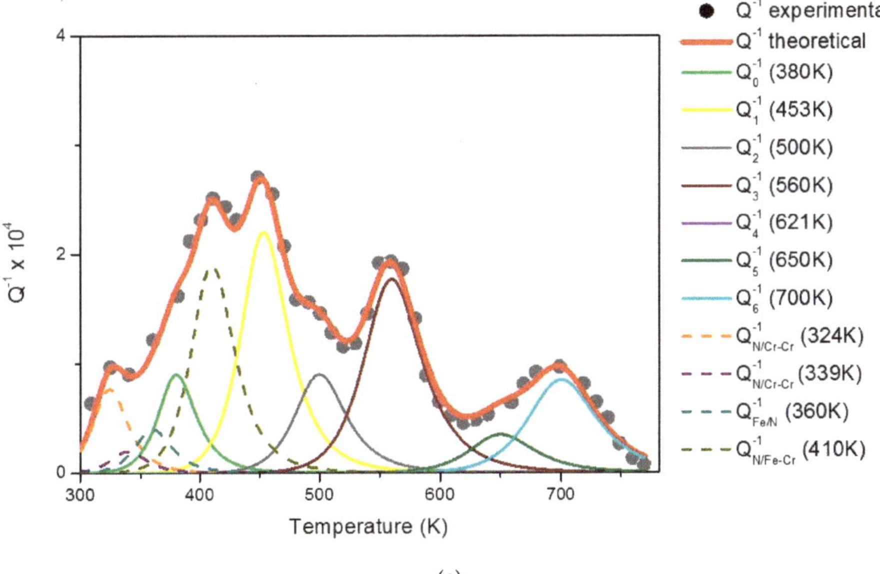

(a)

Figure 10. *Cont.*

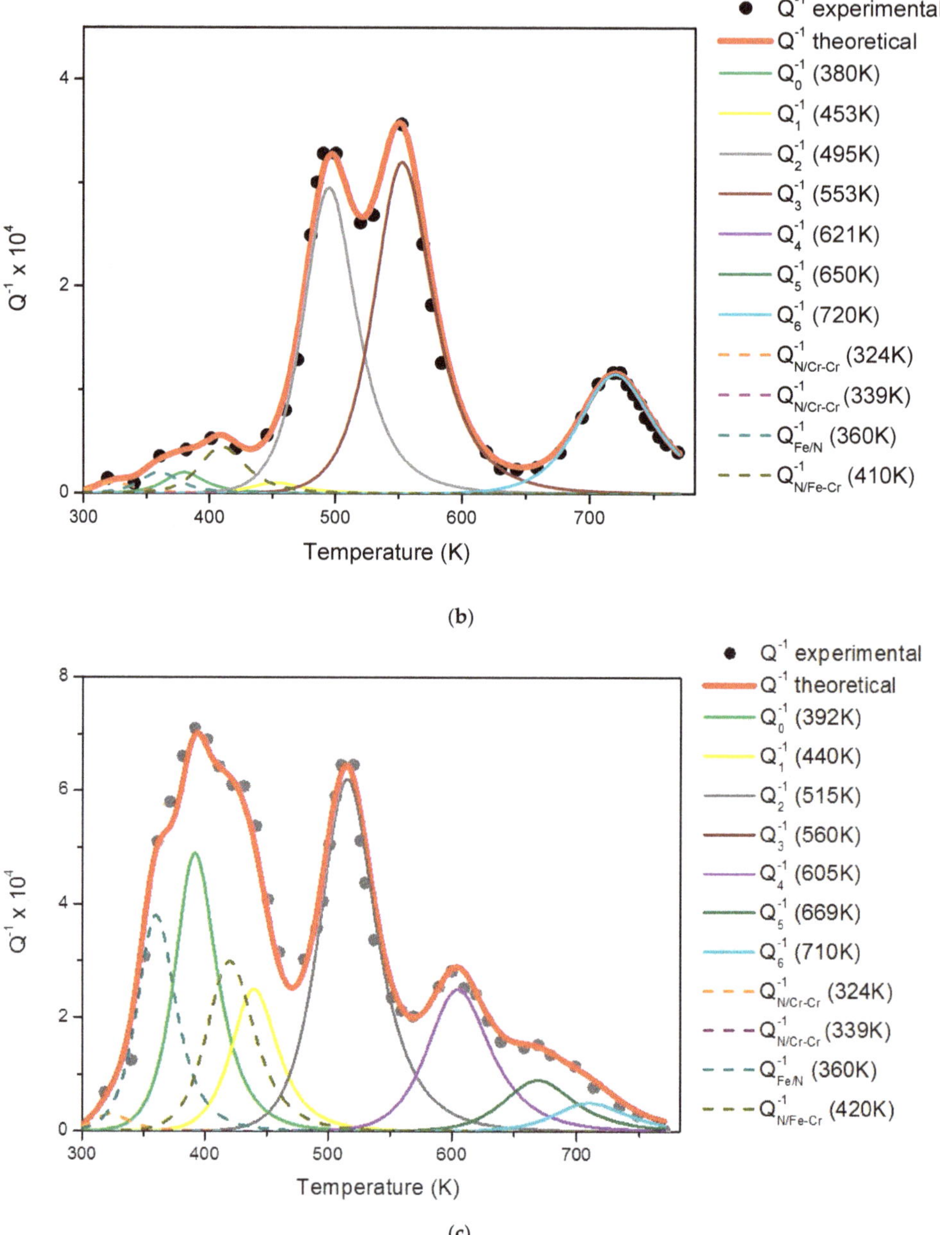

Figure 10. Q^{-1} curves after successive heating steps at 673 K (**a**), 773 K (**b**), and 873 K (**c**). Experimental data are reported after background subtraction.

Table 8. Temperatures T and relaxation strengths Δ of the Q^{-1} peaks used to fit MS experimental data displayed in Figure 10 for the samples heat treated up to 673, 773, and 873 K.

T, Δ	C Peaks							N Peaks			
	C–Cr$_0$	C–Cr$_1$	C–Cr$_2$	C–Cr$_3$	C–Cr$_4$	C–Cr$_5$	C–Cr$_6$	N/Cr–Cr	N/Cr–Cr	Fe/N	N/Fe–Cr
Figure 10a											
T (K)	380	453	500	560	621	650	700	324	339	360	410
Δ × 10^4	0.9	2.2	0.9	1.77	0	0.35	0.85	0.77	0.2	0.4	1.9
Figure 10b											
T (K)	380	453	495	553	621	650	720	324	339	360	410
Δ × 10^4	0.2	0.1	2.95	3.2	0	0	1.15	0.1	0	0.2	0.45
Figure 10c											
T (K)	392	440	515	560	605	669	710	324	339	360	420
Δ × 10^4	4.9	2.5	6.2	0	2.5	0.9	0.5	0.3	0	3.8	3

The results indicate that the distribution of C–Cr associates is altered by treatments at increasing temperature. It is noteworthy that the most intense Q^{-1} peak after the 673 K treatment, namely, that connected with associates with 1 Cr atom, completely disappears after the successive step at 773 K, and reappears after the 873 K one. Such behaviour is clearly related to the precipitation of Cr-rich carbides in this range of temperatures.

EDS results in Table 6 show that the composition of Cr_7C_3 carbides consist of 50% Cr atoms, thus, the ratio between C and Cr atoms is about 1:1. In concomitance with Cr_7C_3 carbides precipitation at 773 K, the Q^{-1} peak due to relaxation of C–Cr associates with 1 Cr atom, which is the most intense after treatment at 673 K, becomes very low, indicating that these associates do not contribution more to relaxation. Therefore, Cr_7C_3 formation seems to occur through a process involving these types of associates. On these grounds, it is supposed that C–Cr$_1$ associates may directly aggregate to form Cr_7C_3 carbides.

The hypothesis of carbide formation through C–Cr$_1$ aggregation is consistent with the Cr_7C_3 structure described by Dyson and Andrews [40]. The positions occupied by metal and C atoms in the Cr_7C_3 carbide cell are displayed in Figure 11: different atomic layers are indicated by specific symbols, and the complex structure consists of distorted octahedra having a C atom at the centre (one of them is indicated).

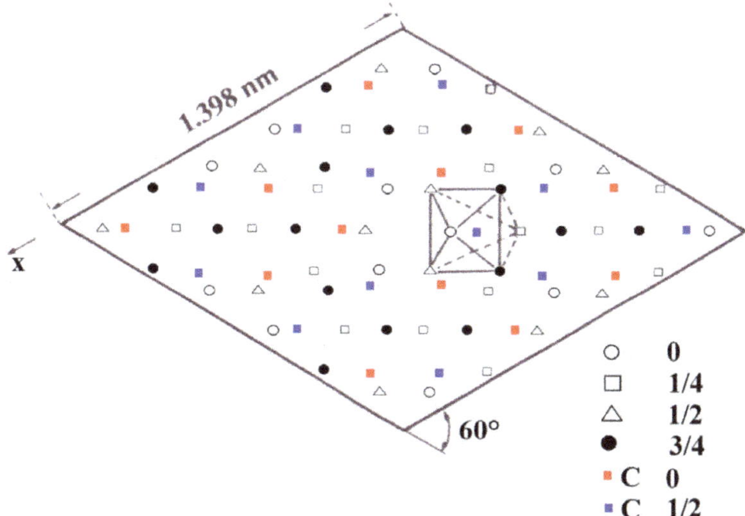

Figure 11. Cr_7C_3 carbide cell according to Dyson and Andrews [40].

Following the 873 K heat step, $Cr_{23}C_6$ carbides form at expenses of Cr_7C_3 carbides, but nucleate on different sites [41]. When Cr_7C_3 carbides solve, C–Cr_1 associates are released in the matrix, can thus participate again to relaxation processes, and the corresponding Q^{-1} peak is observed in the spectrum (Figure 10c).

4. Conclusions

This work presents the results of MS experiments carried out on a Cr martensitic steel. The Q^{-1} vs. temperature curves have been analysed by considering all the relaxation processes involving C and N interstitial atoms. The point defects exhibit different configurations with Cr atoms, which strongly affect the microstructure evolution and mechanical behaviour. The role played by point defects is relevant, and can be summarised as follows:

(i) Cr segregation depends on C–Cr associates distribution in as-quenched material: slow cooling rate from austenitic field involves an unstable distribution which evolves after tempering at 973 K, leading to Cr concentration fluctuations. The phenomenon, which affects the brittle mode of fracture, does not occur if the material is quenched at a higher cooling rate.

(ii) Hydrogen interacts with C–Cr associates, in particular, those which have a high number of Cr atoms. The phenomenon plays a significant role in HA mechanism because H atoms bound by C–Cr associates are not able to diffuse towards lattice defects (grain boundaries, sub-boundaries, and dislocation tangles) where CH_4 bubbles nucleate and successively grow and merge, to form the HA typical cracks.

(iii) C–Cr associates take part in the nucleation mechanism of Cr_7C_3 carbides, and specifically these carbides form by the aggregation of C–Cr associates with 1 Cr atom.

Author Contributions: R.M. planned the experiments; A.V. prepared the samples; R.M. and A.V. performed MS tests; A.F. made the analysis of Q^{-1} spectra; all the authors contributed to analysing the results and writing the manuscript.

Funding: This research received no external funding.

Conflicts of Interest: The authors declare no conflict of interest.

References

1. Schultz, H. Point defects in body-centred cubic transition metals. *Mater. Sci. Eng.* **1968**, *3*, 189–219. [CrossRef]
2. Townsend, J.R.; Di Carlo, J.A.; Nielsen, R.L.; Stabell, D. The elastic bulk effect of point defects in metals. *Acta Metall.* **1969**, *17*, 425–435. [CrossRef]
3. Lutsko, J.F.; Wolf, D.; Phillpot, S.R.; Yip, S. Molecular-dynamics study of lattice-defect-nucleated melting in metals using an embedded-atom-method potential. *Phys. Rev. B* **1989**, *40*, 2841–2855. [CrossRef]
4. Granato, A.V. Interstitialcy model for condensed matter states of face-centered-cubic metals. *Phys. Rev. Lett.* **1992**, *68*, 974–977. [CrossRef] [PubMed]
5. Gordon, C.A.; Granato, A.V. Equilibrium concentration of interstitials in aluminum just below the melting temperature. *Mat. Sci. Eng. A* **2004**, *370*, 83–87. [CrossRef]
6. Montanari, R. Real-time XRD investigations on metallic melts. *Int. J. Mater. Prod. Tech.* **2004**, *20*, 452–463. [CrossRef]
7. Montanari, R.; Varone, A. Synergic role of self-interstitials and vacancies in Indium melting. *Metals* **2015**, *5*, 1061–1072. [CrossRef]
8. Goncharova, E.V.; Makarov, A.S.; Konchakov, R.A.; Kobelev, N.P.; Khonik, V.A. Premelting generation of interstitial defects in polycrystalline indium. *J. Exp. Theor. Phys. Lett.* **2017**, *106*, 35–39. [CrossRef]
9. Safonova, E.V.; Konchakov, R.A.; Mitrofanov, Y.P.; Kobelev, N.P.; Vinogradov, A.Y.; Khonik, V.A. Contribution of interstitial defects and anharmonicity to the premelting increase in the heat capacity of single-crystal aluminum. *J. Exp. Theor. Phys. Lett.* **2016**, *103*, 765–768. [CrossRef]
10. Montanari, R.; Varone, A. Anelastic phenomena preceding the melting of pure metals and alloys. *Mater. Sci. Forum* **2016**, *879*, 66–71. [CrossRef]

11. Montanari, R.; Varone, A. Physical phenomena leading to melting of metals. *Mater. Sci. Forum* **2016**, *884*, 3–17. [CrossRef]
12. Nowick, A.S.; Berry, B.S. *Anelastic Relaxation in Crystalline Solids*, 1st ed.; Academic Press: New York, NY, USA; London, UK, 1972.
13. Brunelli, L.; Gondi, P.; Montanari, R.; Coppola, R. Internal strains after recovery of hardness in tempered martensitic steels for fusion reactors. *J. Nucl. Mater.* **1991**, *179–181*, 675–678. [CrossRef]
14. Capotorto, C.; Coppola, R.; Gondi, P.; Montanari, R.; Tata, M.E. Tempering structures and related ductile to brittle transition in MANET steel. *Fusion Technol.* **1992**, *1*, 1311–1315.
15. Gondi, P.; Montanari, R. On the Cr atom distribution in MANET1 steel. *Phys. Stat. Sol. (a)* **1992**, *131*, 465–480. [CrossRef]
16. Coppola, R.; Gondi, P.; Montanari, R. Effects of C-Cr elementary aggregates on the properties of the MANET steel. *J. Nucl. Mater.* **1993**, *206*, 360–362. [CrossRef]
17. Gondi, P.; Montanari, R. Q^{-1}-spectra connected with C under solute atom interaction. *J. Alloys Compd.* **1994**, *211–212*, 33–36. [CrossRef]
18. Gondi, P.; Montanari, R.; Sili, A.; Coppola, R. Solute Cr atom distribution and fracture behaviour of MANET steel. *J. Nucl. Mater.* **1994**, *212–215*, 564–568. [CrossRef]
19. Coppola, R.; Lukas, P.; Montanari, R.; Rustichelli, F.; Vrana, M. X-ray and neutron line broadening measurements in a martensitic steel for fusion technology. *Mater. Lett.* **1995**, *22*, 17–21. [CrossRef]
20. Brokmeier, H.G.; Coppola, R.; Montanari, R.; Rustichelli, F. Neutron diffraction study of the crystalline texture in a martensitic steel for fusion reactor technology. *Phisica B Condens. Matter* **1995**, *213–214*, 809–811. [CrossRef]
21. Albertini, G.; Ceretti, M.; Coppola, R.; Fiori, F.; Gondi, P.; Montanari, R. Small-angle neutron scattering study of C Cr elementary aggregates in a martensitic steel for fusion-reactor technology. *Phisica B Condens. Matter* **1995**, *213–214*, 812–814. [CrossRef]
22. Gondi, P.; Montanari, R.; Sili, A.; Tata, M.E. Effects of thermal treatments on the ductile to brittle transition of MANET steel. *J. Nucl. Mater.* **1996**, *233–237*, 248–252. [CrossRef]
23. Gondi, P.; Donato, A.; Montanari, R.; Sili, A. A miniaturized test method for the mechanical characterization of structural materials for fusion reactors. *J. Nucl. Mater.* **1996**, *233–237*, 1557–1560. [CrossRef]
24. Gupta, R.; Gondi, P.; Montanari, R.; Principi, G.; Tata, M.E. Internal friction and Mössbauer study of C-Cr associates in MANET steel. *J. Mater. Res.* **1997**, *12*, 296–299.
25. Gondi, P.; Montanari, R.; Tata, M.E. Distribution of C-Cr associates and mechanical stability of Cr martensitic steels. *J. Nucl. Mater.* **1998**, *258–263*, 1167–1172. [CrossRef]
26. Amadori, S.; Campari, E.G.; Fiorini, A.L.; Montanari, R.; Pasquini, L.; Savini, L.; Bonetti, E. Automated resonant vibrating-reed analyzer apparatus for a non-destructive characterization of materials for industrial applications. *Mater. Sci. Eng. A* **2006**, *442*, 543–546. [CrossRef]
27. Gondi, P.; Montanari, R.; Coppola, R. On the statistical distribution of Cr atoms in Fe-Cr alloys with high swelling resistance in NFR. *J. Nucl. Mater.* **1992**, *191–194*, 1274–1278. [CrossRef]
28. Blanter, M.S.; Golovin, I.S.; Neuhäuser, H.; Sinning, H.-R. *Internal Friction in Metallic Materials. A Handbook*; Springer: Berlin, Germany, 2007.
29. Tomilin, I.A.; Sarrak, V.I.; Gorokhova, N.A.; Suvorova, S.O.; Zhukov, L.L. Non-regular carbon atom distribution in iron-chromium alloys. *Phys. Met. Metallogr.* **1983**, *56*, 501–506.
30. Ritchie, I.G.; Rawlings, R. Snoek Peaks in Ternary Iron-Nitrogen Alloys. *Acta Metall.* **1967**, *15*, 491–496. [CrossRef]
31. Tiwari, G.P.; Bose, A.; Chakravartty, J.K.; Wadekar, S.L.; Totlani, M.K.; Arya, R.N.; Fotedar, R.K. A study of internal hydrogen embrittlement of steels. *Mater. Sci. Eng. A* **2000**, *286*, 269–281. [CrossRef]
32. Lu, Q.; Atrens, A. A critical review of the influence of hydrogen on the mechanical properties of medium-strength steels. *Corros. Rev.* **2013**, *31*, 85–103. [CrossRef]
33. Lynch, S.P. Hydrogen embrittlement (HE) phenomenon and mechanisms. In *Stress Corrosion Cracking: Theory and Practice*; Raja, V.S., Shoji, T., Eds.; Woodhead Publishing: Cambridge, UK, 2011; pp. 90–130.
34. Kumar, B.S.; Kain, V.; Singh, M.; Vishwanadh, V. Influence of hydrogen on mechanical properties and fracture of tempered 13 wt % Cr martensitic stainless steel. *Mater. Sci. Eng. A* **2017**, *700*, 140–151. [CrossRef]
35. Turnbull, A.; Carroll, M.W.; Ferriss, D.H. Analysis of hydrogen diffusion and trapping in a 13% chromium martensitic stainless steel. *Acta Metall.* **1989**, *31*, 2039–2046. [CrossRef]

36. Shewmon, P.G. Hydrogen attack of carbon steel. *Metall. Trans. A* **1976**, *7*, 279–286. [CrossRef]
37. Lacombe, P.; Aucouturier, M.; Chene, J. *Hydrogen Embrittlement and Stress Corrosion Cracking*; ASM: Metals Park, OH, USA, 1984; p. 79.
38. Coppola, R.; Gondi, P.; Montanari, R.; Veniali, F. Structure evolution during heat treatments of 12% Cr martensitic steel for NET. *J. Nucl. Mater.* **1988**, *155–157*, 616–619. [CrossRef]
39. *JCPDS-International Centre for Diffraction Data*; Version 2.14; Library of Congress: Newtown Square, PA, USA, 1987–1994.
40. Dyson, D.J.; Andrews, K.W. Carbide M_7C_3 and its formation in alloy steels. *J. Iron Steel Inst.* **1969**, *207*, 208–219.
41. Honeycombe, R.W.K. *Steels Microstructure and Properties*, 3rd ed.; Edward Arnold: London, UK, 1990; p. 158.

© 2018 by the authors. Licensee MDPI, Basel, Switzerland. This article is an open access article distributed under the terms and conditions of the Creative Commons Attribution (CC BY) license (http://creativecommons.org/licenses/by/4.0/).

Article

Modification of Non-Metallic Inclusions in Stainless Steel by Addition of CaSi

Hongying Du [1,*], Andrey Karasev [1], Olle Sundqvist [2] and Pär G. Jönsson [1]

[1] KTH Royal Institute of Technology, Brinellvägen 23, Stockholm 10044, Sweden; karasev@kth.se (A.K.); parj@kth.se (P.G.J.)
[2] Sandvik Materials Technology AB, Sandviken 81181, Sweden; olle.sundqvist@sandvik.com
* Correspondence: hongying@kth.se; Tel.: +46-(0)8-790-8306

Received: 10 December 2018; Accepted: 8 January 2019; Published: 12 January 2019

Abstract: The focus of this study involved comparative investigations of non-metallic inclusions in 316L stainless steel bars without and with Ca treatments. The inclusions were extracted by using electrolytic extraction (EE). After that, the characteristics of the inclusions, such as morphology, size, number, and composition, were investigated by using a scanning electron microscope (SEM) in combination with an energy dispersive X-ray spectroscopy (EDS). The following four types of inclusions were observed in 316L steels: (1) Elongated MnS (Type I), (2) MnS with hard oxide cores (Type II), (3) Undeformed irregular oxides (Type III), and (4) Elongated oxides with a hard oxide core (Type IV). In the reference sample, only a small amount of the Type III oxides (Al_2O_3–MgO–MnO–TiO_x) existed. However, in Ca-treated 316L steel, about 46% of the observed inclusions were oxide inclusions (Types III and IV) correlated to gehlenite and to a mixture of gehlenite and anorthite, which are favorable for the machinability of steel. Furthermore, untransformed oxide cores (Al_2O_3–MgO–MnO) were also found in the inclusions of Type IV. The mechanism leading to different morphologies of oxide inclusions is also discussed.

Keywords: stainless steel; Ca treatment; non-metallic inclusion; electrolytic extraction

1. Introduction

Non-metallic inclusions (NMIs) play a detrimental role with respect to the quality and mechanical properties of steels, causing reduced toughness and shorter fatigue life and potentially increasing the failure risk of the final product [1]. The effect of inclusions on the mechanical properties of steels depends on the chemical composition, density, size, shape, orientation, and distribution of the inclusions. One approach applied for decades in the steelmaking industry is to develop steels with low levels of impurities. However, it is also reported that some inclusions (such as MnS and Ca-modified oxides, among others) can help to improve the steel machinability (for instance by decreasing the wear on the cutting tool, thus extending the tool's life) [2]. Previous studies [3–5] show that inclusions improve machinability primarily by two ways: (1) as a source of stress concentration effects, favoring machinability by reducing the cutting force and increasing the chip breakability during machining, (2) as a lubricant (tool protection layer) in the contact zone of the cutting tool and material (reducing the abrasive and chemical wear of the tool), which is beneficial for the tool life. Thus, modification of NMIs and control of their characteristics in steels during steel manufacturing are very important to balance favorable machinability properties with the desired mechanical property for diverse types of steels.

Ca treatment is one way to modify the inclusions. For instance, Al_2O_3 inclusions in aluminum-killed steels, which can form clusters up to more than 100 μm, can be modified to smaller compact spherical calcium–aluminate (CaO–Al_2O_3) inclusions by calcium addition [6,7]. It is also reported that the suggested content of CaO in modified inclusions after modification should be in the range 25–60 wt % [8]. However, Guo et al. [9] reported that the formation of sulfide enriched by CaS (with quite a high melting temperature) might cause a nozzle clogging problem during casting of free cutting and structural steels with high sulfur contents. Therefore, it is difficult to control sulfide shape by Ca addition. Liu et al. [10] found that inclusion compositions changed from Al_2O_3 to $CaO–Al_2O_3–CaS$ by Ca treatment of X70 pipeline steel.

The modification of oxide inclusions by Ca addition can also improve the machinability of steels [11–15]. For instance, Bletton et al. [11] investigated the inclusion types and machinability properties in three samples of AISI 316L stainless steel with different Ca contents taken from hot-rolled bars. They concluded that the anorthite ($CaO{\cdot}Al_2O_3{\cdot}2SiO_2$) and gehlenite ($2CaO{\cdot}Al_2O_3{\cdot}SiO_2$) inclusions, having relatively low melting temperatures, can be more easily deformed and form a protective lubricant layer between the workpiece and cutting tools during machining. Ånmark et al. [14] investigated the effect of Ca-rich inclusions on tool wear during the hard-turning of a Ca-treated carburizing steel. Ca treatment improved the tool life significantly by the formation of a Ca-enriched slag barrier composed of (Mn,Ca)S and (Ca,Al)(O,S) on the tool. However, systematic investigations of inclusion characteristics and their effect on the final properties of Ca-treated industrial stainless steels have not been sufficient up to now.

Conventionally, NMIs in steels are evaluated in two dimensions (2D) on a polished cross section of a metal sample by using light optical (LO) or scanning electron microscope (SEM). Previous research [16–18] has established significant drawbacks of this 2D conventional method and suggested three-dimensional (3D) investigations based on extraction methods for more reliable measurements regarding the number and size of different inclusions and clusters.

This study is focused on comparative investigations of non-metallic inclusions in industrial 316L Si-killed stainless steels without and with Ca treatment. Here, 3D investigations of inclusions on film filters after electrolytic extraction (EE) were conducted for the evaluation of inclusions characteristics (such as morphology, size, number, and composition) in stainless steel without and with Ca treatment.

2. Materials and Methods

2.1. Materials

Steel samples of industrial 316L stainless steel produced without (reference heat—316R) and with Ca addition (experimental heat—316Ca) were evaluated in this study. The compositions of steels were determined by the steelmaking company (Sandvik Materials Technology AB, Sandviken, Sweden), using usual standard techniques. Details of their compositions are given in Table 1. It should be noted that the contents of sulfur and oxygen in 316Ca steel were higher than in 316R steel.

Table 1. Composition of samples of the reference steel (316R) and the Ca-treated steel (316Ca).

Steel Grade	Ca-treated	C	Si	Mn	Cr	Ni	S [a]	O [a]	Ca [a]	Al [a]
		(mass %)								
316R	No	0.02	0.38	1.60	16.82	11.18	70	20	-	40
316Ca	Yes	0.01	0.46	1.58	16.86	11.14	90	59	28	40

[a] Content of elements given in ppm.

The production route for the reference and experimental heats in the company is shown in Figure 1. It includes scrap melting in an electric arc furnace (EAF), de-carburization in an argon oxygen decarburization (AOD) converter, ladle treatment in a ladle furnace, and continuous casting. Ca treatment is done by adding a CaSi wire at the end of the ladle treatment. A reference sample was produced by using the same process route without Ca treatment. Steel samples for the evaluation of non-metallic inclusions were cut from the middle-radius zone of hot-rolled bars (Ø121 mm) in the longitudinal direction, as shown in Figure 1.

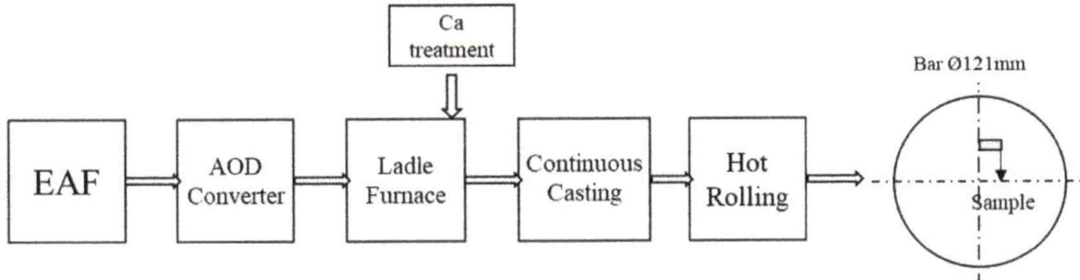

Figure 1. Schematic illustration of the production process of 316L stainless steel, from the melting of scrap and alloys in an Electric Arc Furnace (EAF) to hot rolling.

2.2. Electrolytic Extraction and Investigation of Inclusions

For the evaluation of the characteristics of non-metallic inclusions (such as number, size, composition, and morphology), the steel samples were partially dissolved by using EE. However, inclusions did not dissolve during this EE process. Figure 2a shows a schematic diagram of the EE equipment.

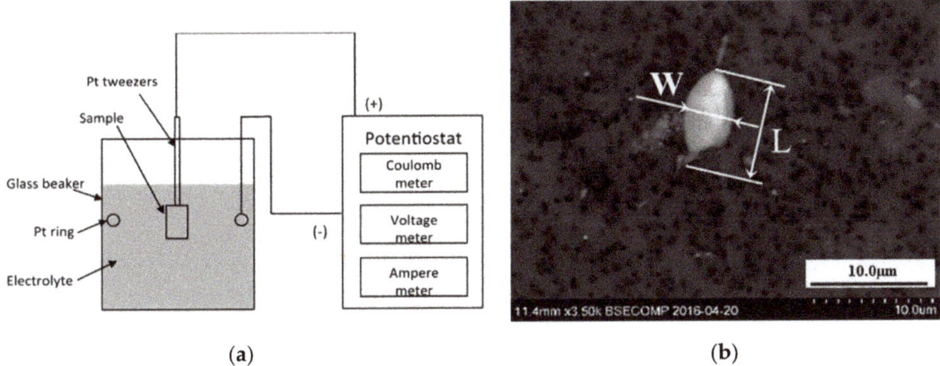

Figure 2. (**a**) Schematic diagram of the equipment used for electrolytic extraction (EE); (**b**) Size measurements of inclusions, where L and W are the length and width of the inclusion.

The electrolytic extraction process was carried out by using two different non-aqueous electrolytes, namely 10% AA (10% acetylacetone–1% tetramethylammonium chloride–methanol) and 2% TEA (2% triethanolamine–1% tetramethylammonium chloride–methanol). The experiments were run with the following electric settings: voltage, 3.2–4.3 V, electric current, 50–70 mA, and electric charge, 500 or 1000 coulombs. The weight of the dissolved metal varied from 0.09 to 0.18 g, depending on the electric charge being used during the extraction.

After extraction, a membrane polycarbonate film filter with an open-pore size of 0.4 μm was used to collect the undissolved non-metallic inclusions. The inclusion investigations were performed by using a scanning electron microscope (SEM, S3700N-Hitachi, Hitachi High-Technologies Corporation, Tokyo, Japan) in the back-scattered electrons (BSE) mode at different magnifications. The compositions of the extracted inclusions were analyzed by using energy dispersive X-ray spectroscopy (EDS, Bruker, Karlsruhe, Germany) included in the SEM equipment. The main parameters of the electrolytic extractions and SEM investigations of non-metallic inclusions in different samples are given in Table 2.

Table 2. Main parameters of the electrolytic extractions and SEM investigations of non-metallic inclusions (NMIs) in different samples.

Steel Grade	Sample	Observed Area, A_{obs} (mm^2)	Dissolved Metal, W_{dis} (g)	Dissolved Depth (μm)	Number of Observed NMIs, n	Size Range (μm)
316R	R	0.898	0.1563	≈188	435	2–98
316Ca	A	0.898	0.0935	≈85	180	3–124

Length (L) and width (W) of different NMIs were measured on the SEM images, as shown in Figure 2b. The equivalent size (D_e) of deformed elongated inclusion was estimated by using the following equation assuming the inclusion is a long ellipsoid:

$$D_e = \sqrt[3]{W^2 \cdot L}, \tag{1}$$

The number of inclusions per unit volume of steel (N_v) and volume fraction (f_v) of different NMIs in steel was estimated by using the equations below:

$$N_v = \frac{n}{V_m}, \tag{2}$$

where n is the number of observed inclusions, and V_m is the volume of dissolved metal that was observed;

$$f_v = \frac{V_{inc}}{V_m} = \frac{\sum_{i=1}^{n} \left(\frac{\pi}{6} \times L_i \times W_i^2\right)}{V_m}, \tag{3}$$

where V_{inc} is the total volume of observed inclusions, L_i is the length of the i-th inclusion, and W_i is the width of the i-th inclusion.

$$V_m = \frac{w_{dis}}{\rho} \times \frac{A_{obs}}{A_{fil}}, \tag{4}$$

where A_{fil} is the whole filtration area of a film filter with collected inclusions (=1200 mm^3), A_{obs} is the total observed area of the film filter used for the image analysis, w_{dis} is the weight loss of the metal sample during EE, and ρ is the metal density (=0.0078 g/mm^3).

3. Results and Discussions

3.1. Characterization of Inclusions after Extraction in Different Electrolytes

It is known that CaO-containing phases in NMIs might be dissolved partially or completely during extraction in strong solutions, otherwise an inaccurate result would be obtained. According to the results reported by Inoue et al. [19], 2% TEA and 10% AA electrolytes can be selected for the extraction of CaO–Al_2O_3 and CaO–SiO_2 inclusions depending on the concentration of CaO in the inclusions. Therefore, both 10% AA and 2% TEA electrolytes were used for the extraction of CaO–Al_2O_3–SiO_2 inclusions from Ca-treated 316L steel and comparison of the obtained results in this study.

The contents of CaO and the ratios of CaO/SiO_2 and CaO/Al_2O_3 in different oxide NMIs extracted by using these two electrolytes are shown in Figure 3, depending on the inclusion length. Dotted lines on this figure correspond to the average values for all analyzed oxide inclusions.

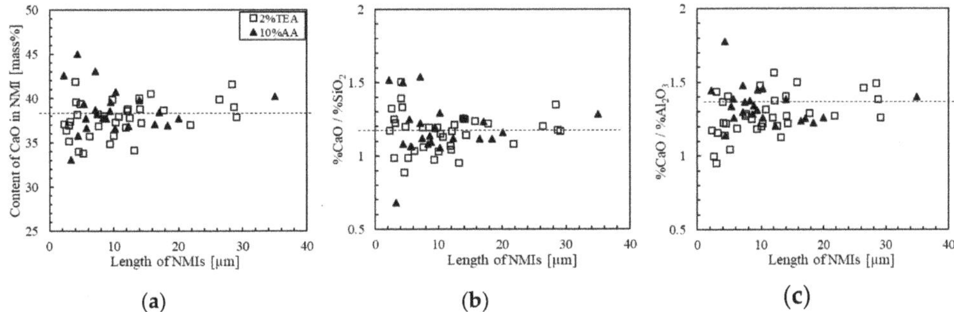

Figure 3. CaO contents (**a**) and ratios of %CaO/%SiO_2 (**b**) and %CaO/%Al_2O_3 (**c**) in oxide inclusions in 316Ca samples after extraction by using 2% TEA and 10% AA electrolytes.

It was found that the content of CaO in all oxide phases of the observed inclusions was lower than 50%. In this case, the compositions of the inclusions extracted by using 2% TEA and 10% AA electrolytes did not show differences. Therefore, it may be safely suggested that the CaO-containing phases (<50% CaO) in NMIs did not dissolve during both extractions. Moreover, the morphology and size of the observed inclusions after extraction in the two different electrolytes were also similar. On the basis of these results, it was concluded that the 10% AA electrolyte was suitable to extract (Ca, Al, Si) oxide. Thus, in the remaining document, only the results from using the 10% AA electrolyte during EE will be discussed.

3.2. Classification of Non-Metallic Inclusions in 316R and 316Ca Steels

Overall, four main types of non-metallic inclusions were observed in 316R and 316Ca steels. Typical photos, size ranges, aspect ratios (AR = L/W), and composition ranges are presented in Table 3.

Table 3. Classification of the inclusions observed in reference (316R) and Ca-treated (316Ca) stainless steels. AR, aspect ratio.

Typical SEM Images	316R			316Ca		
	L (μm)	AR	Composition	L (μm)	AR	Composition
Type I, 10.0μm	5–98	7–30	MnS > 90%	14–124	7–25	MnS 90–100% and (Al, Ca)O < 10%
Type II, 10.0μm, Oxide, MnS	2–36	1–8	MnS 30–90% and (Al, Mg, Mn)O 10–70%	3–33	1–7	MnS 30–90% and (Al, Ca, Si)O 10–70%

Table 3. *Cont.*

Typical SEM Images	316R			316Ca		
	L (μm)	AR	Composition	*L* (μm)	AR	Composition
Type III	2–10	1–3	Al_2O_3 48–75% MgO 2–24% MnO 9–29% TiO_x 1–21%	3–15	1–3	CaO 33–41% Al_2O_3 24–31% SiO_2 30–33% MgO 2–4% TiO_x <1%
Type IV	none	-	-	4–28	2–14	Oxide (CaO 26–39% Al_2O_3 16–30% SiO_2 28–39% MgO 0–5% TiO_x 0–14%) and MnS

Type I inclusions correspond to elongated rod-like MnS inclusions; Type II inclusions are complex oxy-sulfides, containing a hard oxide core or particles and deformed MnS sulfide; Type III inclusions are undeformable irregular oxides and have quite a homogeneous composition. Type III oxide inclusions in 316R steel contained mostly Al_2O_3, MgO, MnO, and TO_x, and their size was in the range from 2 up to 10 μm.

Type III inclusions in the 316Ca steel were mostly undeformable CaO–SiO_2–Al_2O_3–MgO oxides. However, some inclusions had a tail containing a higher concentration of SiO_2 and lower concentrations of CaO and Al_2O_3. This tail was soft and could be deformed during hot rolling, as shown in Figure 4a.

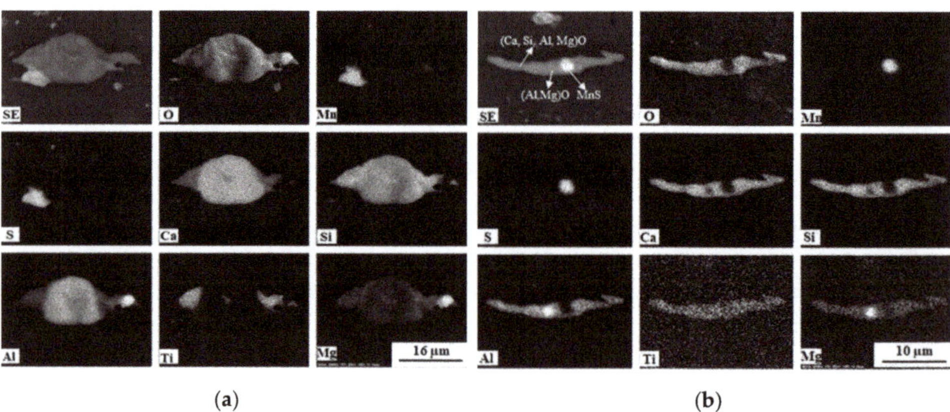

Figure 4. Element mapping of Type III (**a**) and Type IV (**b**) inclusions in Ca-treated steel.

Type IV inclusions in 316R steel were not observed. This was due to all oxide inclusions in the reference sample being Al_2O_3 or Al_2O_3–MnO–MgO phases, which were not deformed during hot rolling. Type IV inclusions in 316Ca steel corresponded to elongated CaO–SiO_2–Al_2O_3 oxides containing a higher concentration of SiO_2 (up to 39%) and lower concentrations of CaO and Al_2O_3. Moreover, these deformed oxides contained also some amounts of TiO_x and MgO. However, line scanning and EDS mapping of some Type IV inclusions detected unmodified Al_2O_3–MgO cores, as shown in Figure 4b. This can explain the quite high Al_2O_3 and MgO contents in some Type IV inclusions determined during EDS investigation.

Similar inclusions, having an Al_2O_3–MgO core covered by a Ca-rich oxide layer, have also been reported by other researchers [20,21]. Furthermore, an undeformed manganese sulfide particle could be found in this inclusion during the investigation. Elongated oxide inclusions containing manganese sulfide cores were also found by Bletton et al. [20]. These deformable lime silicoaluminate inclusions were mainly determined as pseudo-wollastonite, gehlenite, and anorthite [11,20]. They were predominantly bound up with sulfides and were found to form associated inclusions containing manganese sulfide cores.

3.3. Particle Size Distribution of the Inclusions Observed in 316R and 316Ca Steels

The number of inclusions per unit volume (N_v), size, AR, and volume fraction (f_v) of diverse types of inclusions is given in Table 4.

Table 4. Distribution and information about different types of inclusions found in 316Ca steel.

Samples	Inclusion Type	L (μm)	W (μm)	D_e (μm)	AR	$f_v \times 10^4$	N_v (mm^{-3})	Frequency (%)
316R	Type I	33.2 ± 24.8 (5–98)	2.6 ± 1.4 (1–7)	6.0 ± 3.5 (1–16)	7–30	6.7	2667	9.2
	Type II	5.0 ± 3.8 (2–36)	1.8 ± 0.7 (1–7)	2.4 ± 1.1 (1–10)	1–8	3.7	24,933	86.4
	Type III	5.4 ± 2.2 (2–10)	3.6 ± 1.5 (2–8)	4.0 ± 1.6 (2–9)	1–3	0.7	1267	4.4
	Type IV	-	-	-	-		-	-
316Ca	Type I	43.0 ± 32.4 (14–124)	3.7 ± 1.1 (2–5)	8.1 ± 3.2 (4–14)	7–25	4.1	1003	5.0
	Type II	7.0 ± 4.8 (3–33)	2.6 ± 0.8 (1–5)	3.6 ± 1.3 (2–10)	1–7	3.5	9811	48.9
	Type III	6.2 ± 3.0 (3–15)	3.3 ± 0.9 (2–6)	4.1 ± 1.3 (2–8)	1–3	2.1	4460	22.2
	Type IV	10.7 ± 5.2 (4–28)	2.3 ± 0.6 (1–4)	3.8 ± 1.1 (2–7)	2–14	1.7	4794	23.9

D_e, equivalent size of deformed elongated inclusions; f_v, volume fraction; N_v, number of inclusions per unit volume of steel.

It can be seen that the reference 316R sample contained mostly MnS inclusions of Type II (86%) and Type I (9%). Only a small amount of Type III oxides (Al_2O_3–MgO–MnO–TO_x) existed in this sample. Also, almost pure Al_2O_3 inclusions were found in the sample. However, Type IV inclusions were not observed in the reference sample.

Although the content of S in the Ca-treated steel (316Ca) was higher, the number of elongated sulfides of Type I and oxy-sulfides of Type II was about 2.5 times smaller compared to that in the reference 316R steel. However, the length, width, and equivalent size of those inclusions in 316Ca steel were larger than the respective characteristics of the given inclusions in 316R steel. The number of oxide inclusions in the Ca-treated steel increased significantly and corresponded to ≈22% and ≈24% of the total amount of observed inclusions for Type III and Type IV inclusions, respectively. This can be explained by the higher content of oxygen in 316Ca steel compared to the 316R steel. Overall, Ca treatment of 316L stainless steel decreased the total number of inclusions by ≈30%, though the total f_v of the inclusions were similar in both samples.

The size distributions of different types of inclusions obtained by SEM are shown in Figure 5 in terms of N_v with a constant size step of 5 μm. It can be seen that the oxy-sulfide inclusions of Type II were the most numerous. However, Ca treatment of steel decreased the number of small-size oxy-sulfides (L < 10 μm) by about three times.

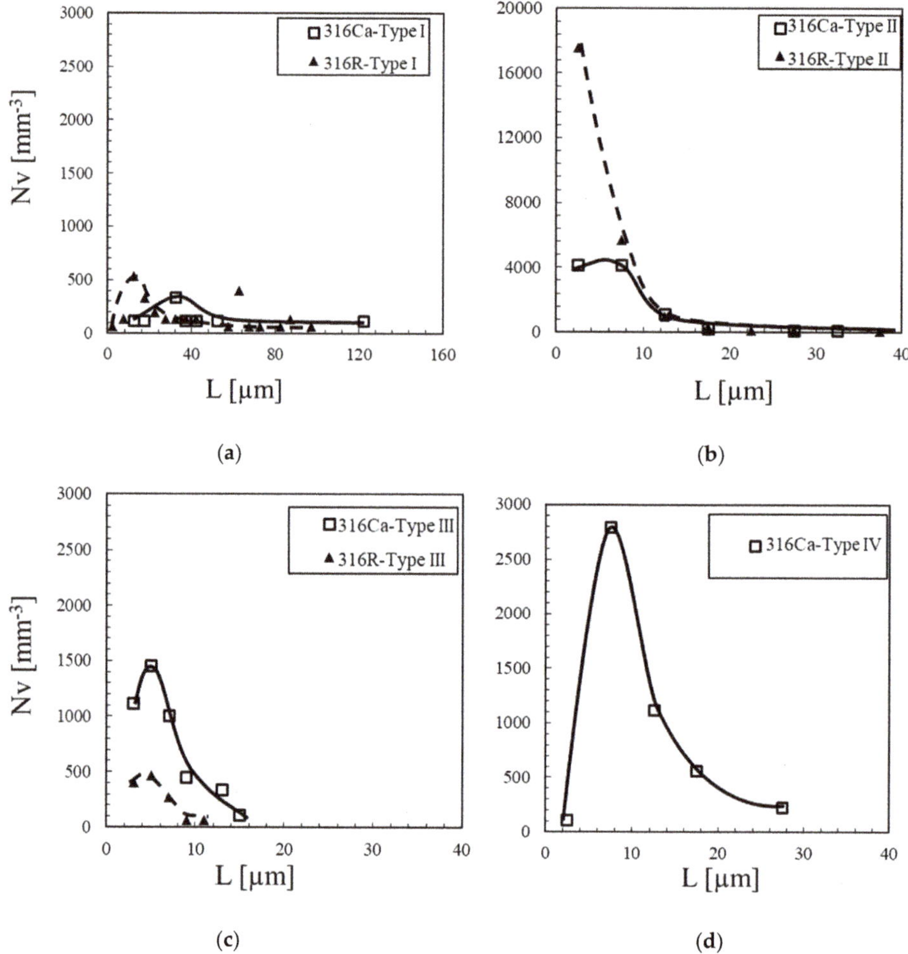

Figure 5. Particle size distributions of different NMI types observed in the metal samples of 316R and 316Ca steel bars.

The contents of the main oxide components (Al_2O_3, SiO_2, and CaO) in the typical Type III and Type IV inclusions in 316Ca steel are shown in Figure 6 as a function of inclusion length. It can be seen that the compositions of large-size Type III oxides ($L > 6$ μm) were more stable and homogeneous because the contents of Al_2O_3, SiO_2, and CaO were similar. However, Type IV complex oxides were heterogeneous due to the high scattering of the composition for inclusions of different size. As shown in Figure 4b, Type IV inclusions contained an untransformed oxide core having very high contents of Al_2O_3 and MgO and small contents of CaO and SiO_2. Furthermore, undeformed MnS particles were detected in some elongated Type IV inclusions, but CaS was not found.

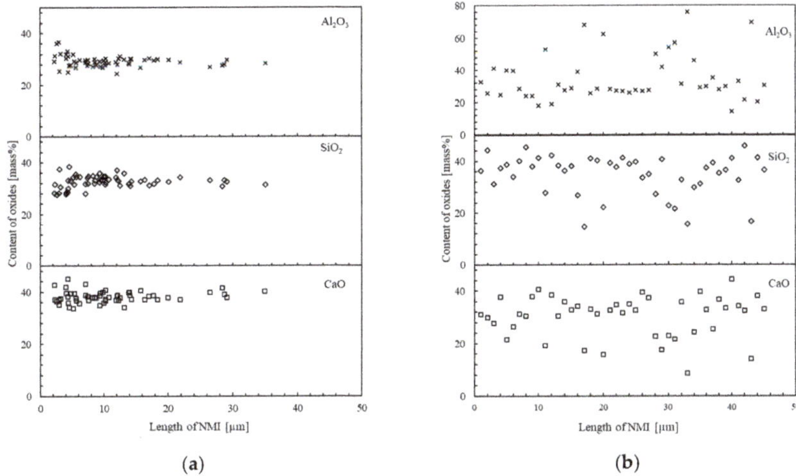

Figure 6. Relationship between the Al_2O_3, SiO_2, and CaO contents and length of Type III (**a**) and Type IV (**b**) oxide inclusions in 316Ca steel.

The triangle $CaO–Al_2O_3–SiO_2$ diagram in Figure 7 shows that most oxides of Type II (black diamond marks) and Type III (green triangle marks) in 316Ca steel were located in the area near gehlenite with a melting point ≥1500 °C. The elongated part of Type IV inclusions had a higher content of SiO_2 and was located in the zone between the anorthite ($CaO{\cdot}Al_2O_3{\cdot}2SiO_2$) and gehlenite ($2CaO{\cdot}Al_2O_3{\cdot}SiO_2$) phases, which corresponds to a lower melting point zone (around 1300–1400 °C). This composition of inclusions corresponds to a liquid phase during steel casting and to a soft oxide during rolling. The undeformed oxide core of Type IV inclusions had a similar ratio of CaO and SiO_2 (around 1:1) as the outer layer but a much higher content of Al_2O_3 and MgO (40%–95%), which corresponded to the oxide phases with higher melting points (>1450 °C). Therefore, it can be concluded that the undeformed cores contained almost pure Al_2O_3 or Al_2O_3–MgO.

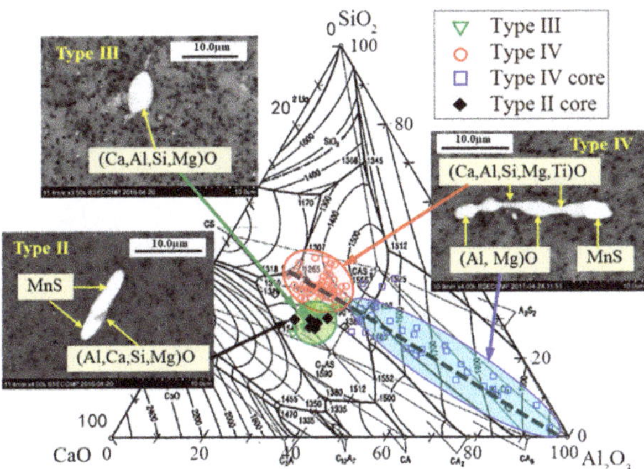

Figure 7. Composition of oxide inclusions in the Ca-treated sample (316Ca) in a $CaO–Al_2O_3–SiO_2$ ternary phase diagram.

3.4. Mechanism of Inclusion Transformation

On the basis of the obtained results, it was shown that two types of oxide inclusions (Type III and Type IV inclusions) can be formed simultaneously during Ca treatment of 316L stainless steel. The presence of these two types of modified inclusions in steel can be explained by some differences between the initial oxide inclusions before CaSi addition. It was found that Type III oxide inclusions in 316R steel could be divided into two groups: (i) Type IIIa—liquid or semi-liquid inclusions with higher content of MnO and TiO_x (17–60% Al_2O_3, <10% MgO, 15–29% MnO, and 5–21% TiO_x) and (ii) Type IIIb—regular/irregular solid oxide inclusions with higher contents of Al_2O_3 and MgO (60–75% Al_2O_3, 10–24% MgO, 9–15% MnO, and 1–10% TiO_x).

After the addition of the CaSi wire, most of MnO and TiO_x, and part of Al_2O_3 in the liquid Type IIIa oxides could be reduced more easily and faster by Ca and Si. Therefore, CaO content was larger in the homogeneous Type III oxides in 316Ca steel. As a result, these obtained inclusions were harder and could not be deformed during hot rolling. A small amount of oxide phases, that were precipitated later on the surface of these solid oxides of Type III, had a smaller CaO content and a larger SiO_2 content due to the fast decrease of the Ca concentration in the melt. These precipitated phases were softer and could be deformed during hot rolling (as shown in by the tails in Figure 4a).

Other solid oxides, having a larger concentration of Al_2O_3 and MgO (Type IIIb), could not be modified quickly and completely due to a limitation of reduction reactions in the solid oxide phases. Therefore, an outer layer of modified oxides was formed slower due to the fast decreasing Ca concentration in the melt. As a result, the outer oxide layer had a smaller CaO content and a larger SiO_2 content compared to the solid oxide of Type III and could be deformed during hot rolling (as shown in Figure 4b). However, these Type IV modified complex inclusions in 316Ca steel could have an undeformed core that contained mostly Al_2O_3 and MgO.

Thus, after Ca treatment of 316L steel, about 46% of the observed inclusions were oxide inclusions (Types III and IV) correlated to gehlenite and to a mixture of gehlenite and anorthite inclusions. Therefore, it is believed that these modified oxide inclusions (especially the Type IV inclusions) can significantly improve the machinability of 316L stainless steel. The effects of different types of inclusions observed in Ca-treated stainless steel will be discussed in a separate article.

4. Conclusions

Non-metallic inclusions in commercial 316L steels without Ca treatment (316R) and with Ca treatment (316Ca) were studied. The characteristics of the inclusions were determined by using 3D investigations after EE. Based upon the presented experimental results, the following conclusions were drawn:

1. Similar compositions of oxide inclusions in 316Ca steel were obtained after EE by using 2% TEA and 10% AA electrolytes.
2. The 316L steels contained four types of inclusions: (1) elongated MnS (Type I), (2) MnS sulfides with hard oxide cores (Type II), (3) undeformed irregular oxides (Type III), and (4) elongated oxides with a hard oxide core (Type IV).
3. In the reference sample, the oxide composition was mainly Al_2O_3–MgO–MnO. However, after Ca treatment of 316L steel, about 46% of the observed inclusions were oxide inclusions (Types III and IV), which correlated to gehlenite and to a mixture of gehlenite and anorthite, which are favorable for the machinability of steel.

Author Contributions: Conceptualization, A.K., O.S., and P.G.J.; formal analysis, H.D., material preparation, O.S.; investigation, H.D.; writing—original draft preparation, H.D.; writing—review and editing, A.K, O.S., and P.G.J.; supervision, A.K. and P.G.J.

Funding: This research received no external funding

Acknowledgments: The VINNOVA, Jernkontoret are acknowledged for the support of this study. H. Du acknowledges the financial support from the China Scholarship Council (CSC).

Conflicts of Interest: The authors declare no conflict of interest.

References

1. Cyril, N.; Fatemi, A.; Cryderman, B. Effects of sulfur level and anisotropy of sulfide inclusions on tensile, impact, and fatigue properties of SAE 4140 steel. *SAE Int. J. Manuf. Mater.* **2008**, *1*, 218–227. [CrossRef]
2. Ånmark, N.; Karasev, A.; Jönsson, P.G. The Influence of Microstructure and Non-Metallic Inclusions on the Machinability of Clean Steels. *Steel Res. Int.* **2016**, *87*, 1–8. [CrossRef]
3. Wang, Y.; Yang, J.; Bao, Y. Effects of Non-metallic Inclusions on Machinability of Free-Cutting Steels Investigated by Nano-Indentation Measurements. *Metall. Mater. Trans. A* **2015**, *46*, 281–292. [CrossRef]
4. Qi, H.S.; Mills, B. On the formation mechanism of adherent layers on a cutting tool. *Wear* **1996**, *198*, 192–196. [CrossRef]
5. Sidjanin, L.; Kovac, P. Fracture mechanisms in chip formation processes. *Mater. Sci. Technol.* **1997**, *13*, 439–444. [CrossRef]
6. Abdelaziz, S.; Megahed, G.; El-Mahallawi, I.; Ahmed, H. Control of Ca addition for improved cleanness of low C, Al killed steel. *Ironmak Steelmak* **2009**, *36*, 432–441. [CrossRef]
7. Verma, N.; Pistorius, P.C.; Fruehan, R.J.; Potter, M.; Lind, M.; Story, S. Transient Inclusion Evolution During Modification of Alumina Inclusions by Calcium in Liquid Steel: Part II. Results and Discussion. *Metall. Mater. Trans. B* **2011**, *42*, 720–729. [CrossRef]
8. Lis, T. Modification of oxygen and sulphur inclusions in steel by calcium treatment. *Metalurgija* **2009**, *48*, 95–98.
9. Guo, Y.; He, S.; Chen, G.; Wang, Q. Thermodynamics of Complex Sulfide Inclusion Formation in Ca-Treated Al-Killed Structural Steel. *Metall. Mater. Trans. B* **2016**, *47*, 2549–2557. [CrossRef]
10. Liu, J.; Wu, H.; Bao, Y.; Wang, M. Inclusion variations and calcium treatment optimization in pipeline steel production. *Int. J. Miner. Metall. Mater.* **2011**, *18*, 527. [CrossRef]
11. Bletton, O.; Duet, R.; Pedarre, P. Influence of oxide nature on the machinability of 316L stainless steels. *Wear* **1990**, *139*, 179–193. [CrossRef]
12. Nordgren, A.; Melander, A. Tool wear and inclusion behaviour during turning of a calcium-treated quenched and tempered steel using coated cemented carbide tools. *Wear* **1990**, *139*, 209–223. [CrossRef]
13. Fang, X.D.; Zhang, D. An investigation of adhering layer formation during tool wear progression in turning of free-cutting stainless steel. *Wear* **1996**, *197*, 169–178. [CrossRef]
14. Ånmark, N.; Björk, T. Effects of the composition of Ca-rich inclusions on tool wear mechanisms during the hard-turning of steels for transmission components. *Wear* **2016**, *368*, 173–182. [CrossRef]
15. Gutnichenko, O.; Bushlya, V.; Zhou, J.M.; Stahl, J.E. Tool wear and machining dynamics when turning high chromium white cast iron with pcBN tools. *Wear* **2017**, *390–391*, 253–269. [CrossRef]
16. Kanbe, Y.; Karasev, A.; Todoroki, H.; Jönsson, P.G. Application of Extreme Value Analysis for Two- and Three-Dimensional Determinations of the Largest Inclusion in Metal Samples. *ISIJ Int.* **2011**, *51*, 593–602. [CrossRef]
17. Karasev, A.; Suito, H. Analysis of size distributions of primary oxide inclusions in Fe-10 mass Pct Ni-M (M = Si, Ti, Al, Zr, and Ce) alloy. *Metall. Mater. Trans. B* **1999**, *30*, 259–270. [CrossRef]
18. Kanbe, Y.; Karasev, A.; Todoroki, H.; Jönsson, P.G. Analysis of Largest Sulfide Inclusions in Low Carbon Steel by Using Statistics of Extreme Values. *Steel Res. Int.* **2011**, *82*, 313–322. [CrossRef]
19. Inoue, R.; Kiyokawa, K.; Tomoda, K.; Ueda, S.; Ariyama, T. Three-dimensional estimation of multi-component inclusion particles in steel. In Proceedings of the 8th International Workshop on Progress in Analytical Chemistry and Materials Characterisation in the Steel and Metal Industries (CETAS'11), Luxembourg, 17–19 May 2011.
20. Bletton, O.; Duet, R.; Henry, M.; Cogne, J.Y. Resulphurized Austenitic Stainless Steel with Improved Machinability. U.S. Patent 5,089,224, 18 February 1992.
21. Todoroki, H.; Mizuno, K. Effect of Silica in Slag on Inclusion Compositions in 304 Stainless Steel Deoxidized with Aluminum. *ISIJ Int.* **2004**, *44*, 1350–1357. [CrossRef]

© 2019 by the authors. Licensee MDPI, Basel, Switzerland. This article is an open access article distributed under the terms and conditions of the Creative Commons Attribution (CC BY) license (http://creativecommons.org/licenses/by/4.0/).

Article

Characteristics and Formation Mechanism of Inclusions in 304L Stainless Steel during the VOD Refining Process

Xingrun Chen [1,2,*], Guoguang Cheng [1,*], Jingyu Li [1], Yuyang Hou [1], Jixiang Pan [2] and Qiang Ruan [2]

1 State Key Laboratory of Advanced Metallurgy, University of Science and Technology Beijing, Beijing 100083, China; b20160493@xs.ustb.edu.cn (J.L.); hyyustb@gmail.com (Y.H.)

2 Hongxing Iron & Steel Co. Ltd., Jiuquan Iron and Steel Group Corporation, Jiayuguan 735100, China; panjixiang@jiugang.com (J.P.); ruanqiang@jiugang.com (Q.R.)

* Correspondence: chenxingrun@jiugang.com (X.C.); chengguoguang@metall.ustb.edu.cn (G.C.); Tel.: +86-106-233-4664 (X.C. & G.C.)

Received: 27 October 2018; Accepted: 29 November 2018; Published: 5 December 2018

Abstract: The formation and characteristics of non-metallic inclusions in 304L stainless steel during the vacuum oxygen decarburization (VOD) refining process were investigated using industrial experiments and thermodynamic calculations. The compositional characteristics indicated that two types of inclusions with different sizes (from 1 μm to 30 μm) existed in 304L stainless steel during the VOD refining process, i.e., CaO-SiO_2-Al_2O_3-MgO external inclusions, and CaO-SiO_2-Al_2O_3-MgO-MnO endogenous inclusions. The calculation results obtained using the FactSage 7.1 software confirmed that the inclusions that were larger than 5 μm were mostly CaO-SiO_2-Al_2O_3-MgO; the similarity in composition to the slag indicated that these inclusions originated from the slag entrapment. The CaO-SiO_2-Al_2O_3-MgO-MnO inclusions that were smaller than 5 μm originated mainly from the oxidation reaction with Ca, Al, Mg, Si, and Mn. The changes in the inclusion composition resulting from changes in the Ca, Al, and O contents, and the temperature during the VOD refining process was larger for the smaller inclusions. Generating mechanisms for the CaO-SiO_2-Al_2O_3-MgO-MnO inclusions in the 304L stainless steel were proposed.

Keywords: 304L stainless steel; non-metallic inclusions; formation mechanism; VOD refining

1. Introduction

In recent years, 304L stainless steel has been rapidly developed, and it is widely used in shipbuilding, offshore drilling platform construction, metal structures for construction of buildings and bridges, containers and cisterns, flux-cored wire, and in industrial transport machinery engineering, petrochemical engineering, nuclear power engineering, etc. [1–4]. Because of its harsh application environment, the requirements for C, N, P, S, and O elements are stringent, the production is difficult, and the added value of the products is high. In addition to the control of the metal elements, the control of inclusions is also the key to improving the quality of 304L stainless steel.

The control of the composition, quantity, and size of the inclusions at the beginning of their formation is likely to become a new and effective method to reduce the harmful influence of the inclusions. Therefore, it is very important to investigate the source and formation mechanism of the inclusions [5–18]. The formation of non-metallic inclusions of Si-killed stainless steel during the GOR (gas oxygen refining) process has been reported by Li et al. [7]. The authors found that the inclusions that were larger than 5 μm and contained more than 30% CaO were attributed to the modification of slag droplets through the oxidation of Si and Al and the collision with deoxidation-type inclusions; in addition, the degree of change was larger for the smaller inclusions. A tracer was used by Kim et al. [8] to determine the source of the large inclusions (larger than 20 μm) in 304 stainless steel, and it was found that these inclusions originated from the slag entrapment. This result was confirmed by the study of Ehara et al. [9], who also pointed out that the oxidation of Si and Al on the surface of slag inclusions can lead to the increase in the SiO_2 and Al_2O_3 contents. Yin et al. [10] reported that after de-oxidation with Si/Mn additions, spherical complex inclusions mainly consisting of calcium silicates were observed. The contents of MgO and Al_2O_3 in these inclusions continuously increased as the steel moved from the argon–oxygen decarburization (AOD) through ladle processing to the tundish. Park et al. [11] reported that the inclusions mainly consisted of Al_2O_3 when the Al content was higher than 0.05%. Ren et al. [12] pointed out that a high-basicity slag improved the cleanness of stainless steel, whereas a low basicity slag lowered the Al_2O_3 content in the inclusions, thereby lowering the melting temperature of the inclusions and improving the deformability of the inclusions. This result was confirmed by the studies of Yan et al. [13,14] and Sakata [15].

However, in previous studies on the formation of inclusions, only a single factor was considered, namely external inclusions or endogenous inclusions. The formation mechanism of both types of inclusions has not been clarified to date. In the present study, the 304L stainless steel was produced by the process route of basic oxygen furnace (BOF)–AOD–vacuum oxygen decarburization (VOD)–ladle furnace (LF)–continuous casting (CC). The characteristics and generating mechanism of the inclusions during the VOD refining process were investigated using industrial experiments and thermodynamic calculations with FactSage 7.1 software; in addition, a formation mechanism for non-metallic inclusions occurring in the VOD refining process was proposed.

2. Experiments

2.1. Experimental Procedure and Sampling

The smelting process route of 304L stainless steel consisted of BOF–AOD–VOD–LF–CC. The molten iron from the blast furnace was used as the raw material, and it was directly poured into the AOD furnace for decarburization and denitrification treatments after the dephosphorization pre-treatment; the target values of the carbon and nitrogen content were 0.30–0.50% and less than 0.08%, respectively and ferrosilicon was used for the Cr reduction in the AOD process. The slag was tapped after the AOD process, and then the ladle (the refractory material was magnesia-calcium) was hung to the VOD for deep decarburization and denitrification; ferrosilicon was used for Cr reduction in the VOD process, and subsequently, the liquid steel was poured into the LF furnace. When the temperature and composition of the liquid steel met the requirements, it was transported to the platform for CC.

The VOD consists of three stages: (i) oxygen blowing: oxygen was blown onto the melt at 1000–1500 m^3/h for 40–60 min. Argon was injected at 200–600 L/min through two porous bricks at the bottom of the ladle. Lime was charged to form a basic slag. The total pressure was maintained at between 2.6 and 16 kPa; (ii) vacuum carbon deoxidation (VCD): the total pressure was reduced to between 26.6 and 66.5 Pa for 10–15 min without any additions of oxygen, fluxes, or ferroalloys. The bottom stirring argon flow rate was increased to 800–1000 L/min because at this stage of very low carbon contents, the decarburization is controlled by the rate of mass transfer of carbon and oxygen in the bath. The oxygen content was 0.015% after the VCD stage; (iii) reduction: ferrosilicon was added as a reducing agent for the chromium oxide in the slag. The aluminum content of the ferrosilicon was 1.3%. The final steel temperature was 1600 °C.

In order to elucidate the formation of the nonmetallic inclusions in the 304L stainless steel, a steel sample was taken at the end of the VOD and was immediately quenched in water. A slag sample was taken after the VOD treatment.

2.2. Composition Analysis and Inclusion Characterization

The chemical composition of the steel samples were determined by a direct reading of the spectrum (ARL4460, Thermo Fisher Scientific, Waltham, MA, USA). The contents of C and S were analyzed by a C/S analyzer (CS-800, ELTRA, Haan, Germany). Cylinders (Φ 5 mm × 5 mm) were machined for the measurement of the total oxygen contents, which were analyzed using the inert gas fusion–infrared absorptiometry method. The acid-soluble Al and Ca contents in the steel were determined using the inductively coupled plasma optical emission spectroscopy method (ICP-OES). The composition of the slags was analyzed by an X-ray fluo-rescence spectrometer (ARL PERFORM'X, Thermo Fisher Scientific, Waltham, MA, USA).

The morphologies of the inclusions in the specimens were observed using scanning electron microscopy (Merlin Compact, Zeiss, Gottingen, Germany) (SEM). The 15 mm × 15 mm × 10 mm samples for the SEM analysis were made by cutting, grinding, and polishing. The chemical compositions of the inclusions were analyzed with an energy dispersive spectrometer (X-Max 80, Oxford Instruments, High Wycombe, UK) (EDS) to determine the inclusion type. A quantitative analysis of the inclusions was performed using the INCA software (Inca Energy 250, Oxford Instruments, High Wycombe, UK) of the scanning electron microscope. To ensure good accuracy for the automated EDS analysis of the inclusions, the size of the inclusions was larger than 1 μm because the interaction volume may spread into the steel and excite electrons from the environment surrounding the inclusions if the diameters are smaller than 1 μm.

3. Results

3.1. Composition of the Molten Steel and Slag

The averages of the chemical compositions of the molten steel and slag after the VOD stage are listed in Tables 1 and 2, respectively. It can be seen that after the VOD treatment, the contents of C and N in steel were 0.008% and 0.015% respectively. The oxygen content and sulfur content were higher because of deep degassing. The S content after VOD reached 30 ppm, indicating that the oxygen potential was higher in the molten steel, which was consistent with the conclusion drawn from the test results of the slag. Conversely, the slag and steel continuously reacted with each other during the VOD refining process, and the slag–steel balance was finally achieved. The contents of Cr_2O_3 and FeO in the slag were much higher after VOD refining, and were 0.43% and 0.15%, respectively.

Table 1. Chemical composition of 304L after vacuum oxygen decarburization (VOD) treatment (wt %).

Stage	C	Si	Mn	P	S	Ni	Cr	Ca	Al	N	Mg	O
VOD	0.008	0.22	1.14	0.015	0.003	8.04	18.00	0.002	0.004	0.015	0.0005	0.007

Table 2. Chemical composition of 304L slag after VOD treatment (wt %).

Stage	CaO	SiO_2	MgO	Al_2O_3	Cr_2O_3	FeO
VOD	59.16	29.12	5.97	1.87	0.43	0.15

3.2. Characterization of the Inclusions

The morphology and the compositions of the inclusions in the molten steel after the VOD process are shown in Figure 1. It was observed that two types of inclusions existed in the 304L steel after the VOD process. The first type of inclusions were spherical CaO-SiO_2-Al_2O_3-MgO with sizes ranging from several to tens of microns at the end of the VOD smelting (Figure 1a). These inclusions contained a small amount of aluminum. The second type consisted of endogenous inclusions with a size smaller than 5 μm, and a different composition. The common types of inclusions were CaO-SiO_2-Al_2O_3-MgO-MnO, which represented the dominant type, as shown in Figure 1b.

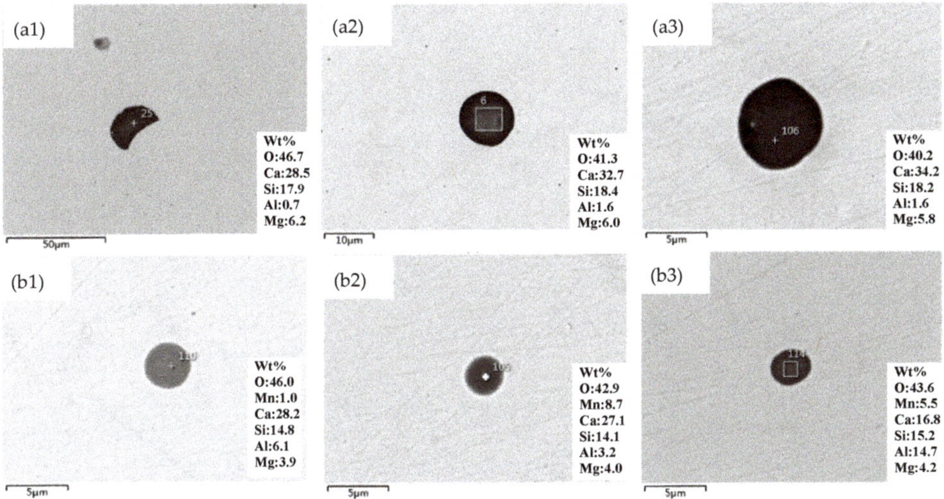

Figure 1. Morphology of typical inclusions encountered in the samples: (**a**) CaO-SiO_2-MgO-Al_2O_3; (**b**) CaO-SiO_2-Al_2O_3-MgO-MnO.

3.3. Corresponding Relation between the Size and Composition of the Inclusions

The mass fractions of CaO, SiO_2, Al_2O_3, MgO, and MnO of the inclusions of various sizes were calculated using the mass fraction derived from the EDS, as shown in Figure 2. It can be clearly seen that the inclusion composition exhibited changes with the increase in size from 1 μm to 30 μm. Most inclusions were smaller than 5 μm. We created two categories based on the relationship between the size and mass percent of composition, i.e., larger than 5 μm and smaller than 5 μm. The inclusions that were larger than 5 μm had almost the same CaO, SiO_2, MgO, and Al_2O_3 contents, but the MnO contents were very low. The composition differed for the inclusions that were smaller than 5 μm. The contents of SiO_2, CaO, and Al_2O_3 fluctuated in a wide range from 0 to more than 30%. The MnO contents increased markedly from 0 to more than 10% as the size of the inclusions decreased. To correlate the compositions of inclusions (size smaller than 5 μm), the inclusions were plotted on the ternary system, as shown in Figure 3. It was shown that the inclusion quantity increased with the inclusion size decreasing. The contents of SiO_2, CaO, and Al_2O_3 fluctuated in a wide range, but the contents of MgO and MnO had smaller variation of range.

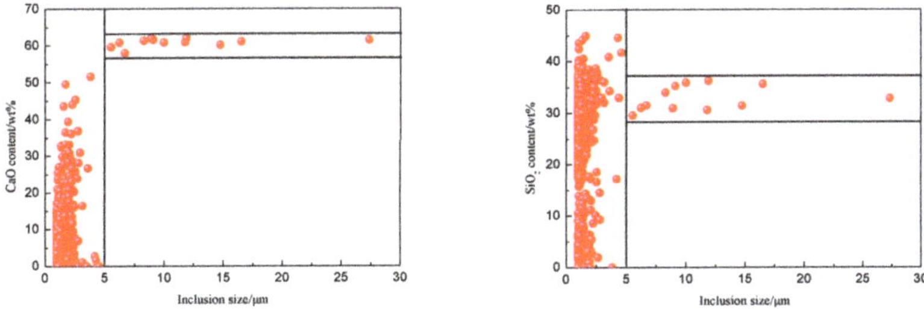

Figure 2. *Cont.*

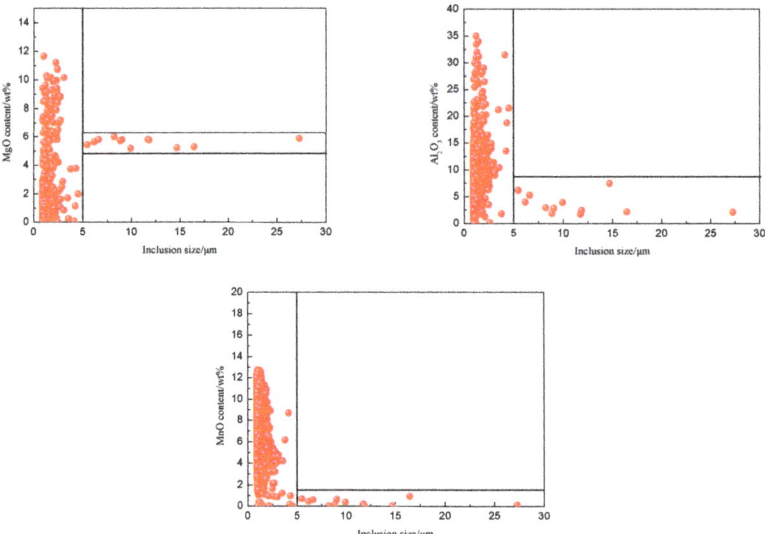

Figure 2. The relationship between inclusion size and inclusion content.

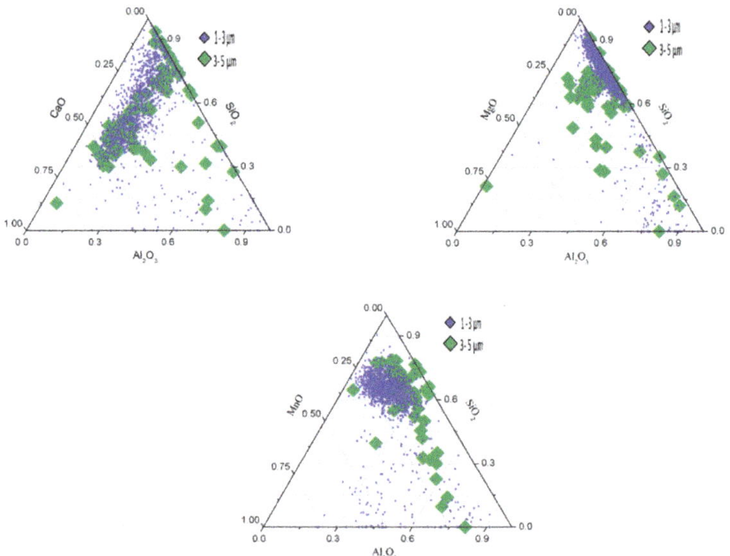

Figure 3. Composition distributions (mass fraction) of inclusions (size smaller than 5 μm).

4. Discussion

The compositional characteristics shown in Figures 2 and 3 indicate that two types of inclusions existed in the 304L stainless steel during the VOD refining process, i.e., the CaO-SiO_2-Al_2O_3-MgO external inclusions that were larger than 5 μm, and the CaO-SiO_2-Al_2O_3-MgO-MnO endogenous inclusions that were smaller than 5 μm. The composition of the inclusions in the same sample showed a large difference. Therefore, the source and the formation process of the inclusions are likely different; this is discussed in detail in the following sections.

4.1. Generating the Mechanism of CaO-SiO_2-MgO-Al_2O_3 Inclusions

Figure 4 shows the CaO-SiO_2-Al_2O_3 phase diagram of the inclusions (larger than 5μm) and their composition during the VOD process. The composition of the slag and the average composition of the inclusions are listed in the figure. It can be clearly seen that the compositions of the inclusions were in good agreement with slag composition. The inclusions consisted mostly of slag components, and they originated from slag entrapment. In addition, the Al_2O_3 contents in the inclusions were very low, and were very close to that of the slag. The MgO contents in the inclusions in the VOD process were also close to that of the slag. Therefore, it can be concluded that the inclusions that were larger than 5 μm during the VOD process were mainly derived from the entrapment of the top slag. The strong argon stirring in the VOD smelting causes the inclusions to be directly or indirectly in equilibrium between the molten steel and slag. Thus, the control of the inclusions can be achieved by controlling the slag composition. The CaO and SiO_2 contents were higher in some inclusions. The main reason was that some inclusions such as $CaSiO_3$ contain a higher content of SiO_2, and Ca_2SiO_4 contains a higher content of CaO precipitated in the cooling process. The oxidations of Si and Al on the surfaces of some inclusions led to increases in the SiO_2 and Al_2O_3 contents.

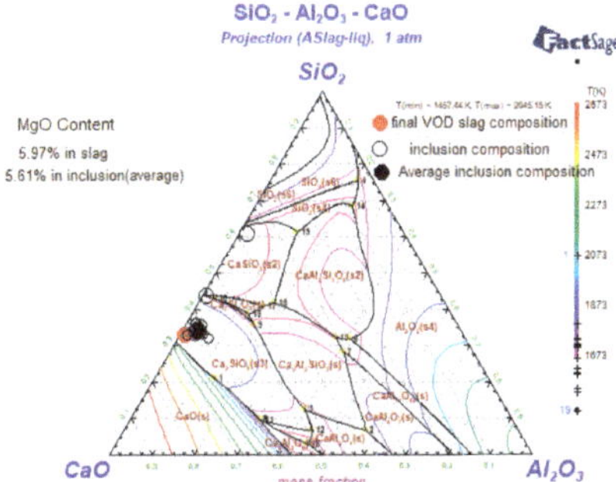

Figure 4. Chemical composition of the inclusions (larger than 5 μm) and the slag during the VOD process.

Qian [19] reported that the flow velocity at the critical interface between the steel and slag was 0.65 m/s. When the flow velocity at the interface was larger than the critical velocity, droplets formed and were entrapped into the molten steel; the average size of the slag droplets from the top slag gradually decreased with increasing velocity at the interface.

The stirring intensity of the argon blowing during the VOD reduction stage was 800–1000 L/min, which was far stronger than the argon blowing intensity of 200–400 L/min during the LF stage [20,21]; therefore, the velocity at the interface between the steel and slag was greater than 2 m/s. The reaction between the slag and steel was very intense during the VOD refining process and the droplets were strongly mixed and changed into very small particles; as a result, the minimum size of the droplets during VOD refining was 5 μm.

4.2. Generating the Mechanism of CaO-SiO_2-Al_2O_3-MgO-MnO Inclusions

Factsage (Factsage^{TM}7.1, Thermfact/CRCT & GTT-Technologies, Aachen, Germany) was used for the thermodynamic calculations of the inclusions in the 304 L stainless steel during the VOD refining process; the databases FTmisc and FToxid were used, and the calculation module was Equilib with 100 g of molten steel. Due to the local non-uniformities of composition and temperature in molten steel during the VOD refining process, the elements with local fluctuations in concentrations in the molten steel had a wide range of concentrations and temperatures compared to the other components determined after the VOD. Figure 5 shows the inclusion composition as a function of the oxygen concentration when the Al content, Ca content, Mg content and the temperature is 0.004%, 0.002%,

0.0005%, and 1600 °C respectively. The CaO concentration exhibits a steady increasing trend with a decrease in the oxygen concentration, whereas the MgO and SiO_2 concentrations exhibit a stable trend, and the MnO concentration decreases. However, the Al_2O_3 concentration remains steady at an oxygen concentration range from 0.015% to 0.011%, and then gradually decreases with decreasing oxygen content at oxygen concentrations of less than 0.011%. The oxygen concentration is at 150 ppm after the VOD refining process, and at this stage, the elements compete with each other for oxidation as ferrosilicon is added for deoxidation. During the alloying deoxidation, strong oxidizing elements such as Ca and Al react with and consume some of the oxygen. However, since the concentration of Si in the molten steel is high, the deoxidation reaction of Si also takes precedence. The Mg concentration is at a low level, resulting in a low concentration of MgO in the inclusion. Moreover, the remaining oxygen reacts with weak oxidizing elements such as Mn, leading to a small amount of MnO in the inclusions. With continuing deoxidation, the Al_2O_3 and MnO concentrations decrease, triggering an increase in the CaO. During the VOD refining process, the oxygen content fluctuates with the deoxidation reaction, causing the fluctuations in each component of the endogenous inclusions that are smaller than 5 μm.

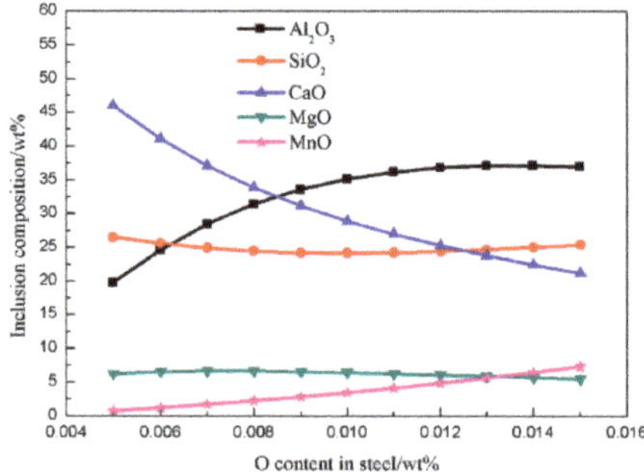

Figure 5. Inclusion composition as a function of the O content.

To determine the effect of Ca on the inclusions, the inclusion composition as a function of increasing Ca content was calculated using the FToxid and FTmisc databases of Factsage^{TM}7.1 when the O content, Al content, Mg content, and the temperature is 0.007%, 0.004%, 0.0005%, and 1600 °C, respectively (Figure 6). It was observed that an increase in the Ca content results in an increase in the CaO and SiO_2 contents, and a decrease in the Al_2O_3 content; however, the MgO and MnO contents did not change much. The Ca was mainly derived from the metal–slag reaction, the metal–refractory reaction, and the Ca from the FeSi alloy during the VOD refining process; this resulted in fluctuations in the Ca content. This is one reason for the fluctuations in the composition of the endogenous inclusions that are smaller than 5 μm.

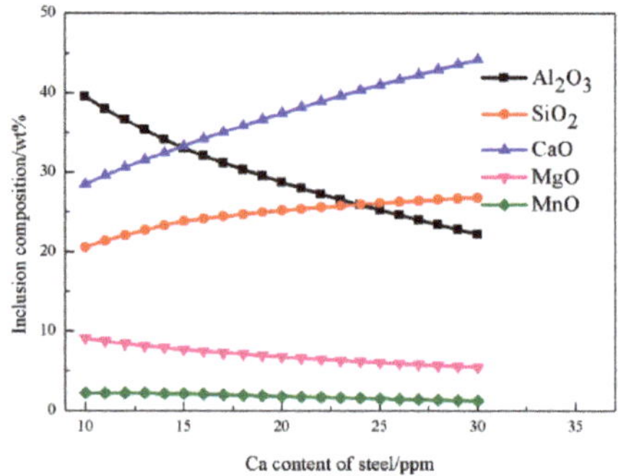

Figure 6. Inclusion composition as a function of the Ca content.

The inclusion composition as a function of the Al content is presented in Figure 7 when the O content, Ca content, Mg content, and the temperatures are 0.007%, 0.002%, 0.0005%, and 1600 °C respectively. It was observed that an increase in the Al content results in an increase in the Al_2O_3 content and a decrease in CaO and SiO_2 contents; however, the MgO and MnO contents did not change much. With regard to the results described by Qian [18], it is worth noting that the Al content comes from the ferrosilicon. A large part of the aluminum reacts with oxygen to form Al_2O_3 inclusions, and the remaining aluminum is involved in the steel–slag reaction, thereby resulting in the fluctuation of the Al content. This is another reason for the fluctuations in the compositions of the endogenous inclusions that are smaller than 5 μm.

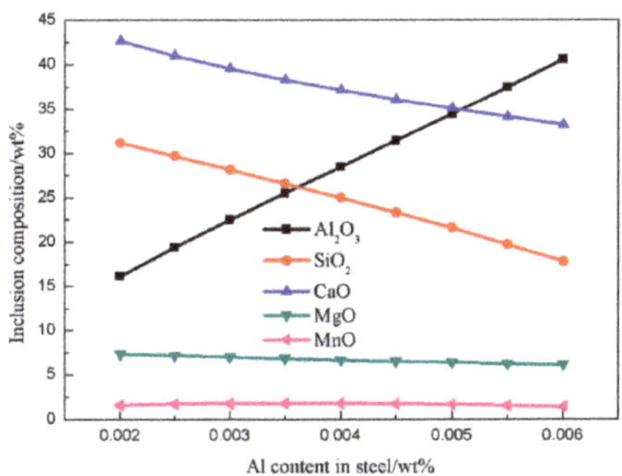

Figure 7. Inclusion composition as a function of the Al content.

The temperature was 1550 °C at the beginning of the VOD process, the highest temperature was 1650 °C at the oxygen blowing stage, and the final steel temperature was 1600 °C; therefore, the temperature changed during the VOD refining process. The relationship between the inclusion composition and the temperature was calculated by using the FToxid and FTmisc databases of Factsage^{TM}7.1 when the O content, Ca content, Mg content, and the Al content were 0.007%, 0.002%, 0.0005% and 0.004%, respectively, and the result are shown in Figure 8. It can be seen that each inclusion composition changes with the changing temperature. The CaO content in the inclusion first increases and then decreases. The Al_2O_3 and MgO contents first decrease from 1650 °C to 1610 °C, and then increase. The change trends of MgO and SiO_2 in the inclusion are just the opposite. The MnO content increases gradually with the decrease in the temperature. The oxidation reaction with Ca, Al, Mg, Si, and Mn are exothermic, so that all of the reactions will occur as the temperature decreases. However, since Ca is a strong oxidizing element, and the concentration of Si in the molten steel is high, the deoxidation reaction of Ca and Si takes precedence. The CaO and SiO_2 contents in the inclusion increase, while the Al_2O_3 and MgO contents in the inclusion decrease. With continuing deoxidation, the CaO and SiO_2 concentrations decrease relatively, triggering an increase in Al_2O_3 and MgO. The change in the temperature during the VOD refining process is one important reason for the fluctuations in the composition of the endogenous inclusions that are smaller than 5 μm.

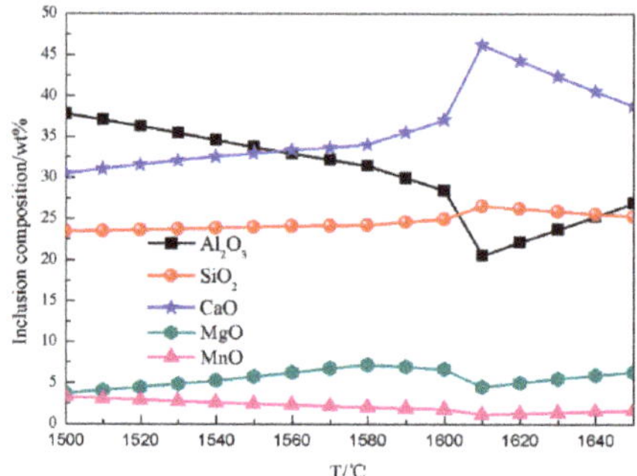

Figure 8. Inclusion composition as a function of the temperature.

The schematic illustration of the formation mechanism of the CaO-SiO_2-Al_2O_3-MgO-MnO inclusions is shown in Figure 9. The oxidation of Si and Al and the collision result in the modification of the smaller-sized inclusions; the reactions are shown in Equations (1)–(3). The source of the total contents of Ca and Mg mainly comes from the addition of FeSi and the reduction of CaO and MgO in the slag or refractory. The appearance of MnO in the inclusions is related to the VOD process; the high oxygen and manganese contents in the liquid steel in part lead to these inclusions.

$$[Si] + 2[O] = SiO_2 \tag{1}$$

$$4[Al] + 3SiO_2 = 2Al_2O_3 + 3[Si] \tag{2}$$

$$2x[Al] + y[Si] + (3x + 2y)[O] = xAl_2O_3 \cdot ySiO_2 \tag{3}$$

$$2[Me] + SiO_2 = 2MeO + [Si] \tag{4}$$

$$3[Me] + Al_2O_3 = 3MeO + 2[Al] \tag{5}$$

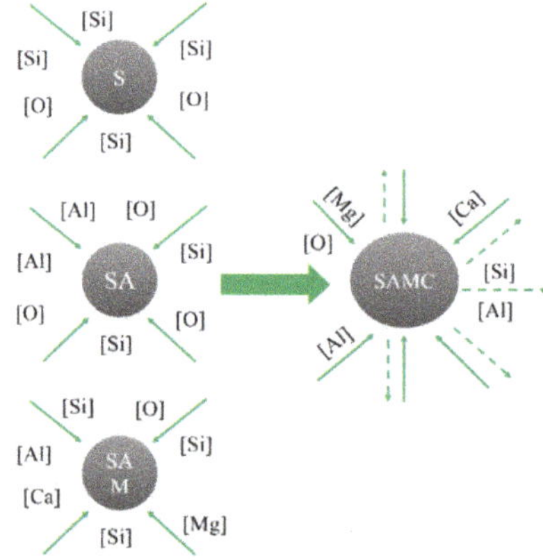

Figure 9. Schematic illustration of the formation mechanism of the CaO-SiO_2-Al_2O_3-MgO-MnO inclusions.

5. Conclusions

(1) The compositional characteristics indicated that two types of inclusions with different sizes existed in the 304L stainless steel during the VOD refining process, namely, CaO-SiO_2-Al_2O_3-MgO external inclusions with sizes ranging from several to tens of microns, and CaO-SiO_2-Al_2O_3-MgO-MnO

endogenous inclusions with sizes smaller than 5 μm. The main inclusion type was CaO-SiO_2-Al_2O_3-MgO-MnO.

(2) The inclusion composition changed with an increasing size of the inclusions from 1 μm to 30 μm. Most of the inclusions were smaller than 5 μm.

(3) The inclusions that were larger than 5 μm were mostly CaO-SiO_2-Al_2O_3-MgO; the similarity in composition to the slag indicated that these inclusions originated from slag entrapment. The CaO-SiO_2-Al_2O_3-MgO-MnO inclusions that were smaller than 5 μm mostly originated from an oxidation reaction with Ca, Al, Mg, Si, and Mn. The changes in the inclusion composition resulting from changes in the Ca, Al, and O contents, and the temperatures during the VOD refining process were larger for the smaller inclusions.

Author Contributions: Conceptualization, G.C.; Methodology, X.C. and J.L.; Software, J.L. and Y.H.; Validation, X.C., J.L. and Y.H.; Formal Analysis, X.C.; Investigation, X.C.; Resources, X.C.; Data Curation, X.C.; Writing-Original Draft Preparation, X.C; Writing-Review & Editing, Y.H.; Visualization, X.C.; Supervision, G.C.; Project Administration, J.P. and Q.R.

Funding: This research received no external funding.

Acknowledgments: The authors gratefully express their appreciation towards the National Nature Science Foundation of China (Grant No. 51374020), the State Key Laboratory of Advanced Metallurgy at University of Science and Technology Beijing (USTB), and Jiuquan Iron and Steel Group Corporation for supporting this work.

Conflicts of Interest: The authors declare no conflict of interest.

References

1. Chen, X.; Pan, J. Analysis on microstructure and inclusions of slab surface layer in 304L stainless steel. *Contin. Casting* **2018**, *43*, 44–48.
2. Momeni, A.; Abbasi, S.M. Repetitive Thermomechanical processing towards ultra fine grain structure in 301, 304 and 304L stainless steels. *J. Mater. Sci. Technol.* **2011**, *27*, 338–343. [CrossRef]
3. Amine, T.; Kriewall, C.S.; Newkirk, J.W. Long-term effects of temperature exposure on SLM 304L stainless steel. *JOM* **2018**, *70*, 384–389. [CrossRef]
4. Padhy, N.; Ningshen, S.; Panigrahi, B.K.; Mudali, U.K. Corrosion behaviour of nitrogen ion implanted AISI type 304L stainless steel in nitric acid medium. *Corros. Sci.* **2010**, *52*, 104–112. [CrossRef]
5. Mizuno, K.; Todoroki, H.; Noda, M.; Tohge, T. Effects of Al and Ca in ferrosilicon alloys for deoxidation on inclusion composition in type 304 stainless steel. *Iron Steelmak.* **2001**, *28*, 93–101.
6. Park, J.H.; Kang, Y.B. Effect of ferrosilicon addition on the composition of inclusions in 16Cr-14Ni-Si stainless steel melts. *Metall. Mater. Trans. B* **2006**, *37*, 791–797. [CrossRef]
7. Li, L.; Cheng, G.; Hu, B.; Wang, C.; Qian, G. Formation of Non-metallic Inclusions of Si-killed Stainless Steel during GOR Refining Process. *High Temp. Mater. Proc.* **2018**, *37*, 521–529. [CrossRef]
8. Kim, J.W.; Kim, S.K.; Kim, D.S.; Lee, Y.D.; Yang, P.K. Formation mechanism of Ca-Si-Al-Mg-Ti-O inclusions in type 304 stainless steel. *ISIJ Int.* **1996**, *36*, S140–S143. [CrossRef]
9. Ehara, Y.; Yokoyama, S.; Kawakami, M. Control of formation of spinel inclusion in type 304 stainless steel by slag composition. *Tetsu-to-Hagané* **2007**, *93*, 475–482. [CrossRef]
10. Yin, X.; Sun, Y.H.; Yang, Y.D.; Bai, X.F.; Deng, X.X.; Barati, M.; McLean, A. Inclusion evolution during refining and continuous casting of 316L stainless steel. *Ironmak. Steelmak.* **2016**, *43*, 533–540. [CrossRef]
11. Park, J.H.; Lee, S.B.; Kim, D.S. Inclusion control of ferritic stainless steel by aluminum deoxidation and calcium treatment. *Metall. Mater. Trans. B* **2005**, *36*, 67–73. [CrossRef]
12. Ren, Y.; Zhang, L.; Fang, W.; Shao, S.; Yang, J.; Mao, W. Effect of Slag Composition on Inclusions in Si-Deoxidized 18Cr-8Ni Stainless Steels. *Metall. Mater. Trans. B* **2016**, *47*, 1024–1034. [CrossRef]
13. Yan, P.; Huang, S.; Pandelaers, L.; Van Dyck, J.; Guo, M.; Blanpain, B. Effect of the CaO-Al_2O_3 based top slag on the cleanliness of stainless steel during secondary metallrugy. *Metall. Mater. Trans. B* **2013**, *44*, 1105–1119. [CrossRef]
14. Yan, P.; Huang, S.; Guo, M.; Blanpain, B. Desulphurisation and inclusion behaviour of stainless steel refining by using CaO-Al_2O_3 based slag at low sulphur levels. *ISIJ Int.* **2014**, *54*, 72–81. [CrossRef]

15. Sakata, K. Technology for production of austenite type clean stainless steel. *ISIJ Int.* **2006**, *46*, 1795–1799. [CrossRef]
16. Park, J.H.; Lee, S.B.; Gaye, H.R. Thermodynamics of the formation of MgO-Al_2O_3-TiO_x inclusions in Ti-stabilized 11Cr ferritic stainless steel. *Metall. Mater. Trans. B* **2008**, *39*, 853–861. [CrossRef]
17. Kang, Y.B.; Lee, H.G. Inclusions chemistry for Mn/Si deoxidized steels: Thermodynamic predictions and experimental confirmations. *ISIJ Int.* **2004**, *44*, 1006–1015. [CrossRef]
18. Qian, G.; Qu, Z.; Cheng, G. Effect of Al in FeSi alloy on continuous casting and surface defect of stainless steel hot-rolled sheet. *Iron Steel* **2016**, *51*, 76–81.
19. Qian, G. The Technologies and Theory of Desulfurization and Inclusion Control for 304 Stainless Steel on GOR Process. Ph.D. Thesis, University of Science and Technology Beijing, Beijing, China, 2015.
20. Swinbourne, D.R.; Kho, T.S.; Langberg, D.; Blanpain, B.; Arnout, S. Understanding stainless steelmaking through computational thermodynamics Part 2–VOD converting. *Miner. Process. Extr. Metall.* **2010**, *119*, 107–115. [CrossRef]
21. Wei, J.H.; Li, Y. Study on Mathematical Modeling of Combined Top and Bottom Blowing VOD Refining Process of Stainless Steel. *Steel Res. Int.* **2015**, *86*, 189–211. [CrossRef]

© 2018 by the authors. Licensee MDPI, Basel, Switzerland. This article is an open access article distributed under the terms and conditions of the Creative Commons Attribution (CC BY) license (http://creativecommons.org/licenses/by/4.0/).

Article

Feasibility of a Two-Stage Forming Process of 316L Austenitic Stainless Steels with Rapid Electrically Assisted Annealing

Viet Tien Luu [1], Thi Anh Nguyet Nguyen [1], Sung-Tae Hong [1,*], Hye-Jin Jeong [2] and Heung Nam Han [2]

[1] School of Mechanical Engineering, University of Ulsan, Ulsan 44610, Korea; tienvietluu@gmail.com (V.T.L.); sweetmoona2@gmail.com (T.A.N.N.)

[2] Department of Materials Science & Engineering and Center for Iron & Steel Research, RIAM, Seoul National University, Seoul 08826, Korea; hj8262@snu.ac.kr (H.-J.J.); hnhan@snu.ac.kr (H.N.H.)

* Correspondence: sthong@ulsan.ac.kr; Tel.: +82-52-259-2129

Received: 27 September 2018; Accepted: 8 October 2018; Published: 11 October 2018

Abstract: The post-annealing mechanical behavior of 316L austenitic stainless steel (SUS316L) after electrically assisted (EA) annealing with a single pulse of electric current is experimentally investigated to evaluate the feasibility of a two-stage forming process of the selected SUS316L with rapid EA annealing. A tensile specimen is deformed to a specific prestrain and then annealed by applying a single pulse of electric current with a short duration less than 1 s. Finally, the specimen is reloaded until fracture. The stress-strain curve during reloading shows that the flow stress of the SUS316L significantly decreases, which indicates the occurrence of EA annealing. The electric current also increases the maximum achievable elongation of the SUS316L during reloading. The stress-strain curve during reloading and the microstructural observation suggest that the effects of EA annealing on the post-annealing mechanical behavior and microstructure strongly depend on both the applied electric current density (electric current per unit cross-sectional area) and the given prestrain. The results of the present study suggest that the EA annealing technique could be effectively used to improve the formability of SUS316L when manufacturing complex parts.

Keywords: electrically assisted annealing; electric current; prestrain; stainless steels

---

1. Introduction

Formability is an important issue for all metallic materials in forming processes, even for austenitic stainless steels (ASSs). ASSs are known to have good formability, excellent corrosion resistance, and good weldability [1–3]. However, in some industrial applications, the good formability of ASSs may still not be satisfactory [4,5]. For example, in deep drawing of consumer electric components using ASSs, the forming process is frequently composed of two (or multiple) forming stages with heat treatment between the forming stages, due to the formability of the ASSs.

Furnace annealing between forming stages is a commercially well-established heat treatment process to control the formability of an ASS in a two- or multiple-stage forming process. A deformed part is heat-treated in a certain temperature range (800–1100 °C) for a specified duration (usually from 30 to 45 min) [4,5] before the next forming stage. While furnace annealing can effectively restore the formability of the product by annihilating strain hardening from the previous forming stage, furnace annealing frequently becomes quite time-consuming and expensive. Thus, a cost-effective and preferably rapid alternative technique to control the formability of ASS is still desirable.

Electrically assisted manufacturing (EAM) is a promising metal forming technique, in which the mechanical property of a metal is controlled by simply applying electricity to the metal during deformation (referred to as electroplasticity) [6,7]. In the investigation of the effect of electric current on the mechanical behavior of metal alloys, several researchers repeatedly applied electric current with a short duration (pulsed electric current) to a metal during deformation. Roth et al. [8] applied a pulsed electric current to aluminum 5754 alloy during tension to achieve maximum elongation close to 400% of the gage length. Salandro et al. [9] applied a pulsed electric current to aluminum 5052 and 5083 alloys during tension. According to Salandro et al. [9], the effectiveness of the electric current on the tensile behavior of selected aluminum alloys depends on both the alloy and the heat treatment. Kim et al. [10] and Roh et al. [11] reported that the formability of aluminum alloys during tension with a pulsed electric current is increased due to electric current induced (or electrically assisted: EA) annealing. It has also been reported that the microstructure of metals can be significantly affected by electric current. Conrad et al. [12] reported that the plasticity and phase transformation of various metals and ceramics were affected by an electric current. Xu et al. [13] reported that recrystallization and grain growth of cold-rolled α-Ti were accelerated by an electric current. More recently, Park et al. [14] reported that the annealing temperature and time for recrystallization of interstitial free (IF) steel and AZ31 magnesium alloy were significantly reduced by electropulsing treatment (EPT) compared to conventional heat treatment with a furnace. It is important to note, however, that for specific metal alloys and experimental conditions, applying an electric current during deformation may induce adverse effects on the formability [15,16]. Magargee et al. [15] observed that the formability of commercially pure titanium was not significantly changed when an electric current was applied during tensile deformation, even though the flow stress substantially decreased. Jeong et al. [16] studied the effect of electric current on the tensile behavior of transformation-induced plasticity (TRIP)-aided steel and reported that the elongation of TRIP-aided steel significantly decreased by applying a pulsed electric current during tensile deformation. They suggested that the effect of electric current on the mechanical behavior can be strongly affected by designing the pulsing pattern of the electric current in accordance with the material's characteristics.

While applying a pulsed electric current to a metal alloy during deformation generally enhances the formability of the metal alloy, it may not be practical to design a commercial metal forming process with a pulsed electric current. The cycle time of the process may be significantly increased and/or the structure of forming machines may become quite complex. Several recent studies on EAM or electroplasticity have reported that the mechanical properties of a metal can be altered by applying a single pulse of electric current of short duration. Kim et al. [17] reported that the springback in U-bending of advanced high strength steel sheets can be reduced or even eliminated by applying a single pulse of electric current prior to removal of the forming load. Thien et al. [18] applied a single pulse of electric current to a complex phase ultra-high strength steel at a specific prestrain and reported that the flow stress significantly decreased and the formability increased during reloading. These previous studies suggest that electrically assisted two- (or multiple-) stage metal forming processes for certain metal alloys may be effectively designed with a single pulse of electric current.

However, studies on the mechanical behavior of metal alloys (including ASSs) under a single pulse of electric current between plastic deformations are still quite limited. In the present study, the effect of a single pulse of electric current on the mechanical behavior and microstructure of a commercially available 316L austenite stainless steel (SUS316L) is reported. Specifically, the goal of the present study is to evaluate the feasibility of a two-stage forming process of the selected SUS316L with rapid EA annealing.

2. Experimental Set-Up

Commercially available 1 mm thick grade SUS316L sheets (Fe-17.06Cr-9.81Ni-1.17Mn-1.97Mo-0.59Si-0.028C in wt.%) were used for the experiment. Typical tensile specimens according to ASTM-E08 with a gage width of 12.5 mm and a gage length of 50 mm were fabricated by laser cutting along the rolling direction of the sheet.

In the present study, electrically assisted (EA) two-stage forming of the selected SUS316L was simply implemented by quasi-static tensile tests with a single pulse of electric current, as schematically described in Figure 1. The quasi-static tensile test was conducted using a universal testing machine. The baseline tensile test was first conducted by simply deforming a specimen by tension until fracture. For the quasi-static tensile test with EA annealing, a specimen was first deformed to a specified prestrain by uniaxial tension with a constant displacement rate of 2.5 mm/min. A single pulse of electric current with a short duration was then applied to the prestrained specimen to induce EA annealing, according to the electric current parameters. Next, the specimen was cooled to room temperature in air and unloaded. Finally, the specimen was reloaded until fracture by uniaxial tension with a constant displacement rate of 2.5 mm/min.

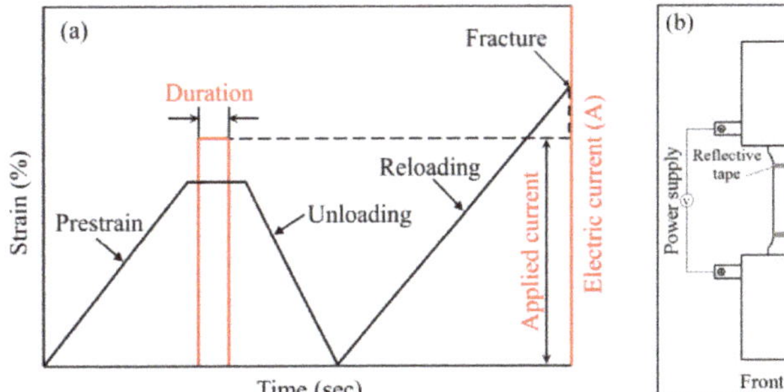

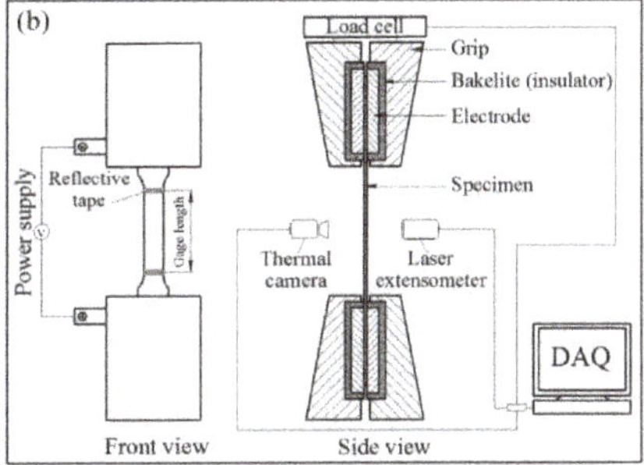

Figure 1. Schematics of (**a**) electrically assisted (EA) annealing with a single pulse of electric current and (**b**) the experimental set-up.

For EA annealing, the electric current was generated by a programmable Vadal SP-1000U power supply (Hyosung, Seoul, Korea). By inserting a set of bakelite insulators between the specimen and grips, the tensile test machine was insulated from the electric current (Figure 1b). The force history during the experiment was measured as a function of time by a load cell using a PC-based data acquisition system (Figure 1b). The displacement history was also measured using a laser extensometer (LX500, MTS, Eden Prairie, MN, USA) by attaching retro-reflective tape to the specimen. Finally, an infrared thermal imaging camera (T621, FLIR, Taby, Sweden) was employed to monitor the temperature change of the specimen throughout the experiment. It should be noted that one side of the specimen was sprayed with black thermal paint to stabilize the emissivity and thus improve the accuracy of the temperature measurement. The emissivity was calibrated by a k-type thermocouple through separate calibration tests.

For the parameter study, two different prestrain levels (20% and 60%) were combined with three different true electric current densities (95, 100, and 105 A/mm^2) and a constant electric current duration of 0.75 s. It should also be noted that the selected prestrains in the present study correspond to 20% and 60% of uniform elongation from the baseline tensile test (without prestrain and electric current), respectively, as indicated in Figure 2. The true electric current density was calculated based on the actual cross-sectional area of the specimen at the given prestrain, as the term "true" indicates. For each parameter set, at least three specimens were tested to verify repeatability of the results.

To analyze the effect of electric current on the microstructure of a prestrained specimen, the EA annealed specimen was removed from the test without reloading to fracture and was prepared for microstructural analysis. The microstructure of the specimen was characterized by an electron probe microanalyzer (EPMA, JXA-8530F, JEOL Ltd., Tokyo, Japan) and a field emission gun scanning electron microscope (FE-SEM, SU70, Hitachi, Tokyo, Japan) equipped with an electron backscatter diffraction system (EBSD, EDAX/TSL, Hikari, Hayward, CA, USA). To evaluate the change of dislocation density by EA annealing, the full width at half maximum (FWHM) of the diffraction peak was measured with an X-ray diffractometer (D8-Advanced, BRUKER MILLER Co., Boston, MA, USA) using a Cu radiation source operating at 50 kV at room temperature. Diffraction patterns were recorded in the scan range of 40–85° with a scan speed of 1°/min. Specimens for microstructural observation were prepared by mechanical grinding followed by electropolishing with an 80 mL perchloric acid, 90 mL distilled water, 100 mL butanol, and 730 mL ethanol solution at 20 V. For EBSD analysis, the accelerating voltage and scan step size were 15 kV and 0.5 μm, respectively. The critical misorientation angle was set to 15° for grain identification.

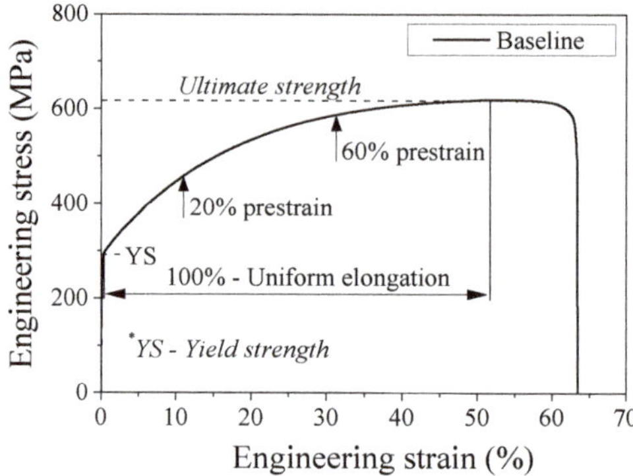

Figure 2. A stress–strain curve of a baseline specimen.

3. Results

During EA annealing, the temperature of the specimen rapidly reached the maximum temperature using resistance heating followed by cooling, as representative temperature histories of the 60% prestrained specimens shown in Figure 3a. It should be noted that the temperature was measured at the center of the specimen, which gave the highest temperature from application of the electric current. Naturally, the peak temperature increased as the current density was increased as shown in Figure 3b. Additionally, there was no significant effect of two different prestrains on the peak temperature.

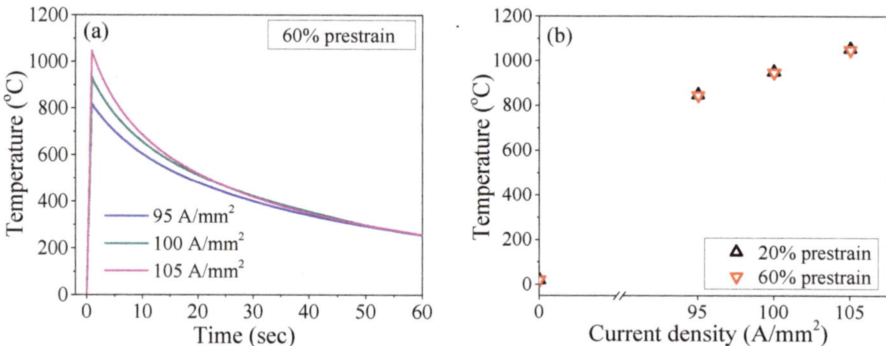

Figure 3. (**a**) Temperature histories of 60% prestrained specimens and (**b**) the peak temperatures as functions of current density during EA annealing for different prestrains.

From the load-displacement history during reloading, engineering stress-strain curves during reloading were constructed based on the gage length and cross-sectional area at each prestrain, as shown in Figure 4. As expected, the yield strengths during reloading (the post-annealing yield strengths) generally showed higher values than the baseline yield strength without prestrain due to strain hardening for both prestrains. For the same reason, the elongations at fracture during reloading (the post-annealing elongation) for both prestrains generally showed lower values than the baseline elongation at fracture without prestrain. As the current density was increased, the engineering stress-strain curves during reloading clearly showed that the flow stress of the SUS316L gradually decreased, which indicates the occurrence of EA annealing [10]. The post-annealing elongation also clearly increased with increasing current density.

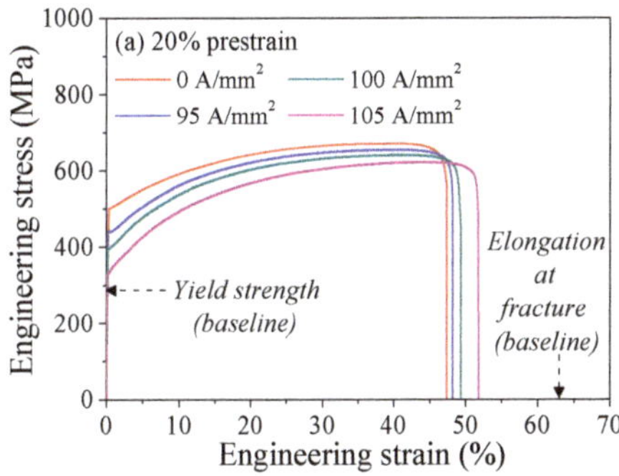

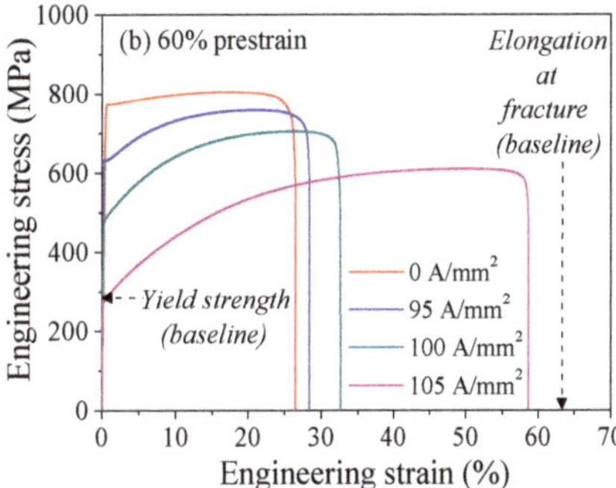

Figure 4. Engineering stress-strain curves after EA annealing with (**a**) 20% prestrain and (**b**) 60% prestrain; the engineering stress-strain curves during reloading was constructed based on the gage length and the cross-sectional area at each prestrain.

The post-annealing yield strength and post-annealing elongation can be plotted as functions of current density (Figure 5a,b, respectively). The results in Figure 5a,b suggest that the effect of EA annealing on the mechanical behavior during reloading is significantly different depending on the magnitude of the prestrain, even with the same true current density and nearly identical peak temperatures. As shown in Figure 5a,b, with the higher prestrain, the effect of EA annealing becomes more pronounced, i.e., the post-annealing yield strength and post-annealing elongation change more rapidly with the higher prestrain, as the current density increases with the constant duration of electric current. The effectiveness of EA annealing on the formability of the SUS316L can be further evaluated by comparing the total achievable displacement or total achievable elongation based on the original gage length of the baseline specimen. The total achievable displacement and total achievable elongation can be simply calculated as:

Total achievable displacement = displacement by prestrain + displacement during reloading

Total achievable elongation = total achievable displacement/the gage length of the baseline specimen

The total achievable elongation as a function of current density, clearly seen in Figure 6a,b, confirms that EA annealing at the higher prestrain is beneficial to improve formability of the given metal alloy. Especially at the highest current density of 105 A/mm^2, the total achievable elongation with the 60% prestrain was approximately 1.5 times higher than that with the 20% prestrain (Figure 6a,b). It is to be noted that the total achievable elongation with the 20% prestrain and current density of 95 A/mm^2 still surpassed that of the baseline tensile test.

For evaluation of the effect of EA annealing on hardening behavior during reloading, the engineering stress-strain curves in Figure 4 were converted to true stress-strain curves up to the engineering strain of uniform elongation, as shown in Figure 7a,b. As expected, strain hardening parameters with the higher prestrain showed lower values in comparison to those with the lower prestrain. However, the strain hardening exponent (Figure 7c) and strength coefficient (Figure 7d) as functions of current density showed that with the higher prestrain, both strain hardening parameters increased more rapidly as the current density was increased. At the current density of 105 A/mm^2, the strain hardening parameters during reloading with the 60% prestrain became nearly identical to those of the baseline curve without prestrain. This finding suggests that the strain hardening by the given prestrain was completely annihilated by the electric current with a duration of 0.75 s, which confirms the benefit of EA annealing as a rapid annealing process for cost-effective two-stage forming. However, it should be noted that with the 20% prestrain, the strain hardening parameters during reloading were not close to those of the baseline curve without prestrain, even at the current density of 105 A/mm^2, which was the highest current density selected in the present study. Therefore, in the design of two-stage forming with EA annealing, the amount of deformation in the first forming stage needs to be carefully decided to optimize the effect of EA annealing in the process.

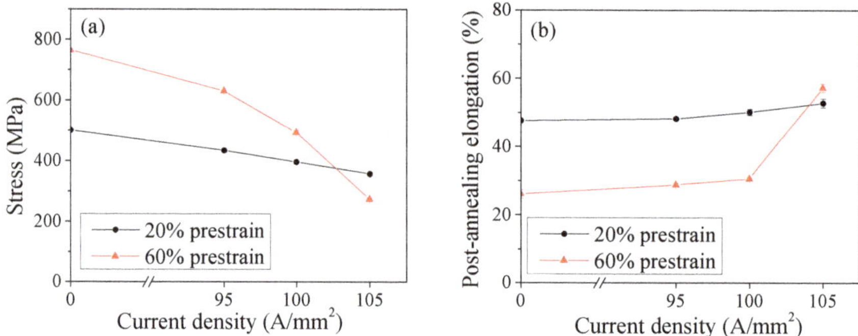

Figure 5. (**a**) Post-annealing yield strength and (**b**) post-annealing elongation of prestrained specimens after EA annealing.

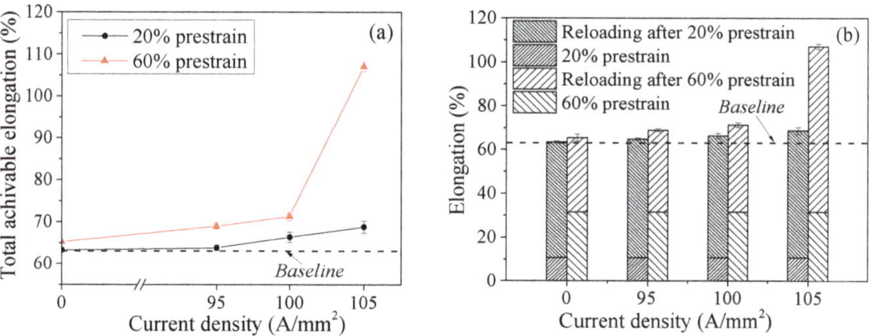

Figure 6. (**a**) Total achievable elongation as a function of current density and (**b**) the effect of prestrain and the current density on the total achievable elongation; the total achievable elongation was calculated based on the original gage length of the baseline specimen.

As presented above, the mechanical behavior of the SUS316L during reloading after a short duration of electric current strongly suggests the occurrence of EA annealing. To confirm this occurrence of EA annealing, FWHM analysis of the diffraction peak profile can be effectively used [19,20]. For crystalline materials, the diffraction peak profile is typically broadened when the crystal lattice is distorted by lattice defects, especially by dislocations [21,22]. As shown in Figure 8a,b, the FWHM values decreased with increasing current density for both prestrains selected in the present study. The results in Figure 8a,b clearly confirm the reduction of dislocation density (annihilation of dislocations) in the prestrained specimen by the applied electric current, which is EA annealing. It should be noted that for the 60% prestrain, the FWHM values show that the specimen was fully annealed to the baseline status without prestrain when the electric current density of 105 A/mm^2 was applied. Therefore, it is speculated that recrystallization occurred under this condition. In contrast, for all of the 20% prestrained specimens, full annealing was not achieved, even though the peak temperatures were nearly the same, as shown in Figure 3b. The results of the FWHM analysis are in good agreement with the observed mechanical behavior during reloading after EA annealing in Figures 4–6.

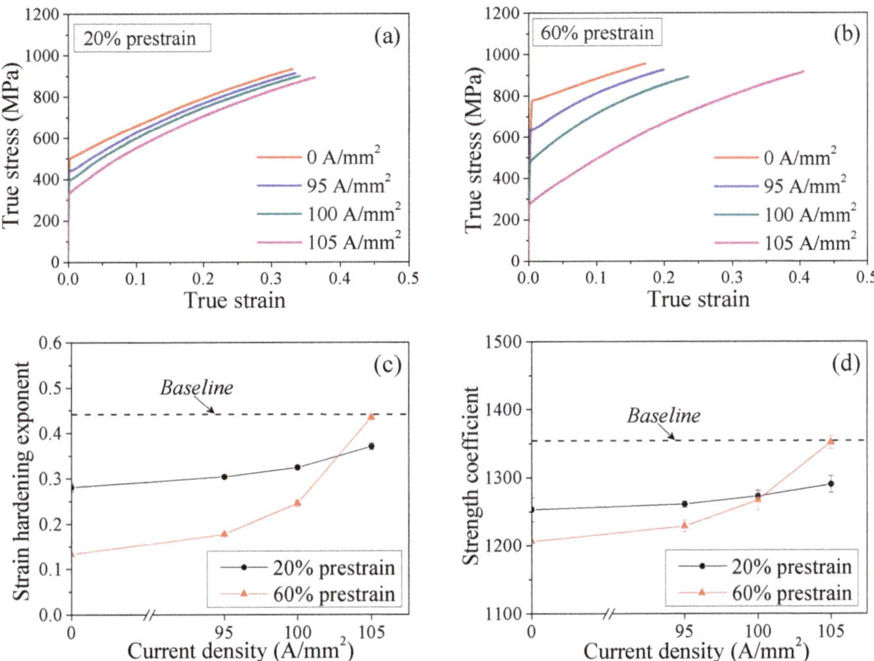

Figure 7. True stress-strain curves after EA annealing with (**a**) 20% prestrain and (**b**) 60% prestrain; (**c**) strain hardening exponent and (**d**) strength coefficient of prestrained specimens after EA annealing.

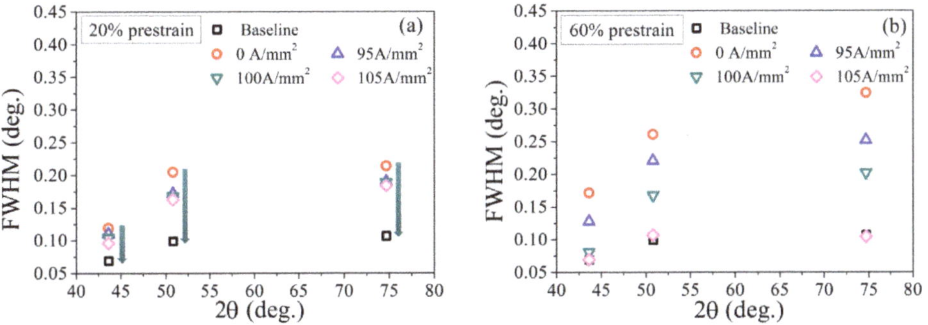

Figure 8. Full width at half maximum (FWHM) profiles of specimens with (**a**) 20% prestrain and (**b**) 60% prestrain.

Figure 9 shows the EBSD inverse pole figures (IPF) with respect to the normal direction (ND) and kernel average misorientation (KAM) maps for each specimen. For KAM maps, the higher degree of misorientation is denoted by a brighter color. The color of the KAM maps is proportional to the amount of dislocation or the accumulated strain energy present in the specimen, as indicated in the color scale bar. The KAM maps display effects, which can be interpreted as dislocations. For EA annealing with 105 A/mm^2, the average grain size of the 60% prestrained specimen was significantly reduced from 17.0 ± 5.9 to 12.1 ± 4.3 μm. The average KAM value of the 60% prestrained specimen was also significantly reduced from 1.10 to 0.33 under the condition of 105 A/mm^2. This result clearly indicates that recrystallization occurred in the 60% prestrained specimen after EA annealing with the current density of 105 A/mm^2. This result coincides with the FWHM value of the 60% prestrained specimen under the electric current density of 105 A/mm^2, as shown in Figure 8b. This result can be explained as the dislocation density induced by the 60% prestrain was high enough to provide the driving force for recrystallization [23] during EA annealing with 105 A/mm^2 for the duration of 0.75 s. This phenomenon might be related to the acceleration of recrystallization kinetics due to the electric current [14]. No significant changes in the grain size were observed for the other combinations of prestrain and electric current density considered in the present study.

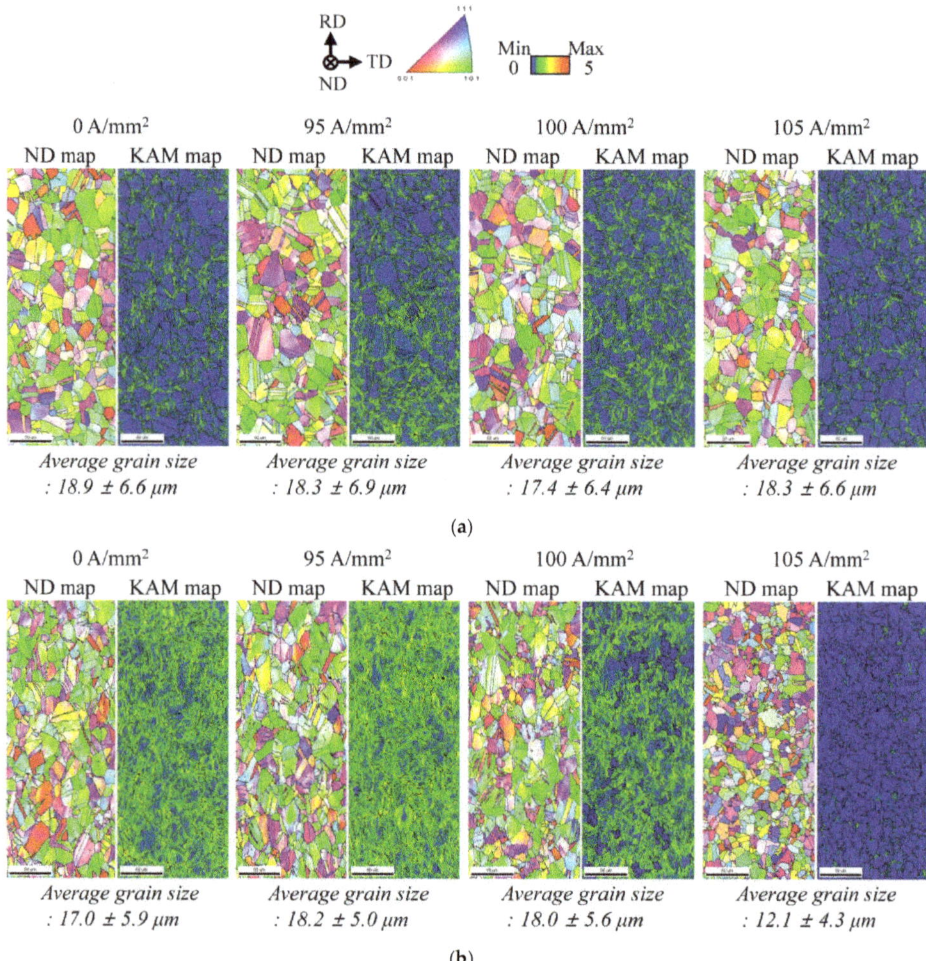

Figure 9. Electron backscatter diffraction (EBSD) inverse pole figures for normal direction (ND) and kernel average misorientation (KAM) maps of specimens with (**a**) 20% prestrain and (**b**) 60% prestrain for various electric current densities.

4. Conclusions

In the present study, the effects of a single pulse of electric current on the mechanical behavior and microstructure of a commercially available SUS316L were investigated to evaluate the feasibility of a two-stage forming process with rapid EA annealing in less than 1 s. The mechanical behavior of the prestrained specimen after applying a single pulse of electric current to the specimen suggests the occurrence of EA annealing. The results of microstructural analysis show the annihilation of dislocations by the electric current, confirming the occurrence of EA annealing. The experimental results show that the effectiveness of EA annealing strongly depends not only on the electric current parameters, but also the magnitude of the prestrain. Depending on the electric current parameters and the magnitude of the prestrain, full annealing with recrystallization was even possible to significantly increase the total maximum achievable elongation of the specimen. The results of the present study suggest that a two-stage forming process of SUS316L can be simplified and expedited by properly implementing rapid EA annealing in the process. As a final remark, in the present feasibility study, the duration of electric current was fixed for simplicity. However, for optimization of a practical dual-stage forming process with rapid EA annealing, a further study with expanded electric current parameter combinations with different electric current durations may be necessary.

Author Contributions: V.T.L. and S.-T.H. conceived and designed the experiments, performed the experiments, analyzed the data and wrote the paper. T.A.N.N. participated in the experimental study. H.-J.J. and H.N.H. carried out and discussed the microstructural study and reviewed the paper.

Funding: The National Research Foundation of Korea (NRF) grant (No. NRF-2015R1A5A1037627); The Encouragement Program for The Industries of Economic Cooperation Region.

Acknowledgments: This research was supported by the Ministry of Trade, Industry & Energy (MOTIE), Korea Institute for Advancement of Technology (KIAT) and Ulsan Institute for Regional Program Evaluation (IRPE) through the Encouragement Program for The Industries of Economic Cooperation Region.

Conflicts of Interest: The authors declare no conflicts of interest.

References

1. Peterson, S.F.; Mataya, M.C.; Matlock, D.K. The formability of austenitic stainless steels. *J. Miner. Met. Mater. Soc.* **1997**, *49*, 54–58. [CrossRef]
2. Andrade, M.S.; Gomes, O.A.; Vilela, J.M.C.; Serrano, A.T.L.; De Moraes, J.M.D. Formability evaluation of two austenitic stainless steels. *J. Braz. Soc. Mech. Sci. Eng.* **2004**, *26*, 47–50. [CrossRef]
3. Acar, M.O.; Fitzpatrick, M.E. Determination of plasticity following deformation and welding of austenitic stainless steel. *Mater. Sci. Eng. A* **2017**, *701*, 203–213. [CrossRef]
4. Anderson, M.; Gholipour, J.; Bridier, F.; Bocher, P.; Jahazi, M.; Savoie, J.; Wanjara, P. Improving the formability of stainless steel 321 through multistep deformation for hydroforming applications. *Trans. Can. Soc. Mech. Eng.* **2013**, *37*, 39–52. [CrossRef]
5. Thanakijkasem, P.; Uthaisangsuk, V.; Pattarangkun, A.; Mahabunphachai, S. Effect of bright annealing on stainless steel 304 formability in tube hydroforming. *Int. J. Adv. Manuf. Technol.* **2014**, *73*, 1341–1349. [CrossRef]
6. Okazaki, K.; Kagawa, M.; Conrad, H. A Study of the electroplastic effect in metals. *Scr. Metall.* **1978**, *12*, 1063–1068. [CrossRef]
7. Conrad, H. Electroplasticity in metals and ceramics. *Mater. Sci. Eng. A* **2000**, *287*, 276–287. [CrossRef]
8. Roth, J.T.; Loker, I.; Mauck, D.; Warner, M.; Golovashchenko, S.F.; Krause, A. Enhanced Formability of 5754 Aluminum Sheet Metal Using Electric Pulsing. In Proceedings of the 36th North American Manufacturing Research Conference, Monterrey, Mexico, 20–23 May 2008.
9. Salandro, W.A.; Jones, J.J.; McNeal, T.A.; Roth, J.T.; Hong, S.-T.; Smith, M.T. Formability of Al 5xxx sheet metals using pulsed current for various heat treatments. *J. Manuf. Sci. Eng.* **2010**, *132*, 1–11. [CrossRef]
10. Kim, M.-J.; Lee, K.; Oh, K.H.; Choi, I.-S.; Yu, H.-H.; Hong, S.-T.; Han, H.N. Electric current-induced annealing during uniaxial tension of aluminum alloy. *Scr. Mater.* **2014**, *75*, 58–61. [CrossRef]
11. Roh, J.-H.; Seo, J.-J.; Hong, S.-T.; Kim, M.-J.; Han, H.N.; Roth, J.T. The mechanical behavior of 5052-H32 aluminum alloys under a pulsed electric current. *Int. J. Plast.* **2014**, *58*, 84–99. [CrossRef]
12. Conrad, H. Effects of electric current on solid state phase transformations in metals. *Mater. Sci. Eng. A* **2000**, *287*, 227–237. [CrossRef]
13. Xu, Z.S.; Lai, Z.H.; Chen, Y.X. Effect of electric current on the recrystallization behavior of cold worked α–Ti. *Scr. Metall.* **1988**, *22*, 187–190. [CrossRef]

14. Park, J.-W.; Jeong, H.-J.; Jin, S.-W.; Kim, M.-J.; Lee, K.; Kim, J.J.; Hong, S.-T.; Han, H.N. Effect of electric current on recrystallization kinetics in interstitial free steel and AZ31 magnesium alloy. *Mater. Charact.* **2017**, *133*, 70–76. [CrossRef]
15. Magargee, J.; Morestin, F.; Cao, J. Characterization of flow stress for commercially pure titanium subjected to electrically assisted deformation. *J. Eng. Mater. Technol.* **2013**, *135*, 041003. [CrossRef]
16. Jeong, H.-J.; Park, J.-W.; Jeong, K.J.; Hwang, N.M.; Hong, S.-T.; Han, H.N. Effect of pulsed electric current on TRIP-aided steel. *Int. J. Precis. Eng. Manuf. Green Technol.* **2018**, in press.
17. Kim, M.-S.; Vinh, N.T.; Yu, H.-H.; Hong, S.-T.; Lee, H.-W.; Kim, M.-J.; Han, H.N.; Roth, J.T. Effect of electric current density on the mechanical property of advanced high strength steels under quasi-static tensile loads. *Int. J. Precis. Eng. Manuf.* **2014**, *15*, 1207–1213. [CrossRef]
18. Thien, N.T.; Jeong, Y.-H.; Hong, S.-T.; Kim, M.-J.; Han, H.N.; Lee, M.-G. Electrically assisted tensile behavior of complex phase ultra-high strength steel. *Int. J. Precis. Eng. Manuf. Green Technol.* **2016**, *3*, 325–333. [CrossRef]
19. Hong, S.-T.; Jeong, Y.-H.; Chowdhury, M.N.; Chun, D.-M.; Kim, M.-J.; Han, H.N. Feasibility of electrically assisted progressive forging of aluminum 6061-T6 alloy. *CIRP Ann.* **2015**, *64*, 227–280. [CrossRef]
20. Kim, M.-J.; Lee, M.-G.; Hariharan, K.; Hong, S.-T.; Choi, I.-S.; Kim, D.; Oh, K.H.; Han, H.N. Electric current-assisted deformation behavior of Al-Mg-Si alloy under uniaxial tension. *Int. J. Plast.* **2017**, *94*, 148–170. [CrossRef]
21. Williamson, G.K.; Hall, W.H. X-ray line broadening from filed aluminum and wolfram. *Acta Metall.* **1953**, *1*, 22–31. [CrossRef]
22. Ungar, T.; Borbely, A. The effect of dislocation contrast on x-ray line broadening: A new approach to line profile analysis. *Appl. Phys. Lett.* **1996**, *69*, 3173–3175. [CrossRef]
23. Cho, S.-H.; Yoo, Y.-C. Static recrystallization kinetics of 304 stainless steels. *J. Mater. Sci.* **2001**, *36*, 4273–4278. [CrossRef]

© 2018 by the authors. Licensee MDPI, Basel, Switzerland. This article is an open access article distributed under the terms and conditions of the Creative Commons Attribution (CC BY) license (http://creativecommons.org/licenses/by/4.0/).

Article

Experimental Data Assessment and Fatigue Design Recommendation for Stainless-Steel Welded Joints

Yang Peng, Jie Chen and Jun Dong *

College of Civil Engineering, Nanjing Tech University, Nanjing 210000, China
* Correspondence: dongjun@njtech.edu.cn; Tel.: +86-136-0140-7837

Received: 12 June 2019; Accepted: 25 June 2019; Published: 26 June 2019

Abstract: Stainless steel possesses outstanding advantages such as good corrosion resistance and long service life. Stainless steel is one of the primary materials used for sustainable structures, and welding is one of the main connection modes of stainless-steel bridges and other structures. Therefore, fatigue damage at welded joints deserves attention. The existing fatigue design codes of stainless-steel structures mainly adopt the design philosophy of structural steel. In order to comprehensively review the published fatigue test data of welded joints in stainless steel, in this paper, the fatigue test data of representative welded joints of stainless steel were summarized comprehensively and the *S–N* curves of six representative stainless-steel welded joints were obtained by statistical evaluation. The comparison of the fatigue strength from existing design codes and fatigue test data was performed, and the results showed that the fatigue strength of welded joints of stainless steel was higher than that of structural-steel welded joints. The flexibility of regression analysis with and without a fixed negative inverse slope was discussed based on the scatter index. It was found that the fatigue test data of stainless-steel welded joints are more consistent with the *S–N* curve regressed by a free negative inverse slope. In this paper, a design proposal for the fatigue strength of representative welded joints of stainless steel is presented based on the *S–N* curve regressed by the free negative inverse slope.

Keywords: stainless-steel structure; welded joints; fatigue strength; *S–N* curves; scatter index

---

1. Introduction

Stainless steel has outstanding corrosion resistance, an elegant surface effect, and good mechanical properties. With the improvement of engineering construction standards, the optimal life cycle benefits are taken into account, as well as the realization of targets to reuse materials for energy conservation and emission reduction. Therefore, stainless steel is one of the preferred materials for a sustainable structure, and has wide development prospects in environments with high durability requirements for materials such as offshore or near the sea [1–3]. The upper tower of Stonecutters Bridge is a composite structure with a stainless-steel skin. The Hong Kong–Zhuhai–Macau Bridge's outer reinforcement parts used duplex stainless steel [4]. The designers of the motorway flyover in Kerensheide, the Netherlands, took the resulting corrosion risks into account by selecting stainless steel. The pillars of the footbridge in Reykjavik, Iceland consist of concrete-filled stainless-steel circular hollow sections. The structure of the Spain Añorga Railway Bridge is completely made from stainless steel [5]. Stockholm Bridge in Sweden used duplex stainless steel in its renovation. Welding is the main method of connection, and fatigue is the main design criterion. The repair of defects is very expensive; moreover, it is difficult to obtain the ideal repair effect [6–9]. Thus, guidelines are needed to avoid fatigue failure of stainless-steel welded joints.

The fatigue properties of welded joints were tested due to difficulties in reaching a consensus on the fatigue design rules. The collection of existing fatigue test data is a common practice in classifying fatigue details. This is, of course, not the most advisable way, given that the costs and the time involved in testing specimens to simulate real bridge details and loads are considerable. Data representative of several countries, steel, and manufacturers may be considered so that the natural scatter based on bridge constructions is included.

The fatigue behavior of stainless-steel welded joints was the subject of substantial research. Niemi et al. [10,11] studied the fatigue strength of welded joints of austenitic and duplex stainless steel. They tested the butt-welded specimens and several types of fillet-welded specimens. The results indicated that the fatigue strength of stainless-steel welded joints was higher than that of the welded joints of structural steel existing in the fatigue design standard [11]. Razmjoo [12] summarized fatigue test data of stainless-steel welded joints in 1995. He compared the data with the 95% confidence limits enclosing the fatigue data obtained from structural-steel welded joints. He found that the fatigue test data of stainless-steel welded joints fell within the scatter band of the structural-steel welded joints. Due to data limitation, he suggested that the *S–N* curves for structural-steel welded joints could be used to conservatively design stainless-steel welded joints. Branco et al. [13] systematically investigated the fatigue behavior of stainless-steel welded joints. They compared the fatigue test data of stainless-steel welded joints with the scatter band of the structural-steel welded joints. The results indicated that the *S–N* curves of the structural-steel welded joints could be applied to stainless-steel welded joints. The International Institute of Welding (IIW) fatigue recommendation [14] *S–N* curves were generally more suitable than those in Eurocode (EC)3 Parts 1–9 [15]. They found that the type of stainless steel did not influence the fatigue strength of stainless-steel welded joints. The specimens made from gas tungsten arc welding (GTAW) have high fatigue strength due to the very favorable weld profiles and low stress concentration. Metrovich et al. [16] studied the fatigue strength of the weld in stainless-steel beams. These beams were made from AL-6XN superaustenitic steel; this stainless steel is popular in the chemical industry. They found that the fatigue strength of the longitudinal fillet weld and bulkhead attachment was the same as a structural-steel weld and the fatigue strength of the transverse groove weld was higher than a structural-steel weld. The National Research Institute for Metals of Japan [17] published fatigue test data of the hot-rolled austenitic stainless-steel and butt weld. Nakamura et al. [18,19] investigated the fatigue strength of the butt weld and the filled weld made from austenitic stainless steel in the air and a corrosive environment. The results illustrated that the negative inverse slope of *S–N* curves (m = 5–15) regressed from the fatigue test data was much larger than the Japanese fatigue design standard [20] for the structural-steel welded joints, in which the negative inverse slope m of the *S–N* curves in the standard is equal to 3. In the corrosive environment, the fatigue strength was reduced. The fatigue strengths of the butt weld and the fillet weld in the corrosive environment were 80% and 66% of the fatigue strength in the air environment, respectively. Singh et al. [21,22] tested the butt weld and the fillet weld of the austenitic stainless steel, and the results indicated that the fatigue test data were consistent with the fatigue design standard, BS 5400 Part 10 [23]. Wu et al. [24] studied the butt weld and the fillet weld made from austenitic stainless steel, duplex stainless steel, and typical structural steel. They found that the fatigue strength of welded joints made from stainless steel was higher than that of the structural-steel welded joints. The *S–N* curves obtained from the fatigue test data of stainless-steel welded joints were quite different from the *S–N* curves from the IIW fatigue design recommendation [14]. The negative inverse slope of the *S–N* curves and the fatigue strength regressed from Wu's fatigue test data were substantially larger than those in the IIW fatigue design recommendation. However, a systematical statistical analysis of the fatigue test data is absent in these studies. Highly conservative design rules were proposed, owing to the fact that experimental evidence was not available to justify more favorable rules. The conclusion is based on the design rules for structural-steel welded joints although the stress–strain curves for stainless steels exhibit no yielding plateau [25], as well as a low proportional limit [25] and low residual stress [26].

The above research focused on the nominal stress approach. Partanen and Niemi [27] proposed *S–N* curves for a hot-spot stress approach based on the fatigue test data of structural-steel and austenitic stainless-steel welded joints. The fatigue strength that resulted in a fatigue life of two million cycles was 107 MPa. This value for the fatigue strength was similar to that of the fatigue classification of the

IIW recommendation [14] and EC3 Parts 1–9 [15]. Jia et al. [28] found that the fatigue strength of duplex stainless-steel welded joints with a fatigue life of two million cycles was equal to 167 MPa. This was much higher than the result of Partanen and Niemi. Feng and Young [29] investigated the stress concentration factor (SCF) of tubular X-joints. They illustrated that the current SCF equations gave unconservative predicted results, and new SCF equations were proposed. The research on hot-spot stress approach for stainless-steel welded joints is not sufficient to suggest reasonable fatigue design rules. Lazzarin et al. [30,31] studied two new approaches for fatigue estimation of stainless-steel welded joints: a local strain energy density approach and a peak stress method. The results showed that the local strain energy density approach was not suitable, and the peak stress method could accurately estimate the fatigue performance of stainless-steel welded joints. Further validation of the peak stress method should be conducted. Fracture mechanics were used to investigate the fatigue strength of stainless-steel welded joints [12,21,22,32–34]. The fatigue crack propagation of stainless steel is influenced by the transformation-induced plasticity (TRIP) effect [35]. The TRIP effect was confirmed to be beneficial to the fatigue crack growth resistance of stainless steels containing austenite as the main phase, such as austenitic stainless steels [35]. This TRIP effect was ignored in the fracture mechanics research for stainless-steel welded joints. After recent developments in the field of stainless-steel construction, it is necessary to conduct a comparative study on the commonly used design standards.

In the fatigue design standards of steel structures, the IIW recommendation [15] and EC3 Parts 1–9 [14] include stainless steel. The IIW recommendation [15] includes austenitic stainless steel and EC3 Parts 1–9 [14] include all grades of stainless steel. The fatigue classifications of the six typical welded joints in EC3 Parts 1–9 [14] and the IIW recommendation [15] are shown in Table 1. There is a slight difference between these standards. The fatigue strength of the plates with longitudinal edge gussets in EC3 Parts 1–9 [14] is lower than that in the IIW recommendation [15].

Table 1. Fatigue strength of typical welded joints in the European Eurocode (EC)3 Parts 1–9 and the International Institute of Welding (IIW) recommendation.

Weld Structure	Additional Provisions	EC3 Codes [14]	IIW Codes [15]
Butt weld	Weld reinforcement less than 10% width	80	80
	Weld reinforcement less than 20% width	90	90
Cruciform joints failing in the weld throat	Failing in the weld throat	36	36
Plates with transverse fillet-welded attachments	/	80	80
Plates with longitudinal fillet-welded attachments	$L \leq 50$ mm	80	80
	50 mm $< L \leq 80$ mm	71	71
	80 mm $< L \leq 100$ mm	63	63
	$L > 100$ mm	56	50
Plates with longitudinal edge gussets	$L < 150$ mm	40	50
Fillet-welded joints with longitudinal round pipes	/	80	80
Base metal	Plates and flats with as-rolled edges	160	160
	Machine gas cut or sheared material with subsequent dressing	140	140
	Material with machine gas cut edges having shallow and regular drag lines or manual gas cut material, subsequently dressed to remove all edge discontinuities	125	125

Note: L stands for weld length.

Thus, it is an opportune time to review the fatigue test data of welded joints. In this paper, the fatigue test data of the welded joints of stainless steel were summarized, and the *S–N* curve and the fatigue strength were obtained using a statistical analysis method. This paper discusses the applicability of a free negative inverse slope regression *S–N* curve and a fixed negative inverse slope regression *S–N* curve using the scatter index, T_{σ}. The fatigue classification of stainless steel followed that of structural steels in Eurocode 3 [14] and the International Institute of Welding recommendations [15]; however, the difference in welding processes and so on between stainless steel and structural steel was not taken into account. By comparing the fatigue classification of Eurocode 3 and IIW recommendations with the fatigue test data of stainless-steel welded joints, the applicability of fatigue classification of structural steel for stainless steel is assessed. In this paper, a design proposal for the fatigue strength of stainless-steel welded joints is presented.

2. Database for the Evaluation

2.1. Data-Pooling of Experimental Results

Fatigue test data of typical welded joints of stainless steel were collected from Japan [17,18], India [21], Europe [11–13], and China [24]. Data including a total of 85 butt welds and 45 cruciform joints failing in the weld throat, 41 plates with transverse fillet-welded attachments, 38 plates with longitudinal fillet-welded attachments, 38 plates with longitudinal edge gussets, and 10 fillet-welded joints with longitudinal round pipes were obtained. The specimen thickness, welding procedure, material type, experimental information, and the fatigue crack initiation position are shown in Table 2.

Table 2. Fatigue test data for stainless steel.

Weld Joint Type	Data Source	Material Trademark	Ultimate Tensile Strength (MPa)	Steel Type	Thickness (mm)	Stress Ratio, R	Frequency, f (Hz)	Welding Procedure	Crack Initiation Position
Butt weld	[17]	SUS304	520	Austenitic	20	−1, 0, 0.5	1–10	–	Weld toe
	[18]	SUS304	520	Austenitic	12	0, 0.5	1–5	–	Weld toe
	[13]	S31803	640	Duplex	10	0.1, 0.5	10–20	GTAW	Weld toe
		304 L	520	Austenitic		0.1, 0.5		GMAW	Weld toe
	[21]	304 L	520	Austenitic	6	0	30	GMAW-GTAW	Weld toe
	[12]	316	620	Austenitic	–	0	–	–	–
		316 L	485		–	0.1	–	–	–
		304 L	485			0			
Load-carrying cruciform joint	[13]	S31803	640	Duplex	10	0.1, 0.5	–	GTAW	Weld root
		304 L	520	Austenitic	10	0.1, 0.5		GMAW	
	[24]	1Cr18Ni9Ti	550	Austenitic	7	0.1	–	SMAW	
Plates with transverse fillet-welded attachments	[13]	S31803	640	Duplex	10	0.1, 0.5	10–20	GTAW	Weld toe
		304 L	520	Austenitic		0.1, 0.5		GMAW	
	[11]	Grade1.4301	625.55	Austenitic	8	0.1–0.41	–	–	Weld toe
		Grade1.4462	640	Duplex					
		Grade1.4436	520	Austenitic	12				
Plates with longitudinal fillet-welded attachments	[13]	S31803	640	Duplex	10	0.1, 0.5	10–20	GTAW	Weld toe
		304 L	520	Austenitic		0.1, 0.5		GMAW	
	[24]	1Cr18Ni9Ti	550	Austenitic	7	0.1	–	SMAW	Weld toe
Plates with longitudinal edge gussets	[13]	S31803	640	Duplex	10	0.1, 0.5	10–20	GTAW	–
		304 L	520	Austenitic		0.1, 0.5		GMAW	
	[24]	SAF2205	620	Duplex	7	0.1	–	GMAW	Weld toe
Fillet-welded joints with longitudinal round pipes	[24]	1Cr18Ni9Ti	550	Austenitic	8	0.1	–	GMAW	Weld toe
Base metal	[18]	SUS304	520	Austenitic	12	0, 0.5	1–5	–	base metal
	[21]	304 L	520	Austenitic	6	0	30	MIG	Base metal

Note: Shielded metal arc welding (SMAW) in the welding process involves manual arc welding, gas metal arc welding (GMAW) involves inert gas shielded arc welding, and gas tungsten arc welding (GTAW) involves non-consumable electrode gas shielded arc welding.

It should be mentioned that the fatigue data from the literature should be verified using the log-normal distribution because of the heterogeneous nature of the database. Statistical examinations for proving the quality of the data being merged were applied to check if each dataset could be regarded as part of the same population. Checking that datasets belonged to the same statistical population was also applied to identify parameters that led to a differentiation in the fatigue category. The details can be found in Section 2.2.

It is well known that, with the increase of plate thickness, the fatigue life of welded joints will decrease [36]. Therefore, most fatigue rules include a correction factor, which can reduce the stress obtained from the design *S–N* curve when the thickness of the plate exceeds a certain reference value. The design curve can be directly applied. In the EC3 Part 9 [14] and IIW recommendations [15], the reference thickness is 25 mm. In Table 1, the thickness of specimens was 6 mm to 12 mm and was much smaller than the reference thickness. The plate thickness correction cannot be studied due to limited data. It is well known that the fatigue strength of welded joints is not related to the material property [36], i.e., the tensile strength. The tensile strength of the duplex stainless steel is much higher than that of the austenitic stainless steel. Maddox [36] reviewed the current research, and the results demonstrated that the influence of the material type on the fatigue strength for stainless-steel welded joints can be neglected. Following the above conclusion and considering the fact that some data did not include the material strength, the influence of material type was not studied in the present paper.

Stainless steel is a rate-sensitive material; however, a load frequency between 3 and 114 Hz has no effect on fatigue behavior when the temperature does not increase and there is no corrosion during the test period [37]. In this paper, the normal distribution of test data under different load frequencies was tested, and it was found that the test data of different load frequencies still satisfied the normal distribution. Therefore, the load frequency indirectly affects the fatigue behavior.

The stress ratio does not influence the fatigue strength of the welded joints. Branco et al. [13] found the same conclusion for stainless-steel welded joints. However, small-scale specimens were tested in their experiments. To the best of the authors' knowledge, no large-scale specimen (welded beam) data could be found in the published literature. These should be researched in the future to reach a solid conclusion.

The different welding procedures produce different weld profiles and imperfections. From a general view, the weld quality represents the weld profile and imperfections [38]. The weld quality has a significant influence on the fatigue strength of the welded joints [38]. Branco et al. [13] reported that there was no difference between the specimens made from gas metal arc welding (GMAW) and from gas tungsten arc welding (GTAW). The weld quality of these two welding procedures can be regarded as being the same.

The crack initiation position is related to the weld profile. This should be illustrated in the fatigue classification category. Except for the load-carrying cruciform joint, the fatigue crack initiation occurs at the weld toe in the other welded joints, as shown in Table 2.

These fatigue test data summarized in Table 1 can be used to determine the fatigue strength of the stainless-steel welded joints. The thickness effect can be studied by using thicker specimens and collecting more beam specimen data in the future.

2.2. Examination of the Log-Normal Distribution

The statistical samples used in regression *S–N* curves had systematic errors that were too large; therefore, the consequences would not be credible. As such, it was necessary to judge whether the systematic error of data could be neglected before regressing the *S–N* curve.

The fatigue life data, $y = \log N$, were prepared for n specimens for probability plotting by ranking the data from minimum to maximum values. Each data item was labeled with an order number, i, as $y_1 \leq y_2 \leq \ldots \leq y_n$. The probability of survival for the ith data item was approximated as follows:

$$p_i = 1 - \frac{i}{n+1}. \tag{1}$$

Statistical tests for verifying the quality of the data being merged were applied to check if each sample set could be seen as part of the same population. The logarithmic fatigue life obeys a normal distribution [39]. As shown in Figure 1, the logarithmic fatigue life was chosen as the abscissa and the survival probability was chosen as the ordinate. When the test data satisfy the normal distribution, the data are reasonably linear on a probability graph [39,40]. An "eyeball" assessment was used to determine whether the logarithmic fatigue life data in the normal probability graph followed a linear trend [40].

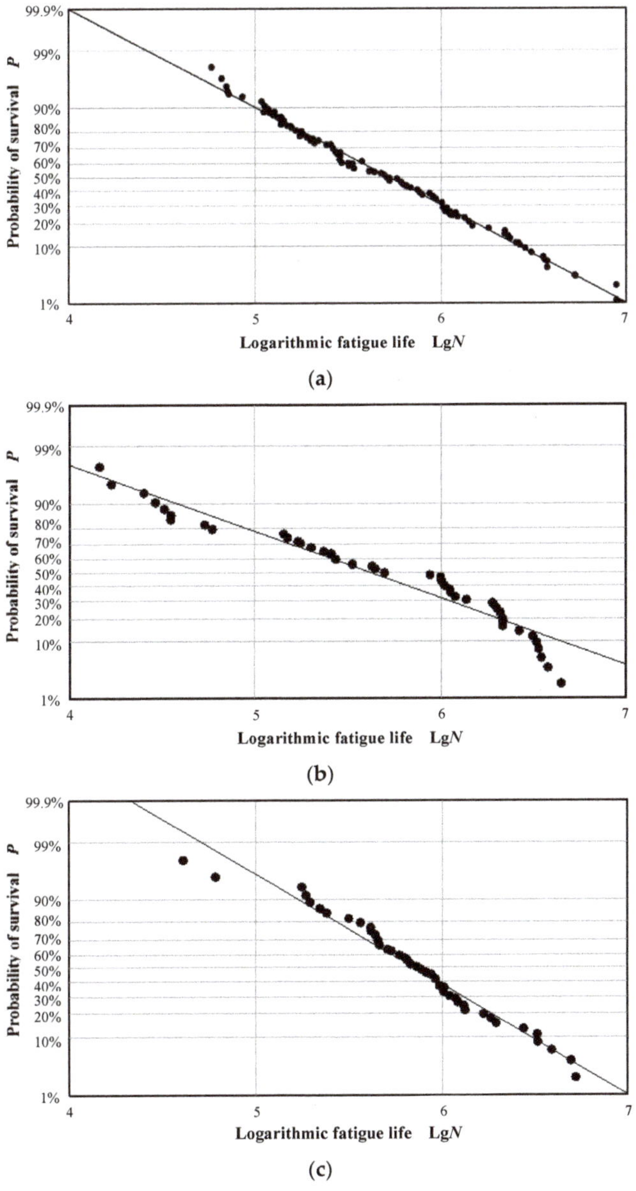

Figure 1. *Cont.*

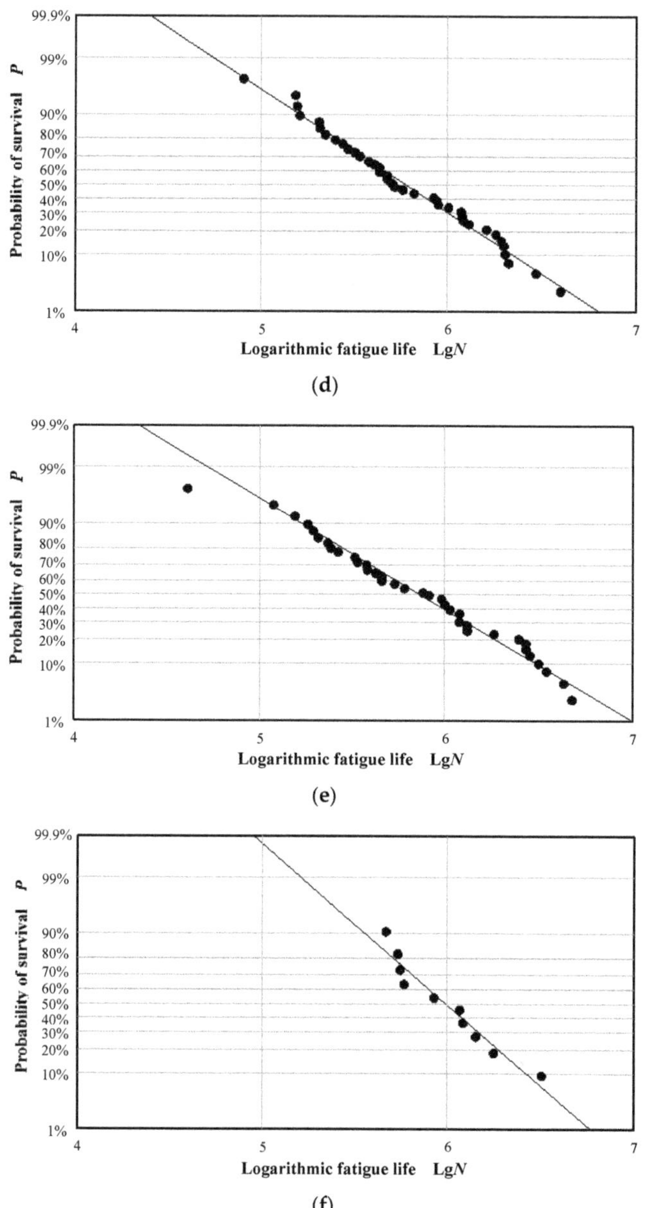

Figure 1. *Cont.*

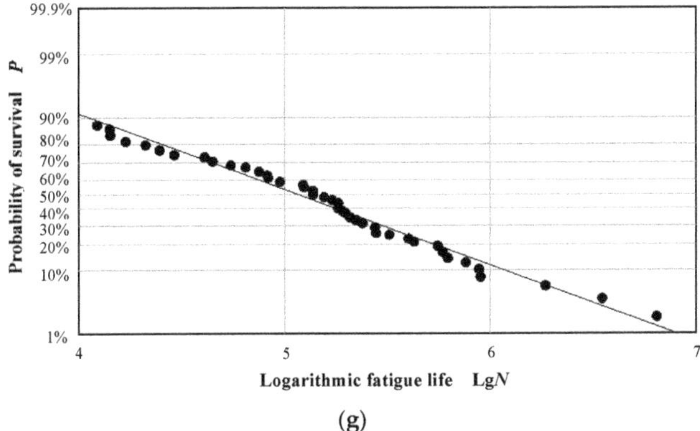

(g)

Figure 1. Verification of the normal distribution of fatigue test data: (**a**) butt weld; (**b**) cruciform joints failing in the weld throat; (**c**) plates with transverse fillet-welded attachments; (**d**) plates with longitudinal fillet-welded attachments; (**e**) plates with longitudinal edge gussets; (**f**) fillet-welded joints with longitudinal round pipes; (**g**) base metal.

The distribution of data points was close to linear, which indicates that the data conformed to a normal distribution. The collected fatigue test data could be regarded as samples from the same statistical matrix and were in accordance with the normal distribution. It was found that the test data of different steel types, thicknesses, stress ratios, load frequencies, and welding procedures still satisfied the normal distribution. Therefore, these factors indirectly affected the fatigue behavior of the test data. The test data collected could be utilized to regress the *S–N* curve.

3. Statistical Re-Analysis, *S–N* Curves, and Fatigue Strength

3.1. *S–N Curve*

Taking the stress range log $\Delta\sigma$ as an independent variable and the number of cycles log *N* as dependent, the mean fatigue *S–N* curve was obtained by estimating both the negative inverse slope *m* and the intercept log *C* as follows:

$$C = \Delta\sigma^m N \Rightarrow log\, N = log\, C \; - \; m\, log\, \Delta\sigma, \tag{2}$$

$$m = \frac{\sum\limits_{i=1}^{n} (y_i - \overline{y})^2}{\sum\limits_{i=1}^{n} (x_i - \overline{x})(y_i - \overline{y})}, \tag{3}$$

$$LogC = \overline{x} + \frac{\sum\limits_{i=1}^{n} (x_i - \overline{x})(y_i - \overline{y})}{\sum\limits_{i=1}^{n} (y_i - \overline{y})^2}\overline{y}, \tag{4}$$

where y represents the logarithmic fatigue life and x represents the logarithmic stress amplitude; $\overline{y}$ represents the mean of logarithmic fatigue life and $\overline{x}$ represents the mean of logarithmic stress amplitude.

$$x_k = x_m - k_{(p,1-\alpha,v)} \cdot \sigma. \tag{5}$$

As shown in Equation (5), the characteristic *S–N* curve was obtained by translating the mean *S–N* curve, and the translation distance was $k{\cdot}\sigma$. $y_{\rm m}$ represents the logarithmic fatigue life of the mean *S–N* curve, and $y_{\rm k}$ represents the logarithmic fatigue life of the characteristic *S–N* curve. σ represents the standard deviation estimator of the logarithmic fatigue life, while k is correlated with the probability of survival p_s, degree of confidence α, and degree of freedom γ. The characteristic value for the fatigue strength was defined with a survival probability of 95% at a one-sided confidence level of 75% at 2×10^6 cycles. The IIW recommendation refers to the logarithmic fatigue life of the average *S–N* curve minus two standard deviations, corresponding to a 97.7% survival probability [15]. The characteristic *S–N* curve obtained at $k = 2$ is lower than the *S–N* curve calculated by Equation (5) [41]. In this paper, the characteristic *S–N* curve was obtained by referring to the method of IIW recommendation.

Figure 2 shows the fatigue test data of six typical welded joints and the fitting characteristics of the *S–N* curve. In order to analyze the rationality of applying the fatigue classification of structural steel to stainless steel, Figure 2 also shows the fatigue classification curves of structural steel. The negative inverse slope of the characteristic *S–N* curve regressed freely was obtained from the test data using the least squares regression method; the negative inverse slope of the fixed slope regression *S–N* curves and fatigue classification curves adopted a value of 3.

The number of fatigue test data points of the six typical welded joints collected in this paper was more than 10, which satisfied the conditions and could be regressed freely. For ease of comparison, the regression *S–N* curves with a fixed slope $m = 3$ are also given in Figure 2.

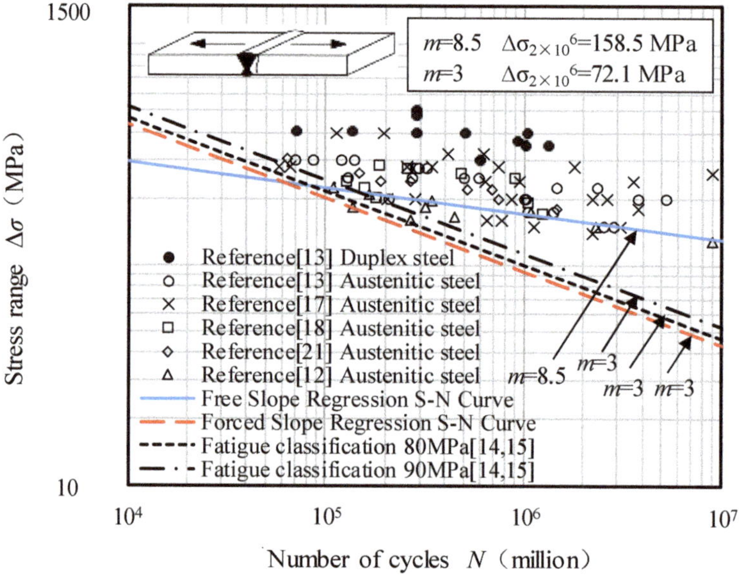

(a)

Figure 2. *Cont.*

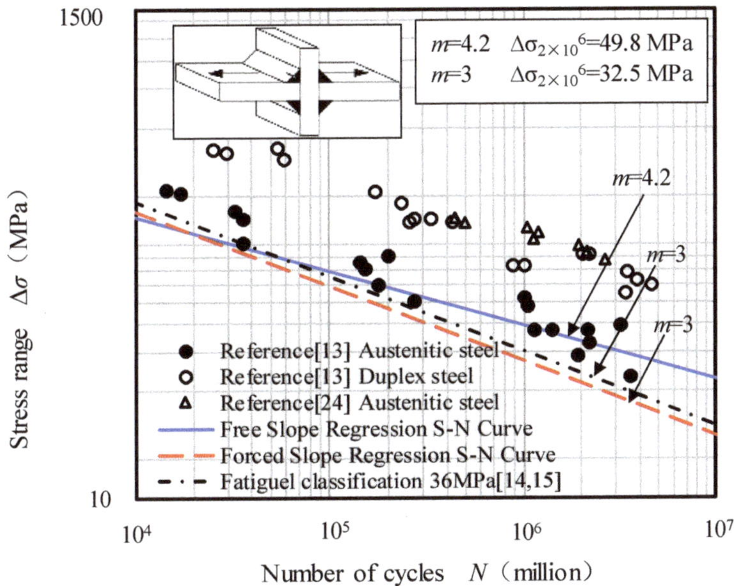

(b)

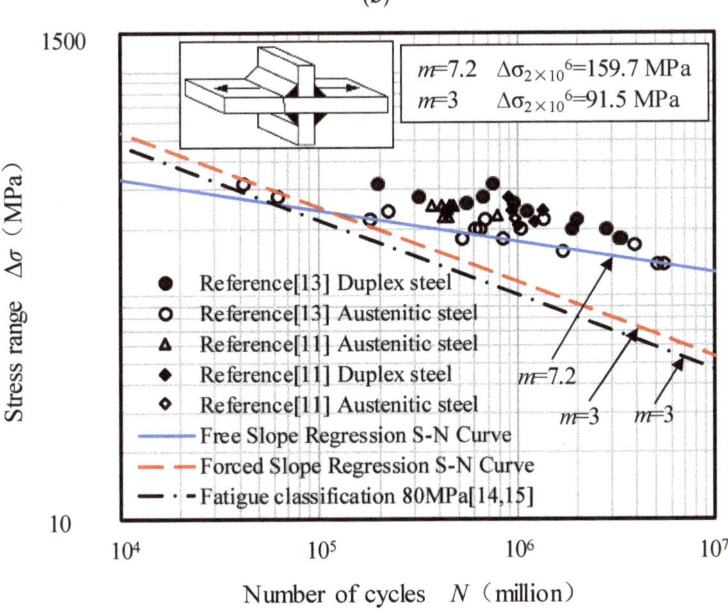

(c)

Figure 2. *Cont.*

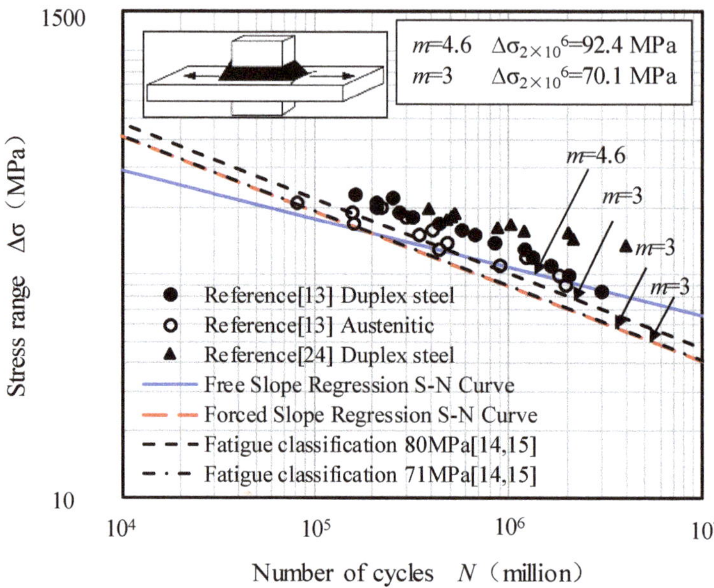

(d)

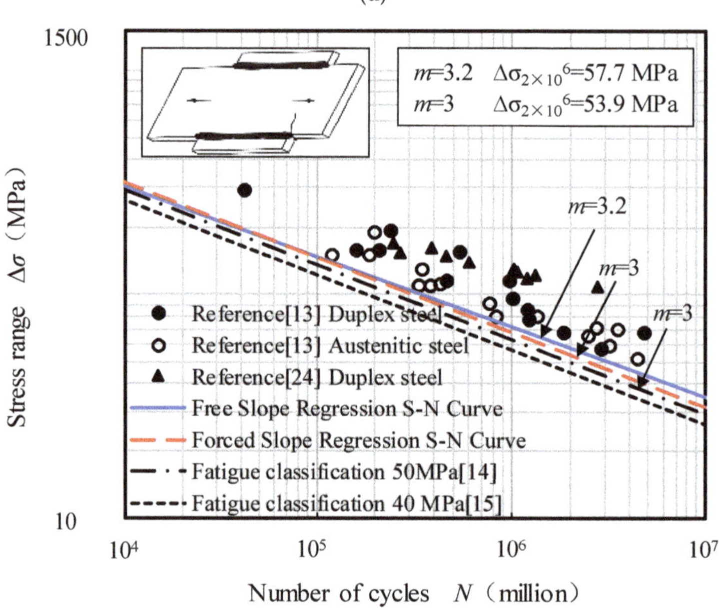

(e)

Figure 2. *Cont.*

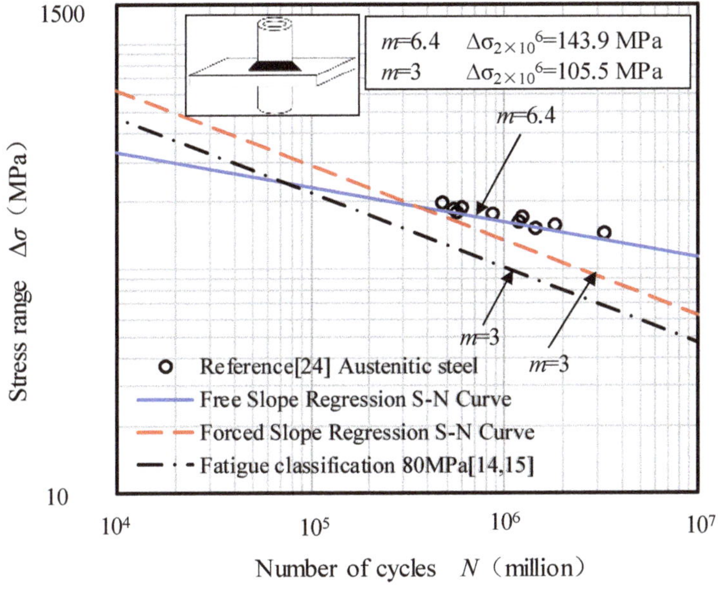

(f)

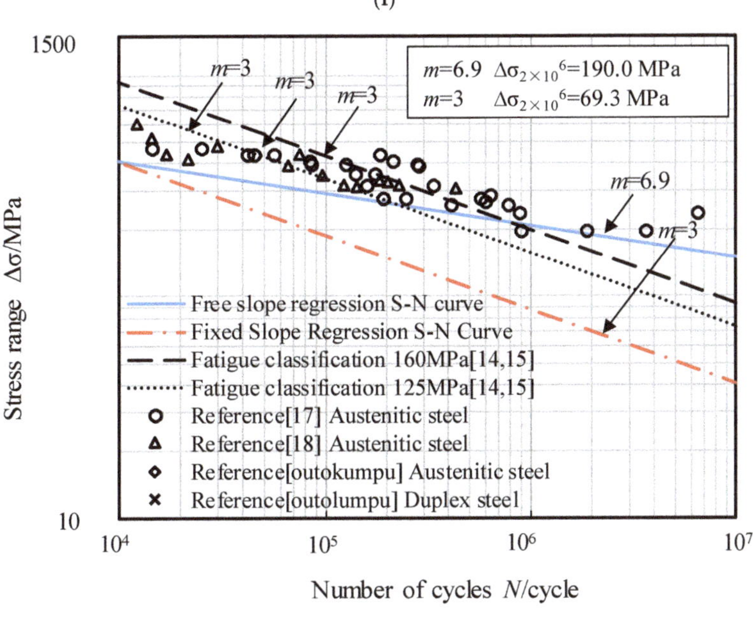

(g)

Figure 2. Characteristic *S–N* curves of stainless-steel welds and classification curves of structural-steel welded joints: (**a**) butt weld; (**b**) cruciform joints failing in the weld throat; (**c**) plates with transverse fillet-welded attachments; (**d**) plates with longitudinal fillet-welded attachments; (**e**) plates with longitudinal edge gussets; (**f**) fillet-welded joints with longitudinal round pipes; (**g**) base metal.

3.2. Scatter Band Analysis

There are inevitable random errors in the fatigue test, which are caused by random changes in the material properties, random errors in the specimen welding process, and random errors in the test loading devices and data testing equipment. The ratio T_{σ} is a parameter that can quantify the correlation degree between fatigue data and the dispersion of the fitting curve. The scatter index of the *S–N* curves is an important indicator of the quality of the assessment [40,42]. The scatter index $1/T_{\sigma}$ was derived from the following equation:

$$1/T_{\sigma} = \frac{\Delta\sigma_{PS=10\%}}{\Delta\sigma_{PS=90\%}} \tag{6}$$

As shown in Figure 3, $\Delta\sigma_{PS=10\%}$ and $\Delta\sigma_{PS=90\%}$ are the stress ranges corresponding to two million cycles of fatigue lives of the *S–N* curves with a probability of survival of 10% and 90%, respectively, and a confidence level of 75% [40]. $\Delta\sigma_{PS=10\%}$ and $\Delta\sigma_{PS=90\%}$ were calculated using Equation (7).

$$\Delta\sigma_{p_s} = \Delta\sigma_{p_s=50\%}\left[\frac{2\cdot 10^6}{10^{\log(2\cdot 10^6)\pm k\cdot\sigma}}\right]^{-m}. \tag{7}$$

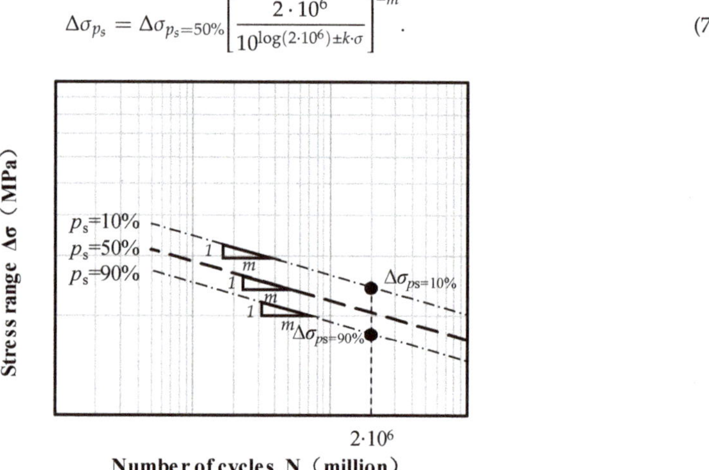

Figure 3. Scatter indexes.

Table 3 shows the scatter indexes of the fatigue test data. The scatter index of the free slope regression was smaller than that of the fixed slope regression. The average scatter index was 1.3 and the standard deviation was 0.5 with free slope regression, while the average scatter index was 1.9 and the standard deviation was 0.7 with fixed slope regression.

Table 3. Scatter indexes of the characteristic *S–N* curve of a typical weld structure.

Weld Structure	Free Slope Regression	Fixed Slope Regression
Butt weld	1.4	2.6
Cruciform joints failing in the weld throat	2.1	3.0
Plates with transverse fillet-welded attachments	1.3	2.0
Plates with longitudinal fillet-welded attachments	1.4	1.6
Plates with longitudinal edge gussets	1.5	1.6
Fillet-welded joints with longitudinal round pipes	1.1	1.4
Base metal	1.3	2.4
Average value	1.3	1.9
Standard deviation	0.5	0.7

When Spindel and Haibach [43] analyzed a large number of fatigue test data points of welded joints of structural steel, the scatter index $1/T_\sigma$ of the free slope regression *S–N* curve was 1.5, which is commonly used as an international standard [44,45].

The above analysis shows that, for typical welded joints of stainless steel, the scatter index obtained by free slope regression was smaller than that obtained by fixed slope regression, and was smaller than the international standard. The free slope regression *S–N* curve can be used as the basis for fatigue design.

3.3. Suggestions on Fatigue Strength

Fatigue test data of duplex stainless steel and austenitic stainless steel were included in all welded joints except for the longitudinal fillet-weld structure. Regression of an *S–N* curve with a fixed slope of 3 was used to compare the fatigue properties of austenitic stainless steel and duplex stainless steel within the matching welded joints. It was noted that the fatigue strength of duplex welded joints of stainless steel was not significantly different from that of austenitic welded joints of stainless steel.

Fatigue classification curves are given in Figure 2 for comparison with the experimental data. As can be seen from Figure 2, for the butt welds, plates with transverse fillet-welded attachments, and fillet-welded joints with longitudinal round pipes, the fatigue test data points were significantly higher than that of the fixed regression *S–N* curve. Fatigue test data of the cruciform joints failing in the weld throat, plates with longitudinal fillet-welded attachments, and a rectangular joint plate were close to the fixed regression *S–N* curve. The fatigue strength obtained by free slope regression was higher than the fatigue classification of structural-steel welded joints. The fatigue strength of stainless-steel welded joints is greater than that of structural-steel welded joints because of the different residual stress. Yuan [26] verified through experiments that the welding residual stress of a welded beam of stainless steel was less than that of structural steel. The linear elastic stage of stainless steel is shorter than that of structural steel, and the tangential modulus decreases rapidly. In materials with non-linear stress–strain curves, the strain required to achieve an equivalent yield stress is higher than that required to reach the unique yield point of traditional structural steel. The residual stress of stainless steel is lower than that of structural steel due to the different constitutive relationship [26]. The heat input of a stainless-steel weld is lower than that of a structural-steel weld due to the higher electrical resistance of stainless steel [26]. This is another reason for the low residual stress of the stainless-steel weld. As demonstrated in the literature [13], the butt weld, fillet-welded joints with longitudinal round pipes, and plates with transverse fillet-welded attachments had small residual stresses. According to the IIW recommendation, the fatigue strength at 2×10^6 cycles should be reduced by 20% to demonstrate the influence of high residual stress. Table 4 shows the reduced fatigue strength. The values are higher than the fatigue strength recommended in the EC 3 Part 1–9 and IIW recommendations. In summary, lower residual stress leads to greater fatigue strength of the stainless-steel welded joints than that of structural steel-welded joints.

Table 4. The fatigue strength of welded joints of stainless steel suggested in this paper.

Weld Structure	Butt Weld	Cruciform Joints Failing in the Weld Throat	Plates with Transverse Fillet-Welded Attachments	Plates with Longitudinal Fillet-Welded Attachments	Plates with Longitudinal Edge Gussets	Fillet-Welded Joints with Longitudinal Round Pipes	Base Metal
The negative inverse slope of *S–N* curve	8.5	4.2	7.2	4.6	3.2	6.4	6.9
Fatigue strength at 2×10^6 cycles (MPa)	155	45	155	90	55	140	190
Lowering Fatigue strength at 2×10^6 cycles by 20% (MPa)	127	-	128	-	-	115	-

In the literature [22], the fatigue life composition of stainless-steel butt welds and plates with transverse fillet-welded attachments was analyzed using the fracture mechanics method, where it was found that the fatigue crack initiation life could not be neglected. Long fatigue crack initiation life will cause the *S–N* curve slope *m* to be much larger than 3 [46,47]. The negative inverse slopes *m* of the *S–N* curve were 8.5, 7.2, and 6.4 for butt welds, plates with transverse fillet-welded attachments, and fillet-welded joints with longitudinal round pipes, respectively, which were much larger than 3. The negative inverse slopes *m* of the *S–N* curve were 4.2, 4.6, 3.2, and 6.9 for cruciform joints failing in the weld throat, plates with longitudinal fillet-welded attachments, plates with longitudinal edge gussets, and base metal, respectively, which were close to 3. The fatigue strengths of the butt welds, plates with transverse fillet-welded attachments, fillet-welded joints with longitudinal round pipes, and base metal were 158.5 MPa, 159.7 MPa, 143.9 MPa, and 190 MPa, respectively. When *S–N* curves were regressed freely, the fatigue limit of base metal at 10^7 cycles obtained from the test data was 150 MPa, and it was within the range of the fatigue limit in the reference [48]. The fatigue strengths of cruciform joints failing in the weld throat, plates with longitudinal fillet-welded attachments, and plates with longitudinal edge gussets were 49.8 MPa, 92.4 MPa, and 57.7 MPa, respectively, when *S–N* curves were regressed freely.

Based on the above analysis, the fatigue strengths of the six typical welded joints of stainless steel are suggested in Table 4.

4. Conclusions

The main conclusions of this study are as follows:

(1) For the same representative welded joints, the fatigue strength of welded joints of stainless steel was greater than that of the structural-steel welded joints, especially for butt welds, plates with transverse fillet-welded attachments, and fillet-welded joints with longitudinal round pipes.

(2) The scatter index $1/T_\sigma$ for the free slope *S–N* curves was 1.3, and the scatter index $1/T_\sigma$ for fixed slope *S–N* curves, $m = 3$, was 1.9. This demonstrates that the scatter band of the free slope S–N curves is much smaller than that of the fixed slope *S–N* curves. The free slope *S–N* curves are more suitable to represent the fatigue performance of the stainless-steel welded joints.

(3) The accuracy of analysis can be increased by adopting the slope of the *S–N* curve according to welding details. Data analysis shows that the slope of the *S–N* curve is between 3.2 and 8.6. With the decrease in the stress concentration, the deviation between the negative inverse slope *m* and 3 is greater. This is why the IIW recommendation and EC 3 design *S–N* curves were used to over-conservatively evaluate the test results.

5. Expectation

The experimental data collected in this paper were all from laboratory specimens, which were different from the actual sizes of the components. It is necessary to supplement these data by verifying full-scale tests under a tangible engineering environment.

Author Contributions: Y.P., J.C., and J.D. prepared the figures and drafted the manuscript. All authors critically revised and approved the final manuscript version.

Funding: This research was funded by the Project of National Natural Science Foundation of China grant number [51408307] and the Project of Nanjing Gongdajianshe Company grant number [2019RD06].

Conflicts of Interest: The authors declare no conflict of interest.

References

1. Wang, Y.; Yuan, H.; Shi, Y.; Gao, B.; Dai, G.X. Application and research status of stainless steel structure. *Steel Struct.* **2010**, *2*, 1–12. (In Chinese)
2. Baddoo, N.R.; Kosmač, A. Sustainable Duplex Stainless Steel Bridges [DB/OL]. 2012. Available online: http://www.worldstainless.org/Files/issf/non-image-files/PDF/Sustainable_Duplex_Stainless_Steel_Bridges.pdf (accessed on 13 March 2018).

3. Beletski, A. Applicability of Stainless Steel in Road Infrastructure Bridges by Applying Life Cycle Costing. Master's Thesis, Helsinki University of Technology, Espoo, Finland, 1 May 2007.
4. Morgenthal, G.; Sham, R.; West, B. Engineering the tower and main span construction of stonecutters bridge. *J. Bridge Eng.* **2010**, *15*, 144–152. [CrossRef]
5. Baddoo, N.; Francis, P. Development of design rules in the AISC design guide for structural stainless steel. *Thin Walled Struct.* **2014**, *83*, 200–208. [CrossRef]
6. Tsakopoulos, P.A.; Fisher, J.W. Full-scale fatigue tests of steel orthotropic decks for the Williamsburg Bridge. *J. Bridge Eng.* **2003**, *8*, 323–333. [CrossRef]
7. Ya, S.; Yamada, K.; Ishikawa, T. Fatigue evaluation of rib-todeck welded joints of orthotropic steel bridge deck. *J. Bridge Eng.* **2010**, *16*, 492–499. [CrossRef]
8. Aygul, M.; Al-Emrani, M.; Urushadze, S. Modelling and fatigue life assessment of orthotropic bridge deck details using FEM. *Int. J. Fatigue* **2012**, *40*, 129–142. [CrossRef]
9. Sim, H.B.; Uang, C.M.; Sikorsky, C. Effects of fabrication procedures on fatigue resistance of welded joints in steel orthotropic decks. *J. Bridge Eng.* **2009**, *14*, 366–373. [CrossRef]
10. Lahti, K.E.; Hänninen, H.; Niemi, E. Nominal stress range fatigue of stainless steel fillet welds—The effect of weld size. *J. Constr. Steel Res.* **2000**, *54*, 161–172. [CrossRef]
11. Burgan, B.; Baddoo, N.; Gardner, L.; Way, L.; Johansson, B.; Olssen, A.; Sélen, E.; Viherma, R.; Kouhi, J.; Talja, A.; et al. *Development of the Use of Stainless Steel in Construction*; European Commission: Brussels, Belgium, 1999.
12. Razmjoo, G.R. Design Guidance on Fatigue of Welded Stainless Steel Joints. In Proceedings of the International Conference on Offshore Mechanics and Arctic Engineering, Copenhagen, Denmark, 18–22 June 1995.
13. Branco, C.M.; Maddox, S.J.; Sonsino, C.M. *Fatigue Design of Welded Stainless Steels*; European Commission: Brussels, Belgium, 1998.
14. Hobbacher, A.F. *Recommendations for Fatigue Design of Welded Joints and Components*, 2nd ed.; IIW document IIW-2259-I5 ex XIII-2460-13/XV-1440-13; International Institute of Welding: Cambridge, UK, 2016.
15. *Eurocode 3, Design of Steel Structures—Part 1–9: Fatigue*; EN 1993-1-9; European Committee for Standardization: London, UK, 2005.
16. Metrovich, B.; Fisher, J.W.; Yen, B.T.; Kaufmann, E.J.; Cheng, X.; Ma, Z. Fatigue strength of welded AL-6XN superaustenitic stainless steel. *Int. J. Fatigue* **2003**, *25*, 1309–1315. [CrossRef]
17. NIMS. Data Sheets on Fatigue Crack Propagation Properties for Butt Welded Joints SUS304-HP(18Cr-9Ni) Hot Rolled Stainless Steel Plate. 1986. Available online: http://smds.nims.go.jp/openTest/en/content/fatiguelist.html (accessed on 3 September 2018).
18. Nakamura, Y.; Chihiro, I.; Mimura, H. Research on for Fatigue Strength of Austenitic Stainless Steel (Effects of Stress Ratio and Corrositive Environment). *Kou Kouzou Rombunshuu* **2009**, *16*, 55–64. (In Japanese)
19. Nakamura, Y.; Chihiro, I.; Hiroaki, M. Study on fatigue strength of austenitic stainless steel. *J. Struct. Constr. Eng. (Trans. AIJ)* **2010**, *16*, 55–64. (In Japanese)
20. Japanese Society of Steel Construction. *Fatigue Design Recommendation for Steel Structures*; Japanese Society of Steel Construction: Tokyo, Japan, 1995.
21. Singh, P.J.; Guha, B.; Achar, D.G.R. Fatigue life prediction for AISI 304L butt welded joints having different bead geometry using local stress approach. *Sci. Technol. Weld. Join.* **2003**, *8*, 303–308. [CrossRef]
22. Singh, P.J.; Achar, D.G.R.; Guha, B. Fatigue life prediction of gas tungsten arc welded AISI 304L cruciform joints different LOP sizes. *Int. J. Fatigue* **2003**, *25*, 1–7. [CrossRef]
23. BS 5400-10:1980. *Steel Concrete and Composite Bridges, Code of Practice for Fatigue*; British Standards Institution: London, UK, 1980.
24. Wu, B.; Yang, X.; Jia, F.; Huo, L. Experimental study on fatigue strength of welded joint of stainless steel. *Mech. Strength* **2004**, *26*, 321–325. (In Chinese)
25. Yu, W.W.; LaBoube, R.A. *Cold-Formed Steel Design*, 4th ed.; John Wiley & Sons, Inc.: Hoboken, NJ, USA, 2010.
26. Yuan, H.X.; Wang, Y.Q.; Shi, Y.J.; Gardner, L. Residual stress distributions in welded stainless steel sections. *Thin Walled Struct.* **2014**, *79*, 38–51. [CrossRef]
27. Partanen, T.; Niemi, E. Hot spot stress approach to fatigue strength analysis of welded components: Fatigue test data for steel plate thicknesses up to 10 mm. *Fatigue Fract. Eng. Mater. Struct.* **1996**, *19*, 709–722. [CrossRef]

28. Jia, F.-Y.; Huo, L.-X.; Wu, B.; Yang, X.-Q.; Jin, X.-J. Evaluation on fatigue strength of duplex stainless steel welded joints by hot spot stress. *J. Tianjin Univ.* **2004**, *37*, 1014–1017. (In Chinese)
29. Feng, R.; Young, B. Stress Concentration Factors of Cold-formed Stainless Steel Tubular X-joints. *J. Constr. Steel Res.* **2013**, *91*, 26–41. [CrossRef]
30. Livieri, P.; Lazzarin, P. Fatigue strength of steel and aluminium welded joints based on generalised stress intensity factors and local strain energy values. *Int. J. Fract.* **2005**, *133*, 247–276. [CrossRef]
31. Meneghetti, G.; Lazzarin, P. The peak stress method for fatigue strength assessment of welded joints with weld toe or weld root failures. *Weld. World* **2011**, *55*, 22–29. [CrossRef]
32. Infante, V.; Branco, C.M.; Martins, R. a fracture mechanics analysis on the fatigue behaviour of sruciform joints of duplex stainless steel. *Fatigue Fract. Eng. Mater. Struct.* **2003**, *26*, 791–810. [CrossRef]
33. Singh, P.J.; Guha, B.; Achar, D.R.G. Fatigue tests and estimation of crack initiation and propagation lives in AISI 304L butt-welds with reinforcement intact. *Eng. Fail. Anal.* **2003**, *10*, 383–393. [CrossRef]
34. Singh, P.J.; Guha, B.; Achar, D.R.G.; Nordberg, H. Fatigue life prediction improvement of AISI 304L cruciform welded joints by cryogenic treatment. *Eng. Fail. Anal.* **2003**, *10*, 1–12. [CrossRef]
35. Trudel, A.; Lévesque, M.; Brochu, M. Microstructural effects on the fatigue crack growth resistance of a stainless steel CA6NM weld. *Eng. Fract. Mech.* **2014**, *115*, 60–72. [CrossRef]
36. Maddox, S.J. Fatigue design rules for welded structures. *Prog. Struct. Eng. Mater.* **2015**, *2*, 102–109. [CrossRef]
37. Gurney, T.R.; Saunders, H. *Fatigue of Welded Structures*, 2nd ed.; Cambridge University Press: Cambridge, UK, 1979.
38. Maddox, S.J. Status review on fatigue performance of fillet welds. *J. Offsh. Mech. Arct. Eng.* **2008**, *130*, 1469–1476. [CrossRef]
39. ISO 12107-2003. *Metallic Materials—Fatigue Testing—Statistical Planning and Analysis of Data*; ISO: Geneva, Switzerland, 2003.
40. Schneider, C.R.A.; Maddox, S.J. *Best Practice Guide on Statistical Analysis of Fatigue Data*; Comission: XIII-WG1-114-03; Internatinal Institute of Welding: Cambridge, UK, 2003.
41. Baptista, C.; Reis, A.; Nussbaumer, A. Probabilistic S-N curves for constant and variable amplitude. *Int. J. Fatigue* **2017**, *101*, 312–327. [CrossRef]
42. Zhang, Y. Research on Mechanical Properties and Design Method of Stainless Steel Weld Joints. Ph.D. Thesis, Beijing University of Technology, Beijing, China, 20 June 2016. (In Chinese)
43. Spindel, J.E.; Haibach, E. The method of maximum likelihood applied to the statistical analysis of fatigue data including run-outs. *Int. J. Fatigue* **1979**, *1*, 81–88. [CrossRef]
44. Knight, J.W. *Some Basic Fatigue Data for Various Types of Fillet Welded Joints in Structured Steel*; Welding Institute Members Report 9/1976/E; The Welding Institute: Cambridge, UK, 1976.
45. Maddox, S.J. Assessing the significance of flaws in welds subjects to fatigue. *Weld. J. Weld. Res. Suppl.* **1974**, *53*, 401–409.
46. Atzori, B.; Lazzarin, P.; Meneghetti, G.; Ricotta, M. Fatigue design of complex welded structures. *Int. J. Fatigue* **2009**, *31*, 59–69. [CrossRef]
47. Peng, Y. Fatigue Behavior of Butt Welded Joints between Cast Steel and Rolled Steel Based on Experimental Study and Fracture Mechanics Analysis. Ph.D. Thesis, Tongji University, Shanghai, China, 20 December 2012. (In Chinese)
48. De Finis, R.; Palumbo, D.; Ancona, F.; Galietti, U. Fatigue limit evaluation of various martensitic stainless steels with new robust thermographic data analysis. *Int. J. Fatigue* **2015**, *74*, 88–96. [CrossRef]

© 2019 by the authors. Licensee MDPI, Basel, Switzerland. This article is an open access article distributed under the terms and conditions of the Creative Commons Attribution (CC BY) license (http://creativecommons.org/licenses/by/4.0/).

Article

New Ferritic Stainless Steel for Service Temperatures up to 1050 °C Utilizing Intermetallic Phase Transformation

Timo Juuti [1,*], Timo Manninen [2], Sampo Uusikallio [1], Jukka Kömi [1] and David Porter [1]

[1] Materials and Mechanical Engineering, Faculty of Technology, University of Oulu, P.O. Box 4200, 90014 Oulu, Finland; sampo.uusikallio@oulu.fi (S.U.); jukka.komi@oulu.fi (J.K.); David.porter@oulu.fi (D.P.)
[2] Outokumpu, Process R&D, 95490 Tornio, Finland; timo.manninen@outokumpu.com
* Correspondence: timo.juuti@oulu.fi; Tel.: +358-50-337-3377

Received: 25 April 2019; Accepted: 5 June 2019; Published: 7 June 2019

Abstract: A large number of thermodynamic simulations has been used to design a new Nb-Ti dual stabilized ferritic stainless steel with excellent creep resistance at 1050 °C through an optimal volume fraction of Laves (η) phase stabilized by the alloying elements Nb, Si and Mo. By raising the dissolution temperature of the phase, which also corresponds to the onset of rapid grain growth, the steel will better maintain the mechanical properties at higher service temperature. Laves phase precipitates can also improve creep resistance through precipitation strengthening and grain boundary pinning depending on the dominant creep mechanism. Sag tests at high temperatures for the designed steel showed significantly better results compared to other ferritic stainless steels typically used in high temperature applications at present.

Keywords: simulations; phase diagrams; iron alloys; intermetallics; phase transformation

---

1. Introduction

Stainless steel is used for many different automobile components due to its excellent corrosion and heat resistance, good appearance and mechanical properties. The use of stainless steel in automobile manufacture began with decorative trims, but has since spread into more functional components, such as exhaust systems, where stainless steel accounts for more than half of all the stainless steel used in automotive applications [1–3]. The increased use of stainless steel in exhaust systems has been result of tightening exhaust gas regulations and producer warranties together with the desire to decrease fuel consumption and vehicle weight. Those parts of the exhaust system closest to the engine are exposed to the highest service temperatures and are therefore the most demanding as regards properties. For the hot end of the exhaust system, high-Cr ferritic stainless steels have been introduced instead of more commonly used AISI 409 or 304 steels. However, these steels lack the properties needed to be used above 850 °C, i.e. nearest to the engine, where the material is required to have excellent high-temperature strength, together with resistance to thermal fatigue, oxidation and corrosion. Such a component is for example the exhaust manifold, which is directly attached to the engine and susceptible to temperatures up to 1050 °C [1–7]. The austenitic stainless steel AISI 304 has excellent high-temperature strength but weak cyclic oxidation resistance, while the ferritic stainless steel AISI 409 has inferior high-temperature strength. To replace these, a Nb-Ti-stabilized ferritic stainless steel with 18% chromium, i.e. AISI 441, was introduced and is currently used because of its better high temperature properties compared to AISI 409. AISI 441 can be used at temperatures up to 950 °C, but still does not meet the full requirements of modern exhaust systems [5]. In the current study, the authors show how a new type of ferritic stainless steel with 21% chromium and a relatively high Nb content has been designed with the aim of meeting these increasing requirements by being usable even at 1050 °C with the aid of precipitation. Potential precipitate phases stable above 700 °C are Laves (η) and Chi (χ), which are typically encountered in high-temperature applications [8]. Provided the precipitates are small enough, these phases can increase both the ambient and high-temperature strength of the steel via precipitation strengthening [9]. In addition, such precipitates may increase creep resistance as is common in martensitic pipeline steels [10–12]. However, in ferritic stainless steels the precipitation of intermetallic phases to improve creep resistance is not widely exploited [13–16].

In this research, the authors design a new dual stabilized ferritic stainless steel with enhanced creep resistance compared to conventional High-Cr ferritic stainless steels used in high temperature applications. The improvement of creep resistance is achieved by utilizing the precipitation of η-phase and by controlling the amount, nucleation site and the solvus temperature of the phase. η-phase is an intermetallic AB_2 type phase such as Fe_2Nb and Fe_2Ti, but sometimes including other alloying elements, for e.g. Si, Cr or Mo. η-phase has a hexagonal crystal structure with c/a ratio of approximately 1.633 with unit cells in the range of a = 0.473–0.495 nm and c = 0.770–0.815 nm [17,18]. The distribution of the η-phase in the microstructure can influence the high temperature mechanical properties of the steel. The creep resistance can be improved whether the η-phase is on the grain boundaries or within the grains depending on the creep mechanism and η-phase amount and particle size. The nucleation of η-phase in α-Fe based alloys has been observed to first occur at the grain boundaries, then on dislocations and finally in the α-Fe matrix [19]. The η-phase amount and solvus temperature limits the service temperature of the steel but can be controlled by carefully balancing the alloying elements included in η-phase, such as Nb, assuming there is no nucleation of interfering Nb containing precipitates [16]. In addition, the nucleation site of the η-phase can be controlled by applying precise annealing parameters such as the peak temperature and holding time.

2. Methodology

2.1. Thermo-Calc and TC-Prisma

To help in the selection of an optimal composition for a trial steel a comprehensive set of thermodynamic simulations has been carried out using TC-Prisma software together with the Thermo-Calc thermodynamic database for Fe-based alloys TCFE9 and the atomic mobility database for Fe-based alloys MOBFE2 [20]. TC-Prisma simulates the diffusion controlled nucleation, growth and dissolution of precipitates using the Langer-Schwartz theory and the Kampmann-Wagner numerical approach in conjunction with Thermo-Calc and its diffusion controlled phase transformation module (DICTRA). In TC-Prisma the classic nucleation theory has been extended to model nucleation in multicomponent alloy systems [21]. The classic nucleation theory describes the time dependent nucleation rate defined as:

$$J(t) = J_s \exp\left(-\frac{\tau}{t}\right), \tag{1}$$

where J_s is the steady state nucleation rate, τ is the incubation time for establishing steady state nucleation conditions, and t is the isothermal reaction time. The steady state nucleation rate is given by:

$$J_s = Z\beta^* N_0 \exp\left(-\frac{\Delta G^*}{kT}\right), \tag{2}$$

where Z is the Zeldovich factor, a measure of the probability with which an embryo with a radius slightly larger than the critical radius passes back across the free energy barrier thereby dissolving in the matrix, β^* is the rate at which atoms attach to the critical nucleus, N_0 is the number of nucleation sites per unit volume, ΔG^* is the Gibbs energy for the formation of a critical nucleus, k is Boltzmann's constant and T is the absolute temperature. The number of nucleation sites depends on the shape and size of the grains. Assuming all grains are of equal size tetrakaidekahedral, the density for the grain boundary area per unit volume is calculated as:

$$\rho = \frac{6\sqrt{1+2A^2}+1+2A}{4A}D^{-1}, \tag{3}$$

where A is the aspect ratio (1 by default) and D is the distance between one pair of hexagonal faces. Assuming boundary thickness as one atomic layer, the number of nucleation sites is:

$$N = \rho\left(\frac{N_A}{V_m^\alpha}\right)^{2/3}, \tag{4}$$

where ${V_m}^\alpha$ is the molar volume of the matrix phase and N_A is the Avogadro number. The Gibbs energy required for the formation of critical nucleus is expressed as:

$$\Delta G^* = \frac{16\pi\sigma^3}{3\left(\Delta G_m^{\alpha\rightarrow\beta} / V_m^\beta\right)^2}, \tag{5}$$

where σ is the interfacial energy, ${\Delta G_m}^{\alpha\rightarrow\beta}$ is the molar driving force for the formation of a β precipitate from the α matrix and ${V_m}^\beta$ is the molar volume of the β precipitate phase. Interfacial energy σ is calculated with an extended Becker's model using thermodynamic data from the existing CALPHAD thermodynamic database and can be described as:

$$\sigma = \frac{n_s z_s}{n_A z_l}\Delta E_s, \tag{6}$$

where n_s is the number of atoms per unit area at the interface, z_s is the number of bonds per atom crossing the interface, z_l is the coordination number of an atom within the bulk crystal lattice, and ΔE_s is the energy of solution in a multicomponent system involving the two phases being considered [21].

2.2. Sag Tests

Bars with dimensions of 150 × 15 × 2 mm were cut for a simple high-temperature sag tests as illustrated in Figure 1. This is a simple test to evaluate high-temperature creep resistance. The bars were placed horizontally in a furnace with both ends supported 20 mm above the bottom of the furnace. The distance between the supports was L = 140 mm. After holding the sample in the furnace at temperatures in the range 800 to 1050 °C for 20 or 100 h, the deflection of the bars is measured. The creep deformation is caused by the sample's own weight and is heavily reliant on the bar dimensions. Hence, the dimensions, furnace and other conditions are kept constant throughout the test for all samples. The active creep mechanism in the sag test can be estimated using the deformation map in which the different creep mechanisms are located in normalized stress - temperature space.

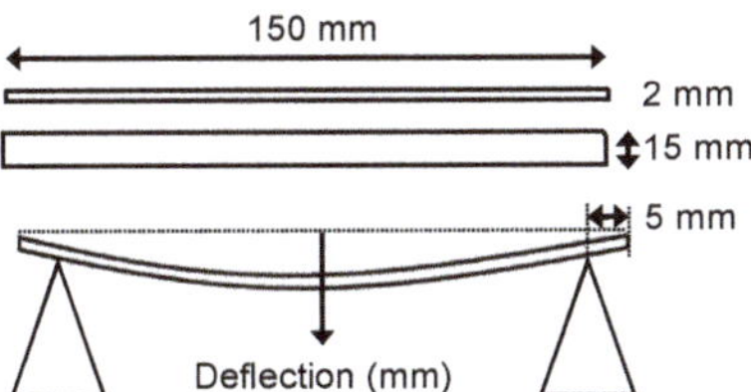

Figure 1. Schematic illustration of the Sag test and sample dimensions.

The creep mechanisms considered are power-law or dislocation creep, and the diffusional creep mechanisms known as Coble creep and Nabarro-Herring creep [22,23]. The stress state in the sag test can be estimated using classical beam theory. The specimen is then modelled as a simply supported beam. The problem is statically determinate, and therefore the distribution of bending moment can be determined directly based on equilibrium equations. In the beginning of the test, the cumulated inelastic strain is negligible, and the distribution of axial normal stress may be assumed linear over the thickness of the specimen. The maximum shear stress exists in the center point between the supports, on the upper and lower surfaces, and is given by:

$$\tau_{\max} = \left(\frac{3}{8}\right)\frac{\rho g L^2}{t}, \tag{7}$$

where ρ is the density of the material, g acceleration of gravity, L distance between the supporting points and t the thickness of the specimen. The origin of the coordinate system is positioned in the middle plane of the specimen on the leftmost supporting point. The maximum shear stress has the value of τ_{max} = 0.28 MPa. The shear modulus as a function of temperature can be approximated as:

$$\mu(T) = \mu_o - \frac{d\mu}{dT}\Delta T, \tag{8}$$

where $d\mu/dT$ is −37.3 [MPa/K] and the shear modulus at 300 K (μ_0) is 78 GPa [24,25]. Therefore, the shear stress normalized with respect to the shear modulus is 7.0 × 10^{-6} at 1050 °C and the homologous temperature, T/T_m, where T_m is the absolute melting temperature, is 0.7.

3. Results

3.1. Thermodynamic Simulations

TC-Prisma and Thermo-Calc was used to simulate solvus temperature, equilibrium volume fraction and precipitation kinetics of η-phase in fixed base composition (0.02C-0.02N-21Cr-0.32Mn-0.2Ti wt.%) with Nb content in a range of between 0.2 and 1.0 wt.%, Si between 0.2 and 0.8 wt.%, and Mo between 0.03 wt.% and 2.0 wt.%. The interfacial energy for the simulations was calculated with the modified Becker's model to be between 0.14 and 0.16 J/m^2 in the temperature range of 400 °C to 1000 °C. Nucleation was assumed to take place at the grain boundaries with grain size of 50 μm (5.7 ASTM) used for the simulations. According to the simulation results shown in Figure 2, alloying with Nb increases the solvus temperature of η-phase and therefore the temperature in which the grain growth begins, ultimately affecting the service temperature of the steel as shown by Nabiran et al. in the case of Fe_2Nb type η-phase [26].

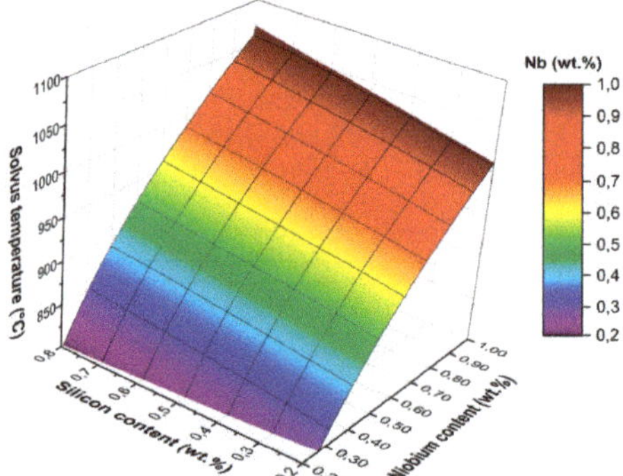

Figure 2. The simulated solvus temperature of η-phase as a function of Si and Nb contents (wt.%) in 0.02C-0.02N-21Cr-0.32Mn-0.2Ti base composition.

A similar effect to that achieved with Nb is also achieved by alloying with Si, however, to a significantly smaller degree. According to the thermodynamic simulations, increasing the Si content from 0.4 to 0.8 wt.% increases the solvus temperature of η-phase by approximately 30 °C, whereas increasing the Nb content by 0.4 wt.% would result in an increase of 100 to 150 °C within the range of 0.35 and 1 wt.% of Nb, as shown in Figure 2. Ferritic stainless steels typically contain significantly larger amounts of Mo than Nb and Si, for e.g. type AISI 444 contains 2 wt.% of Mo. The thermodynamic simulation results suggest that Mo alloying does not have a significant effect on the η-phase solvus temperature as seen in Figure 3. Therefore, it was decided to omit Mo alloying. However, Mo can increase the volume fraction of η-phase and Mo alloying could be considered if the amount of η-phase in the microstructure is insufficient to increase creep resistance.

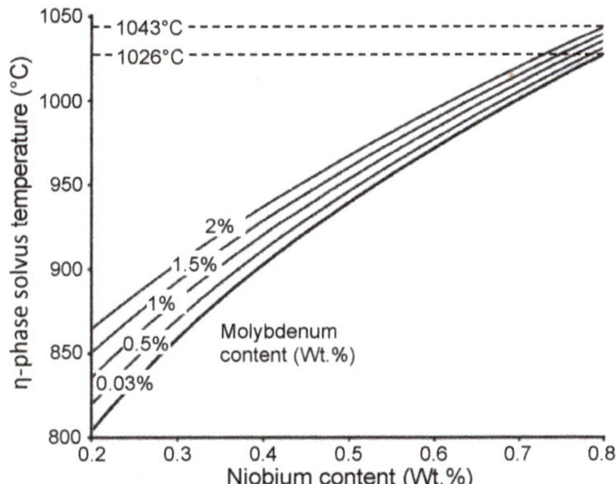

Figure 3. The simulated solvus temperature of η-phase as a function of Nb with various Mo contents (wt.%) in 0.02C-0.02N-21Cr-0.32Mn-0.2Ti base composition.

Concerning the optimization of Ti in the steel, the precipitation of Ti containing carbonitrides at very high temperatures during casting must be considered. The amount of Ti should be high enough to deplete the microstructure of free C and N so no Nb containing carbonitrides form and Nb is isolated for η-phase precipitation. The ideal amount of Ti alloying is achieved when a tiny amount is precipitated in η-phase, the carbon and nitrogen level are kept as low as possible and the amount of Ti carbonitrides is high enough to deplete all free interstitials from the matrix. However, excess alloying of Ti can result in surface defects and is typically kept at 0.2 wt.% or below [27,28]. When alloying 0.2 wt.% Ti to a steel containing 0.02C-0.02N-0.8Nb-0.8Si, according to Thermo-Calc the equilibrium composition of η-phase in temperatures above 850 °C is 46Nb-39Fe-11Cr-3Si-0.4Ti in wt.%, which means most of the Ti has precipitated as carbonitrides. Therefore 0.2 wt.% Ti can be considered very close to optimal.

3.2. *Annealing Microstructures*

Based on the results of the thermodynamic simulations and the criteria given above, a trial steel with the chemical composition 0.02C-0.02N-0.8Nb-0.8Si-0.2Ti (wt.%) was prepared as a 65 kg laboratory cast ingot. The cast ingot was reheated to 1100 °C and held for 75 min which is typical for ferritic stainless steels, and then hot rolled with a starting temperature of 1160 °C in seven passes to an end thickness of 6 mm. This was followed by cold rolling with a 30% reduction to a thickness of 4.2 mm. The cold rolled strips were then pre-annealed and final cold rolled with 48% reduction to a final thickness of 2 mm following final annealing to achieve optimal microstructure to the finished product. Pre-annealing and final annealing were done to obtain a fully recrystallized microstructure with a low volume fraction of η-phase as described below.

The pre-annealing was simulated with Gleeble 3800 thermomechanical simulator using parameters that match typical plant-scale annealing as regards heating and cooling rates as seen in Figure 4. The temperature range of 950 °C to 1200 °C was selected for pre-annealing simulations. The isothermal precipitation diagram calculated with TC-Prisma in Figure 5 shows that 1% of η-phase is predicted to precipitate in approximately ten seconds in the temperature window 700 to 800 °C. The precipitation slows down significantly at higher temperatures due to the reduction of the driving force ($\Delta G_m{}^{\alpha\rightarrow\beta}$) and high interfacial energy (σ) of the η-phase. Comparing the TC-Prisma predictions with the Gleeble simulation heating curves presented in Figure 4, it can be seen that a significant amount of η-phase can precipitate during the pre-annealing and some nucleation will likely occur during the cooling as well.

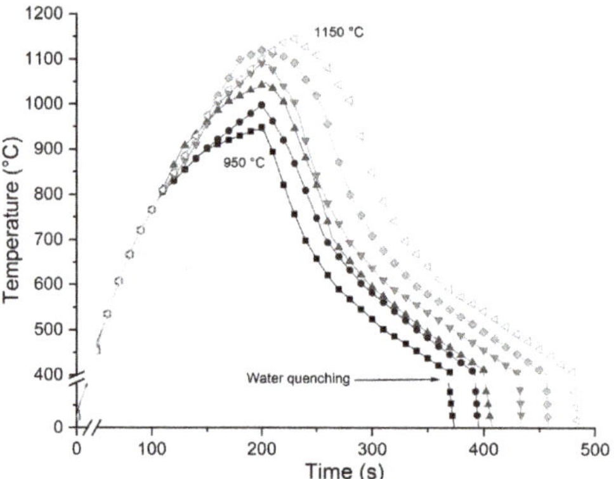

Figure 4. Thermal cycles of the Gleeble annealing simulations (peak temperatures 950–1150 °C).

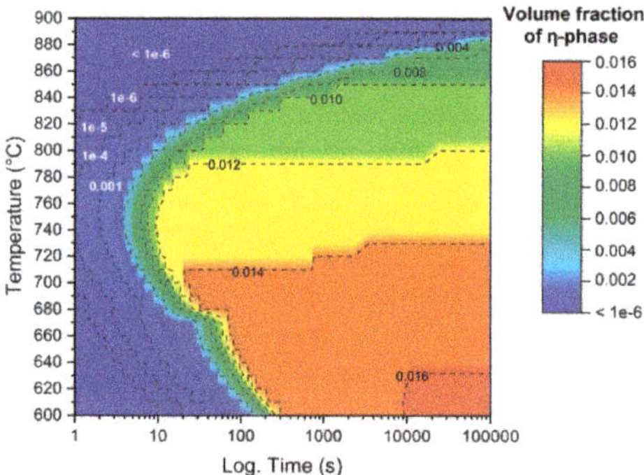

Figure 5. Isothermal precipitation diagram of η-phase in the trial High-Nb-SS steel.

There are two things that need to be considered when selecting the pre-annealing peak temperature; first, the microstructure should be fully recrystallized after the annealing, and secondly, any previously precipitated η-phase should be dissolved so that it can precipitate after recrystallization for good creep resistance. The microstructure of the steel is ferritic from ambient to solvus temperature so no phase transformation apart from the precipitation will occur at any temperature.

The thermodynamic and Gleeble simulations suggest that the recrystallization temperature is in fact, also dependent on the dissolution temperature of η-phase. This dependency can be related to the pinning effect of η-phase particles at grain boundaries or to solute Nb and its decreasing effect on grain boundary mobility; however, confirmation of this is not covered in present study [29,30]. The pre-annealing temperature had to be set high enough to obtain full recrystallization and complete or almost complete η-phase dissolution. The grain size should be kept reasonably low (around ASTM

6 or smaller) by avoiding rapid grain growth in order to meet the requirements for good formability during manufacturing as well as low and high temperature strength, although larger grain size would likely increase the creep resistance. However, the grain size after pre-annealing is not significant, as the material will be later cold rolled and final annealed. Microstructures after the simulated annealing cycles were analyzed using a ZEISS Ultra+ Field Emission Scanning Electron Microscope (FESEM) and typical microstructures from each stage of the cold rolling and annealing process can be seen in Figure 6. After pre-annealing at 950 °C, the microstructure was not fully recrystallized. However, after pre-annealing at 1050 °C (Figure 6) the microstructure was recrystallized, but a high volume fraction of η-phase was present and partly distributed in rows, indicating that the pre-annealing temperature had to be even higher. The microstructure of a sample pre-annealed at 1150 °C contained no η-phase and showed that full recrystallization had occurred. However, the grain size was increased compared to that before annealing. As mentioned earlier the grain size after pre-annealing is not as significant as it is in the final, so 1150 °C was selected as the pre-annealing temperature.

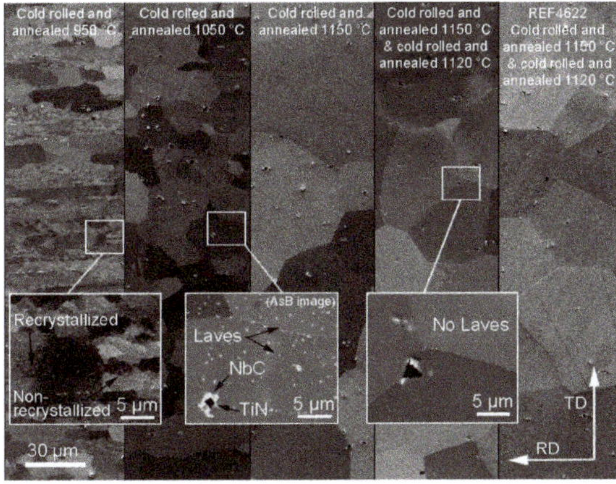

Figure 6. FESEM secondary electron and backscatter images of samples after each process phase.

The parameters for final annealing were determined using a similar approach. To get smaller grain size after final annealing, the temperature window has to be smaller compared to pre-annealing. The lowest peak temperature that resulted in full recrystallization was 1120 °C, so this was chosen for the final annealing, although some grain growth was observed, and some η-phase was still left but in very small amounts. The obtained grain size 5.0 ASTM was slightly larger than the target of 6 ASTM.

3.3. Creep Resistance

Figure 7 shows the results of the sag tests for typical stainless steel grades used in high temperature applications, the high-Nb-SS trial test material and reference material REF4622 with composition close to that of EN 1.4622 but manufactured using similar process parameters those employed here for the High-Nb-SS steel. The compositions of the steels, grain size at room temperature and solvus temperatures of η-phase can be seen in Table 1.

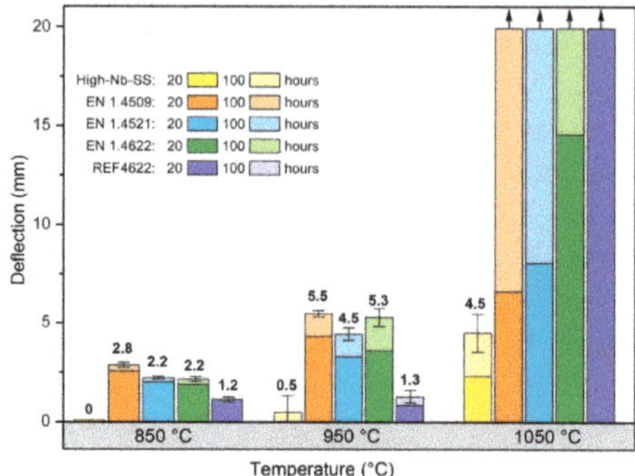

Figure 7. Deflection measured and standard deviation after sag test at 850 °C, 950 °C and 1050 °C for the trial steel and conventional ferritic stainless steels typically used in high-temperature applications.

Table 1. Chemical compositions, grain size (GS) at room temperature (RT) and after Sag test in 950 °C for 100 h, and simulated solvus temperatures of η-phase in the High-Nb-SS trial and reference steels, and existing steels used in high temperature applications produced by Outokumpu Stainless.

Steel Product	C	N	Cr	Ti	Mo	Nb	Si	Mn	GS in RT (ASTM)	GS in Sag (ASTM)	η-phase solvus T
EN 1.4509	0.01	0.02	18.0	0.12	0.03	0.36	0.51	0.53	6.0	2.0	851
EN 1.4521	0.01	0.02	17.9	0.13	2.1	0.39	0.48	0.49	7.8	<0	906
EN 1.4622	0.02	0.02	20.8	0.17	0.03	0.36	0.48	0.40	8.8	4.3	846
REF4622	0.02	0.02	21.0	0.10	0.12	0.40	0.43	0.33	6.1	4.5	831
High-Nb-SS	0.02	0.02	21.0	0.20	0.02	0.80	0.80	0.32	5.0	4.0	1026

The High-Nb-SS steel showed significantly better sag test results, i.e. higher creep resistance in the absence of external stress, than the reference material or other stainless steel grades. At 850 °C there was no measurable sag (i.e. < 0.1 mm) for the High-Nb-SS steel during the 100 h test, and only 0.5 mm deflection was measured after testing at 950 °C. When tested at 1050 °C the High-Nb-SS steel exhibited similar deflection to that shown by the other steels at 850 °C and 950 °C.

4. Discussion

According to the deformation maps for ferritic steels by Ashby [23] and Zinkle et al. [31] and considering the calculated shear stress and homologous temperature, the dominant creep mechanism during the sag test would be Coble creep, which typically occurs at low stresses and high temperatures and is controlled by grain boundary diffusion. The creep rate in Coble creep is proportional to the stress. The linear distribution of axial normal stress will therefore prevail provided that the curvature of the specimen remains moderate. The maximum shear stress will therefore be approximately constant and Cobble creep will remain as the active deformation mechanism throughout the test in this particular case. The creep rate in Coble creep is inversely proportional to the cube of the grain size [22,32]. Another high-temperature creep mechanism in which finer grain size has negative effect is grain boundary sliding. Grain boundary sliding however is not an independent mechanism but accommodated by other creep modes and its effect can be mitigated if the grain boundaries are effectively covered with precipitates [32,33]. The overlapping of the mechanisms is discussed by Hertzberg et al. (1989) who suggest that Coble creep is dependent on grain boundary sliding [32]. Therefore, suppression of both mechanisms should be considered when designing steel with good creep resistance. The negative effect of a fine grain size on grain boundary sliding can be reduced if the grains are sufficiently covered with η-phase precipitates that act as pinning points resisting grain boundary sliding [11,34].

Figure 8 shows the η-phase precipitated along the grain boundaries and within the grains of the microstructure of a sag sample after 100 h at 950 °C as expected. The larger grain size of the trial steel compared to other steels (Table 1) can have a positive effect on creep resistance because of Coble creep mechanism [9], however, during the sag test the grain size increases in other steels resulting in larger grain size. This is result of the considerably higher solvus temperature of η-phase in High-Nb-SS, which correlates to simulated equilibrium solvus temperatures shown in Table 1. The grain growth was further observed by measuring the grain size from each sample after sag test at 950 °C for 100 h. The grain size (ASTM) in High-Nb-SS, REF4622 and EN 1.4622 was approximately 4–5, 2 in EN 1.4509 and too large to be measured in EN 1.4521 (i.e. <0 ASTM) as shown in Table 1. It seems that the negative effect of a finer grain size in High-Nb-SS steel is mitigated by the precipitation of η-phase on the grain boundaries, as previously explained.

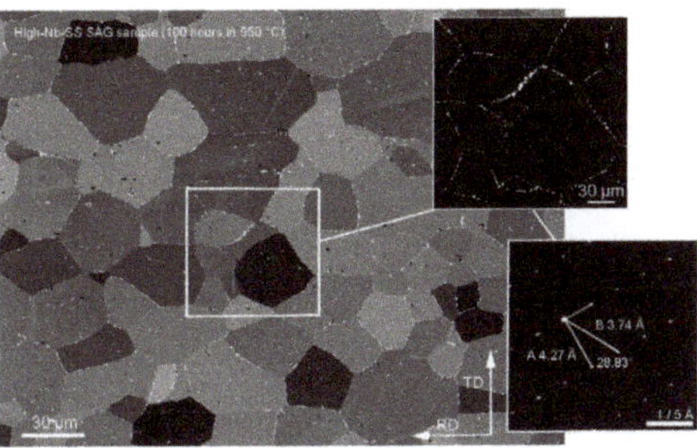

Figure 8. FESEM backscatter image of sag test sample after 100 h in 950 °C. In-set shows η-phase only on grain boundaries (=0.77 wt.%) and typical TEM diffraction image of η-phase particle.

According to Thermo-Calc simulations for equilibrium at 950 °C, the volume fractions of η, TiN and NbC are 0.6%, 0.2% and 0.1%, respectively. With simple image analysis carried out for sag tested sample using ImageJ software with 1.64 mm^2 surface area and hundreds of particles, a volume fraction of 1.124% for all particles visible in a backscattered electron image was measured, i.e. η and NbC excluding TiN. FESEM image of a part of the analyzed area is shown in Figure 8. η-phase was identified with TEM diffraction image (Figure 8) to have a hexagonal C14 crystal structure with lattice parameters a = 4.27 Å and c = 7.77 Å and composition of 53Fe-25Nb-11Cr-9Si-2Ti (wt.%) measured with EDS [8,17]. The results correlate well with the Thermo-Calc predictions for the volume fraction at equilibrium state. The equilibrium composition of η-phase according to Thermo-Calc is 49Fe-32Nb-11Cr-7Si-1Ti which also correlates well with the EDS analysis.

When particles inside the grains are removed from the image (Figure 8), the volume fraction of η-phase left at grain boundaries was calculated 0.77% using the ImageJ software. The amount could be high enough to retain good creep resistance in High-Nb-SS steel even with its finer grain size under low stress–high temperature conditions. The finer grain size could be achieved by further restraining the rapid grain growth during annealing with a more optimal peak temperature. However, this would require more simulations and tests. In the sag test the dominant creep mechanism involves diffusion along grain boundaries, however, the presence of η-phase inside the grains implies that good creep resistance is also expected for higher stresses where other creep mechanisms are dominant. This needs to be examined with comprehensive creep tests. The formation of η-phase can be expected to have some influence on the mechanical properties of the steel in room temperature based on the Author's previous studies on the effect η-phase on strengthening in AISI 444 type ferritic stainless steel [9].

5. Conclusions

1) When alloying 0.8 wt.% of Nb and Si, the equilibrium solvus temperature of the η-phase in High-Nb-SS was calculated as 1026 °C. The solvus temperature is 175 °C higher than in EN1.4509 which is a commonly used steel in high temperature applications. When exposed to temperatures

above the solvus temperature, the rapid grain growth begins, limiting the service temperature of the steel.

2) The dominant creep mechanism during the sag test is Cobble creep, which is related to the grain boundary diffusion and can therefore be suppressed by grain boundary precipitation.
3) Grain boundary precipitation of η-phase resulted in significantly lower deflections in Sag test, i.e. enhanced the creep resistance of High-Nb-SS compared to commercially available steels used in high temperatures.
4) Creep rate in the Cobble mechanism is inversely proportional to the cube of the grain size. However, the η-phase precipitation mitigated the negative effect of a finer grain size in High-Nb-SS steel. Avoiding grain growth at higher temperature will better maintain other mechanical properties of the steel.

6. Patents

The work reported in this manuscript resulted in Patent no. 18215480.7-1103

Author Contributions: Conceptualization, T.J.; methodology, T.J. and S.U.; software, T.J.; validation, T.J. and T.M.; formal analysis, T.J., S.U. and T.M.; investigation, T.J., D.P.; resources, T.J. and T.M.; data curation, T.J.; writing—original draft preparation, T.J.; writing—review and editing, T.J., T.M. and D.P.; visualization, T.J.; supervision, D.P.; project administration, J.K.; funding acquisition, T.J., T.M. and J.K.

Funding: This research was funded by Companhia Brasileira de Metalurgia e Mineração and the Academy of Finland through the project #311943.

Acknowledgments: The authors wish to acknowledge Juha Uusitalo and Aarne Pohjanen for their assistance with Gleeble thermomechanical simulations and stress calculations.

Conflicts of Interest: The authors declare no conflict of interest.

References

1. Hua, M.; Garcia, C.I.; DeArdo, A.J.; Tither, G. Dual-Stabilized Ferritic Stainless Steels for Demanding Applications Such as Automotive Exhaust Systems. *Iron Steelmak.* **1997**, *24*, 41–44.
2. Inoue, Y.; Kikuchi, M. Present and future trends of stainless steel for automotive exhaust system. *Nippon Steel Tech. Rep.* **2003**, *28*, 62–69.
3. Ali-Löytty, H.; Jussila, P.; Juuti, T.; Karjalainen, L.P.; Zakharov, A.A.; Valden, M. Influence of precipitation on initial high-temperature oxidation of Ti-Nb stabilized ferritic stainless steel SOFC interconnect alloy. *Int. J. Hydrogen Energy* **2012**, *37*, 14528–14535. [CrossRef]
4. Ishii, K.; Miyazaki, A.; Satoh, S. Stainless Steels for Automotive Exhaust Systems. *Kawasaki Steel Tech. Rep.* **1999**, *40*, 39–41.
5. Fujita, N.; Ohmura, K.; Sato, E.; Yamamoto, A. Development of Ferritic Stainless Steel YUS 450 with High Heat Resistance for Automotive Exhaust System Components. *Nippon Steel Tech. Rep.* **1996**, *71*, 25–30.
6. Fujita, N. New Ferritic stainless steels in automotive exhaust system. *Nippon Steel Tech. Rep.* **2000**, *81*, 29–33.
7. Yan, H.; Bi, H.; Li, X.; Xu, Z. Effect of two-step cold rolling and annealing on texture, grain boundary character distribution and r-value of Nb + Ti stabilized ferritic stainless steel. *Mater. Charact.* **2009**, *60*, 65–68. [CrossRef]
8. Juuti, T.; Rovatti, L.; Mäkelä, A.; Karjalainen, L.P.; Porter, D. Influence of Long Heat Treatments on the Laves Phase Nucleation in a type 444 Ferritic Stainless Steel. *J. Alloys Compd.* **2014**, *616*, 250–256. [CrossRef]
9. Juuti, T.; Rovatti, L.; Porter, D.; Angella, G.; Kömi, J. Factors controlling ambient and high temperature yield strength of ferritic stainless steel susceptible to intermetallic phase formation. *Mater. Sci. Eng. A* **2018**, *726*, 45–55. [CrossRef]
10. Hald, J. Microstructure and long-term creep properties of 9–12% Cr steels. *Int. J. Press. Vessel. Pip.* **2008**, *85*, 30–37. [CrossRef]
11. Abe, F. Effect of fine precipitation and subsequent coarsening of Fe2W laves phase on the creep deformation behavior of tempered martensitic 9Cr-W steels. *Metall. Mater. Trans. A* **2005**, *36*, 321–332. [CrossRef]
12. Li, Q. Precipitation of Fe2W laves phase and modeling of its direct influence on the strength of a 12Cr-2W steel. *Metall. Mater. Trans. A* **2006**, *37*, 89–97. [CrossRef]

13. Miyazaki, A.; Hirasawa, J.; Furukimi, O.; Kobayashi, M.; Kihara, S. Development of High Heat—Resistant Ferritic Stainless Steel with High Formability, "RMH-1", for Automotive Exhaust Gas Systems. *Mater. Jpn.* **2003**, *42*, 157–159. [CrossRef]
14. Sim, G.M.; Ahn, J.C.; Hong, S.C.; Lee, K.J.; Lee, K.S. Effect of Nb Precipitate Coarsening on the High Temperature Strength in Nb Containing Ferritic Stainless Steels. *Mater. Sci. Eng. A* **2005**, *396*, 159–165. [CrossRef]
15. Fujita, N.; Ohmura, K.; Kikuchi, M.; Suzuki, T.; Funaki, S.; Hiroshige, I. Effect of Nb on High-temperature Properties for Ferritic Stainless Steel. *Scr. Mater.* **1996**, *35*, 705–710. [CrossRef]
16. Froitzheim, J.; Meier, G.H.; Niewolak, L.; Ennis, P.J.; Hattendorf, H.; Singheiser, L.; Quadakkers, W.J. Development of high strength ferritic steel for interconnect application in SOFCs. *J. Power Sources* **2008**, *178*, 163–173. [CrossRef]
17. Andrews, K.W.; Dyson, D.J.; Keown, S.R. *Interpretation of Electron. Diffraction Patterns*; Springer US: Boston, MA, USA, 1967; ISBN 978-1-4899-6228-7.
18. Brown, E.L.; Burnett, M.E.; Purtscher, P.T.; Krauss, G. Intermetallic phase formation in 25Cr-3Mo-4Ni ferritic stainless steel. *Metall. Trans. A* **1983**, *14*, 791–800. [CrossRef]
19. Speich, G. Precipitation of Laves Phases from Iron-niobium (Columbium) and Iron-Titanium Solid Solutions. *Trans. Met. Soc. AIME* **1962**, *224*, 850–858.
20. Chen, Q.; Jou, H.J.; Sterner, G. *Thermodynamic and Mobility Databases Overview*; Thermo-Calc Software AB: Stockholm, Sweden, 2018; pp. 1–20.
21. Chen, Q.; Jou, H.J.; Sterner, G. *TC-PRISMA User's Guide*; Thermo-Calc Software AB: Stockholm, Sweden, 2015.
22. Frost, H.J.; Ashby, M.F. *Deformation-Mechanism Maps: The Plasticity and Creep of Metals and Ceramics*, 1st ed.; Pergamon Press: Amsterdam, The Netherlands, 1982.
23. Frost, H.J.; Ashby, M.F. Deformation-Mechanism Maps for Pure Iron, Two Austenitic Stainless Steels, and a Low-Alloy Ferritic Steel. In *Fundamental Aspects of Structural Alloy Design*; Jaffee, R.I., Wilcox, B.A., Eds.; Springer US: Boston, MA, USA, 1977; pp. 27–65. ISBN 978-1-4684-2421-8.
24. Puchi-Cabrera, E.S.; Guérin, J.D.; Barbier, D.; Dubar, M.; Lesage, J. Plastic Deformation of Structural Steels Under Hot-working Conditions. *Mater. Sci. Eng. A* **2013**, *559*, 268–275. [CrossRef]
25. Schneibel, J.H.; Heilmaier, M. Hall & Petch Breakdown at Elevated Temperatures. *Mater. Trans.* **2014**, *55*, 44–51.
26. Nabiran, N.; Klein, S.; Weber, S.; Theisen, W. Evolution of the Laves Phase in Ferritic Heat-Resistant Steels During Long-term Annealing and its Influence on the High-Temperature Strength. *Metall. Mater. Trans. A* **2014**, *46*, 102–114. [CrossRef]
27. Grubb, J.; Wrigth, R.; Farrar, P.J. *Toughness of Ferritic Stainless Steels*; Lula, R., Ed.; ASTM International: West Conshohocken, PA, USA, 1982; ISBN 978-0-8031-0792-2.
28. Redmond, J. Toughness of 18Cr-2Mo Stainless Steel. In *Toughness of Ferritic Stainless Steels*; ASTM International: West Conshohocken, PA, USA, 1980; pp. 123–144.
29. Suehiro, M.; Liu, Z.-K.; Ågren, J. Effect of niobium on massive transformation in ultra low carbon steels: A solute drag treatment. *Acta Mater.* **1996**, *44*, 4241–4251. [CrossRef]
30. Suehiro, M. Recrystallization and Related Phenomena. An Analysis of the Solute Drag Effect of Nb on Recrystallization of Ultra Low Carbon Steel. *ISIJ Int.* **1998**, *38*, 547–552. [CrossRef]
31. Zinkle, S.J.; Lucas, G.E. *Deformation and Fracture Mechanisms in Irradiated FCC and BCC Metals*; Oak Ridge National Laboratory: Oak Ridge, TN, USA, 2003.
32. Hertzberg, R.W. *Deformation and Fracture Mechanics of Engineering Materials*, 3rd ed.; John Wiley & Sons: Hoboken, NJ, USA, 1989.
33. Čadek, J. *Creep in Metallic Materials*; Elsevier: Amsterdam, The Netherlands, 1988; ISBN 0444989161.
34. Erneman, J.; Schwind, M.; Andren, H.; Nilsson, J.; Wilson, A.; Agren, J. The evolution of primary and secondary niobium carbonitrides in AISI 347 stainless steel during manufacturing and long-term ageing. *Acta Mater.* **2006**, *54*, 67–76. [CrossRef]

© 2019 by the authors. Licensee MDPI, Basel, Switzerland. This article is an open access article distributed under the terms and conditions of the Creative Commons Attribution (CC BY) license (http://creativecommons.org/licenses/by/4.0/).

Article

On the Evaluation of Surface Fatigue Strength of a Stainless-Steel Aeronautical Component

Filippo Cianetti [1,*], Moreno Ciotti [2], Massimiliano Palmieri [1] and Guido Zucca [3]

1 Department of Engineering, University of Perugia, 06125 Perugia, Italy; massimiliano.palmieri@studenti.unipg.it
2 UmbraGroup S.p.A, Via V. Baldaccini 1, 06034 Foligno (PG), Italy; MCiotti@umbragroup.com
3 Italian AirForce, Flight Test Center, Technology Material for Aeronautics and Space Department, Military Airport M. De Bernardi, via Pratica di Mare, 000040 Pomezia (RM), Italy; guido.zucca@aeronautica.difesa.it
* Correspondence: filippo.cianetti@unipg.it; Tel.: +39-75-585-3728

Received: 8 March 2019; Accepted: 15 April 2019; Published: 17 April 2019

Abstract: In this paper, a novel method for the evaluation of the surface fatigue strength of a stainless-steel component is proposed. The use of stainless steel is necessary indeed, whenever a component has to work in a particularly aggressive environment that may cause an oxidation of the component itself. One of the major problems that affect stainless-steel components is the possible wear of the antioxidant film that reduces the antioxidant properties of the component itself. One of the main causes that can lead to wear is related to the surface corrosion that occurs every time two evolving bodies are forced to work against each other. If the antioxidant film is affected by surface fatigue problems, such as pitting or spalling, the antioxidant capacities of this type of steel may be lost. In this context, it is, therefore, necessary to verify, at least, by calculation that no corrosion problems exist. The method proposed in this activity is a hybrid method, numerical-theoretical, which allows to estimate the surface fatigue strength in a very short time without having to resort to finite element models that often are so complex to be in contrast with industrial purposes.

Keywords: stainless steel; structural dynamics; finite element explicit analysis; Hertz theory

1. Introduction

The use of stainless steels is spreading in all those sectors where the working conditions of various mechanical components are quite critical and could trigger structural problems of the component.

Following their good strength/ductility combination coupled with their excellent corrosion resistance [1], stainless steels are adopted in many applications including automotive [2], construction and building [3], oil and gas [4], aeronautical [5], medical [6,7], and food [8].

In the abovementioned sectors, the mechanical components have to work in environments that can be saturated with aggressive substances, risking undermining the mechanical properties of the material and thus inducing possible problems of strength [9]. It is, therefore, evident that, in this context, the use of stainless materials becomes essential and, therefore, is increasingly widespread.

Stainless steel owes its resistance to corrosion to a simple chemical reaction: the combination of chrome in steel and the oxygen in the air or water on the surface form a very thin passivation film. This film protects the component against all aggressive substances. If this film is damaged, however, it spontaneously reforms from the steel matrix [10–13]. For this reason, stainless steels are generally called self-passivation steel. The main actor in the self-passivation phenomenon is chrome. When the oxygen content is above a certain level, chrome reacts with this, forming a layer of chromium oxide or hydroxide. This thin film (with a thickness of the order of a few atoms, $3 \div 5 \times 10^{-7}$ mm) protects the underlying metal from the action of external agents. The corrosion resistance also depends on the surface morphology: the more the surface is smooth and homogeneous, the greater its resistance to corrosion will be.

On the basis of this, many researchers of different scientific backgrounds are actively working, studying the properties of stainless steels both from a chemical [11–16] and from a structural point of view [17–22]. The first category is a majority focus on studying how the properties of these materials vary according to their chemical composition and electrochemical behaviour, while the second category is mainly focus on the consequences generated by a possible mechanical degradation of the component surface.

It is known how mechanical components working in contact are subjected to degradation issues. The two most common phenomena that induced the degradation of a mechanical component are pitting and spalling [23]. According to the American Society for Metals (ASM) handbook, "Usually, pitting is the result of surface cracks caused by metal-to-metal contact of asperities or defects due to low lubricant film thickness" [24]. Even if the starting point of pitting phenomena is still under investigation by the scientific community, the link between bodies working in contact and pitting issues is real and approved without doubts [25].

Given the reduction of the antioxidant properties that occur in the case of a degradation of the surface layer of components, an efficient tool for predicting and evaluating contact fatigue may help the design engineer when they have to face this kind of problem.

To solve mechanical contact issues, engineers must generally choice between a kinematic-theoretical approach and a numerical approach. The first exploits the kinematic laws to determine impact forces and then applies the Hertz theory [26,27] to calculate stresses. This approach requires, however, an assessment of the initial condition of contact dynamics. This aspect it is often so complex to make that this method is not applicable in a real complex model. The second approach is based on the explicit finite element method (FEM), that allows a more accurate assessment of both the impact forces and surface pressures. The major drawback of this type of approach is the computational time: To obtain reliable results in terms of stresses, very complex models and, so, impracticable times of the simulations are generally required [28].

The present work seeks to propose a simple procedure for the evaluation of pressures and contact stresses in mechanical components that may result in surface resistance (pitting) issues. The aim of the work is that of evaluating, by calculation, the possible occurrence of pitting phenomena, trying to avoid such phenomenon at the beginning of the design stage [29–31].

The attempt of the authors is to find a trade-off between the efficiency requirement on one side and the effectiveness of the other one, typical in the design scenario and in contrast with the time and economic constraints of industries.

For this reason, the authors developed a hybrid approach, theoretical-numerical, which allows the calculation effort to speed up obtaining, at the same time, reliable results. Developing a finite element model not voted to contact stress calculation but only to contact forces evaluation involves a drastic reduction in computational efforts. Once contact forces are known, the classical Hertz theory [27] is used to estimate the stresses on the component surface.

The fully theoretical and numerical procedures and the hybrid one proposed in this activity are devoted to evaluating contact stress due to bodies in contact. The evaluation of fatigue life is a shared step of all three methodologies, and it is always based on the assessment of the fatigue curve and on the damage computational model. In this case, the linear accumulation law of Palmgren-Miner [32] is used. The proposed hybrid method was validated on a simple case of sphere-plane contact. At the end, a more complex case study was used to validate the effectiveness of the proposed approach. The mechanical component used as benchmark is a stainless-steel recirculation mechanism of a ball screw developed for aeronautical applications by a leader company in this sector.

As a matter of fact, ball screws are used successfully in all those applications, both industrial and aeronautical, in which high-speed and high-accuracy movements of components are required. Ball screws in aeronautical field are generally used for moving mobile surfaces (flaps and slats) and the stabilizer. For this reason, they are exposed to atmospheric conditions that are sufficiently critical to require the use of stainless steels.

Due to the high speed with which these systems have to work, one of the most frequently encountered problem by ball-screw manufacturers is the premature failure of the evolving components in contact [26]. The areas most affected by surface fatigue problems in a ball screw are essentially two: the contact zone between the screw tracks and the balls themselves during normal operation and the impact zone in the recirculation mechanism where the balls are forced to abruptly change direction [33,34]. The high number of impacts to which a recirculation system is generally subject tend to generate a plastic deformation of the recirculation system itself. This modifies the geometry of the system, causing a rapid increase in the maximum contact pressure on the surface. This aspect is often shown in a wear on the surface of the recirculation mechanism (pitting) which leads to a reduction of the antioxidant properties of the stainless steel used and consequently to a drastic reduction of the fatigue strength of the component.

Since this phenomenon is often rediscovered, numerous studies have been carried out trying to better understand the phenomenon of the ball's impact during the recirculation phase and to propose different approaches for the estimation of the life of ball screws. Jiang et al. [35] studied the dynamics and impact forces of spheres in a recirculation mechanism, using a multi-body code and trying to optimize the geometry of the recirculation mechanism itself. Braccesi et al. [36] presented a relationship between the Hertz equivalent elastic stress limit with typical design parameters such as the allowable material stress, the geometric characteristics in contact areas, the impact angles, and the rotational speed for different types of materials, using an elastic-plastic model for a curved body in contact. Zhang et al. [37] studied the influence of various factors on the impact forces generated between the spheres and recirculation mechanism for different types of materials and for different recirculation geometries.

Although the available methods for the calculation of surface fatigue strength are multiple, the one proposed in the present activity is aimed at being user-friendly for all those realities that have not enough resources to devote to the development of complex models that are multi-body model or finite elements model. What is immediately highlighted by the proposed approach is that, remaining linked to the theory and exploiting the numerical simulation, it is possible to obtain a more accurate estimation of the fatigue life than would be obtained using a completely kinematic-theoretic approach. At the same time, the solution is very close to that obtainable from a completely numerical approach but, obviously, with a drastic reduction of the computational effort.

2. Theoretical Approach

In order to obtain a clear exposition, this section describes the theoretical approach that consists in two consecutive steps:

1. The determination of the contact force starting from the laws of classical mechanics, i.e., the conservation of energy and momentum;
2. The evaluation of contact stresses from the Hertz theory on contact.

The last paragraph of this sections is dedicated to the evaluation of fatigue: as declared in the previous paragraph, the third step is represented by the fatigue evaluation that is "shared" with the numerical method.

2.1. *From Impact Dynamics to Contact Force: The Relative Approach*

To understand the relationship between impact dynamics and contact forces, two simplifying assumptions are necessary. The first one foresees that the only component of the velocity during the impact is perpendicular to the surface; the second is that the impact time is much longer than the fundamental vibration period of the bodies. The first hypothesis allows the impact to be treated as a centred impact, while the second one instead allows an application of the principle of energy and momentum conservation. Under these hypothesis, impact waves are exchanged between two bodies and there is no energy dissipation [38]. Adopting these assumptions and adopting the formulation proposed by Braccesi et al. [36], contact forces can be easily expressed as a function of the second derivative of the mutual deformation of the two bodies at the time of contact: the approach acceleration $\ddot{\alpha}$. With reference to Figure 1, the approach acceleration $\ddot{\alpha}$ can be written as shown in Equation (1):

$$\ddot{\alpha} = -P\frac{m_b + m_{sp}}{m_b \cdot m_{sp}}, \tag{1}$$

where P is the force that presses the two bodies against each other and m_{sp} and m_b are the masses of the two bodies, respectively. If a mass is much greater than the other $\left(m_b \gg m_{sp}\right)$, it is possible to rewrite Equation (1) as follows:

$$\ddot{\alpha} = \frac{-P}{m_{sp}}, \tag{2}$$

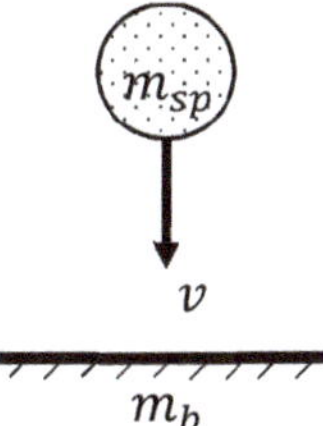

Figure 1. A Hertzian distribution of contact pressure.

In order to integrate Equation (2), the Hertz theory, which expresses the contact force as a function of the relative approach, must be introduced. The contact between machine components, bearings, gears, cams, etc., is a problem known since more than a century and a half, and it was accurately addressed by Hertz [27] in an analytical-nonlinear theory. In general, the problem consists in having very small contact surfaces, theoretically degenerating in a point or in a line, through which the bodies exchange forces. As a direct result, there are very intense pressure distributions surrounded by sharp gradients. In particular, with reference to the scheme of Figure 2, two general curved geometries were placed in contact each other by the action of the force P.

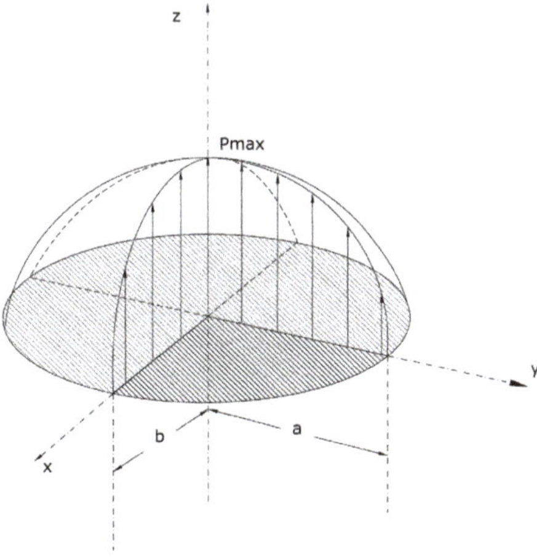

Figure 2. A Hertzian distribution of contact pressure.

In the general case, there are four different curvature radii, and the contact surface is an ellipse of semiaxes a (the major) and b (the minor) (see Figure 2). The pressure $p(x,y)$ will have a semielliptic distribution of maximum $p_{\max}$. Such quantities can easily be derived from the Hertz theory as

$$p(x,y) = p_{\max}\sqrt{1-\left(\frac{x}{a}\right)^2-\left(\frac{y}{b}\right)^2}, \tag{3}$$

$$p_{\max} = \frac{3\cdot P}{2\cdot\pi\cdot a\cdot b}, \tag{4}$$

The Hertz theory [27] leads to an accurate modeling of reality, in particular in all those cases, where no plastic deformation occurs. If yield occurs, it is possible to define the contact yield pressure as

$$P_Y = \frac{2\cdot E^*\cdot a_Y}{\pi\cdot R^*}, \tag{5}$$

which indicates the limit to not be exceeded. In Equation (5), a_y is the contact radius when a yield occurs. Let α be the relative approach between two bodies at the contact point due to the presence of force P. Introducing:

$$\frac{1}{E^*} = \sum_{i=1}^{2}\frac{1-\nu_i^2}{E_i}, \tag{6}$$

$$\frac{1}{R^*} = \sum_{i,j=1}^{2}\frac{1}{R_{i,j}}, \tag{7}$$

$$A = \frac{2\pi}{3\cdot k}\left(\frac{2\cdot L(e)}{K^3(e)}\right)^{\frac{1}{2}}, \tag{8}$$

where ν_i and E_i are the Poisson and Young's modulus of ith body, $R_{i,j}$ the radius of the jth curvature and $K(e)$ and $L(e)$ are elliptic integrals defined in Reference [36]. The relation between the relative approach and the contact force [36] is shown in Equation (9).

$$P = A\cdot E^*\cdot R^{*\frac{1}{2}}\cdot\alpha^{\frac{3}{2}} \tag{9}$$

Equation (9) allows the determination of the contact force in case of elastic behaviour. If yield occurs, i.e., in case of the plasticization of the component, Equation (8) is not still valid and Thornton's hypotheses [26] must be introduced.

2.1.1. 1st Thornton's Hypothesis

The pressure distribution presents a cutoff to the yield stress (Figure 3); the contact force is expressed in the following equation:

$$P = P_e - 2\pi\int_0^{a_p}(\sigma(r)-\sigma_Y)r dr, \tag{10}$$

where a_p is the radius of the plasticized area, P_e is the force resulting from the perfectly elastic behavior and σ_y is the yield contact stress.

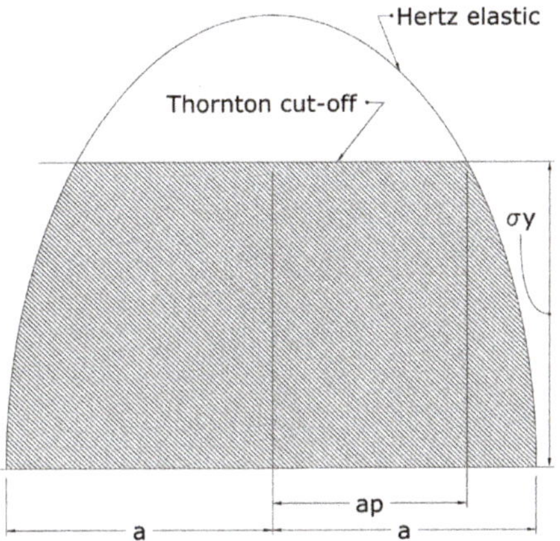

Figure 3. The Thornton cutoff at yielding stress.

Integrating Equation (10) and considering Equation (5), it is possible to evaluate the contact load as shown in Equation (11).

$$P_l = P_Y + \pi \cdot \sigma_Y \cdot R^* \cdot (\alpha - \alpha_Y), \tag{11}$$

Equation (11) describes the elastic-plastic behavior of the material after yielding an identification of the linear relation with the relative approach. In Equation (11), α_y represents the relative approach when a yield occurs. The material will assume a hysteretic behavior, described in Figure 4. In this way, when the maximum value P_u is reached, the material will follow the unloading curve shown in Equation (12).

$$P_u = \frac{4}{3} \cdot E^* \cdot R_p^{\frac{1}{2}} \cdot \left(\alpha - \alpha_p\right), \tag{12}$$

where α_p represents the permanent deformation.

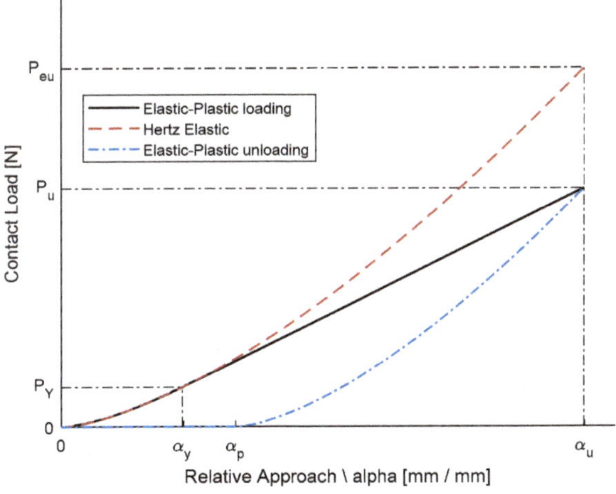

Figure 4. The elastic vs. perfect plastic curve.

2.1.2. 2nd Thornton's Hypothesis

Identified with $(P_u; \alpha_u)$, the conditions at the time of discharge, the contact area developed by the contact force, and the reduced curvature (caused by the plastic deformation) will be equal to that developed under elastic conditions by an equivalent force P_{eu} and by the curvature contact. By Equation (13), it is possible to determine the reduced radius by plastic deformations:

$$R_p = \frac{\alpha^*}{\alpha^* - \alpha_p} R^*, \tag{13}$$

Recalling what previously stated that the general elastic plastic loading and unloading curves are defined in Equation (14) and in Equation (15) respectively:

$$P_l = P_Y + \frac{3}{2} \cdot A \cdot E^* \cdot R_H^{\frac{1}{2}} \cdot \alpha^{\frac{1}{2}} \cdot (\alpha - \alpha_Y), \tag{14}$$

$$P_u = A \cdot E^* \cdot R_H^{\frac{1}{2}} \cdot \alpha^{\frac{1}{2}} \cdot \left(\alpha - \alpha_p\right). \tag{15}$$

Replacing Equation (9) or Equations (14) and (15) in Equation (1) and imposing a zero approach velocity ($\dot{\alpha} = 0$), it is possible to obtain the equivalent approach as shown in Equation (16)

$$\alpha = \left(\frac{5}{4} \cdot \frac{m_1 v^2}{AE^* R^{*\frac{1}{2}}}\right)^{\frac{2}{5}}, \tag{16}$$

In Equation (16), v is the approach velocity of the two bodies at the beginning of the impact. Using the obtained equivalent approach value α, it is possible to determine the impact force and the corresponding maximum pressure value.

2.2. From Contact Force to Contact Stress

Setting $R_{1,1} = R_{1,2}$, and $R_{2,1} = R_{2,2}$, the problem is simplified to the contact between two spheres. In this case stress components can be calculated analytically. Considering the variable z as from the reference system of Figure 2, the contact stresses can be computed as in Equations (17)–(19) [27].

$$\sigma_{1,2} = \sigma_{x,y} = -p_{\max}\left[\left(1 - \left|\frac{z}{a}\right| \cdot \tan^{-1}\left|\frac{1}{\frac{a}{z}}\right|\right)(1 + \nu) - \frac{1}{2 \cdot \left(1 + \frac{z^2}{a^2}\right)}\right], \tag{17}$$

$$\sigma_3 = \sigma_z = -p_{\max} \cdot \frac{1}{\left(1 + \frac{z^2}{a^2}\right)}, \tag{18}$$

$$\tau_{\max} = \frac{\sigma_{1,2} - \sigma_3}{2}, \tag{19}$$

The stress components are shown as a function of the distance from the surface (see Figure 5). As can be seen, the tangential stress reaches its maximum at a depth equal to about half of the contact area radius. The rolling through the latter generates an inversion of the stress giving rise to a purely alternating main component responsible for surface fatigue.

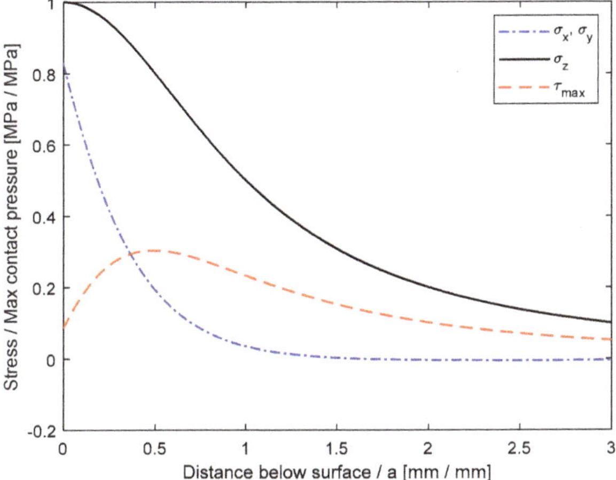

Figure 5. Subsurface stresses distributions.

2.3. Method for Surface Fatigue Damage Evaluation

Contact fatigue or surface fatigue is a particular phenomenon of fatigue that presents itself with characteristics different from the traditional ones and leads to the early and sometimes catastrophic wear and tear of the affected components. It is characterized by the formation of cavities on the surface, generally small, which can, however, grow, resulting in a characterization of the surface itself. Surface fatigue breaks derive from the repeated application of loads that produce stresses on the surfaces in contact and below them. The forms of damage due to superficial fatigue are pitting or spalling.

Surface fatigue resistance depends mostly on the surface resistance of the material, and therefore, it is necessary to readapt formulas generally used in classical fatigue before using them also for the evaluation of surface fatigue life.

The most used law for calculating the fatigue damage is the linear accumulation law of Palmgren-Miner [32,39,40]. This theory states that the component fatigue failure occurs when the cumulative damage is equal to one, i.e., when 100% of available life has been "consumed". The Palmgren-Miner rule is expressed as follows:

$$D = \sum_{i=1}^{n} \frac{n_i}{N_i}, \tag{20}$$

where n_i is the number of cycles to which the component is subject to a certain value of alternating stress and N_i is the number of cycles at which, given the value of alternating stress, the component is able to resist. This value is uniquely determined by the Wöhler curve (fatigue strength curve) which is described by the following Equation:

$$S_f = a \cdot N^b, \tag{21}$$

Parameters a and b respectively represent the intercept and slope of the above curve. For the calculation of the damage, however, it is convenient to rearrange Equation (21) as follows:

$$N_i = \left(\frac{\sigma_{a,i}}{a}\right)^{1/b}, \tag{22}$$

where $\sigma_{a,i}$ is the alternating stress value to which the component is subjected, obtainable by a cycle counting algorithms. Substituting Equation (22) in Equation (20), it is possible to obtain the following formula that is most useful for calculating the fatigue damage:

$$D = \sum_{i=1}^{N} \frac{n_i}{\left(\frac{\sigma_{a,i}}{a}\right)^{1/b}}, \tag{23}$$

Equation (23) is the most used formula for calculating the classical fatigue life of a component subjected to fluctuating loads over time. With some consideration, it is possible to use both the fatigue curve formulation and the damage formula to surface fatigue problems.

Considering what was stated in Section 2.2, when two bodies are forced to roll against each other, an inversion of the tangential stress occurs. This fact makes possible the hypothesis that the bodies are subjected to a purely alternate stress state (Figure 6).

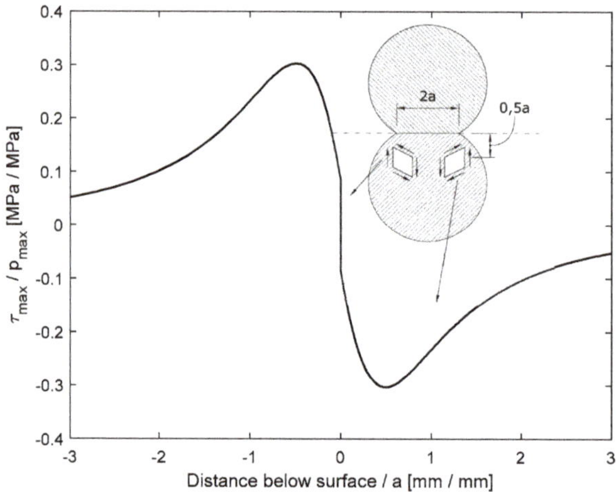

Figure 6. The tangential stress after and before rolling contact.

Recalling the Tresca criterion [41,42], which states that the surface stress is twice the maximum tangential stress and taking advantage of the previous hypothesis, it is possible to affirm that even the surface stress is purely alternate, and therefore, it is the alternating stress of Equations (22) and (23) ($\sigma_{id} = 2{\cdot}\tau_{\max}$). The previous equations are indeed still valid. In Equation (23), n_i is the number of cycles to which the component is subjected and $\sigma_{a,i} = \sigma_{id}$ is the alternating stress at the ith cycles.

Recalling what was stated in Section 2.2, i.e., that the maximum tangential stress is 0.3 times the maximum pressure ($\tau_{\max} = 0.3{\cdot}p_{\max}$), it is possible to rewrite Equations (21) and (23) as a function of the contact pressure. In such a way, it is possible to evaluate the fatigue life of the component directly from the computed contact pressures.

The fatigue strength curve, expressed as a function of the contact pressure, is therefore described by the following equation:

$$p_f = a_s{\cdot}N^b, \tag{24}$$

where p_f is the fatigue limit in terms of pressure and $a_s = \frac{a}{0.6}$. As a consequence, Equation (23) can also be rearranged as well as a function of the contact pressure, as shown in Equation (25).

$$D = \sum_{i=1}^{N} \frac{n_i}{\left(\frac{p_i}{a_s}\right)^{1/b}}, \tag{25}$$

As in Equation (23), n_i is the number of impulses to which the component is subject and p_i is the pressure generated by the ith impulse.

With the previous formulas, it is possible to write the fatigue strength curve of the material according to both the surface stress and the contact pressure. The same is valid for the formula used to calculate the surface fatigue damage.

3. Theoretical vs. Numerical: The Birth of a Hybrid Method

In the previous paragraph, the theoretical approach for contact stress evaluation and contact fatigue prediction was briefly explained. The theoretical method exploits the kinematical laws [43] and the Hertz theory [27], determining the surface fatigue life of the component. A second option that can be followed by the mechanical designer is a completely numerical approach that, exploiting the potentialities of explicit dynamic analysis codes, allows the very accurate determination of both the forces and the contact pressures. To demonstrate the potentials and limitations of both approaches, a simple test case has been analyzed, i.e., a sphere that falls from a 26-mm height on a plane, determining both impact forces and contact pressures. To this aim, the inputs shown in Table 1 were considered.

Table 1. The geometrical and material parameters used in the simulation.

	Description	Symbol	Value	Unit
Sphere (Mat. 100 Cr6)	Sphere diameter	r	6.3	mm
	Curvature radius (First direction)	r_{11}	3.15	mm
	Curvature radius (Second direction)	r_{12}	3.15	mm
	Weight	m	0.001	Kg
	Density	ρ	7.85×10^{-6}	$\frac{Kg}{mm^3}$
	Young modulus	E_1	205,000	MPa
	Poisson coefficient	ν_1	0.3	-
Plane (Mat. 17–4 PH)	Curvature radius (First direction)	r_{21}	$+\infty$	mm
	Curvature radius (Second direction)	r_{22}	$+\infty$	mm
	Young modulus	E_2	196,500	MPa
	Poisson coefficient	ν_2	0.27	-

It has been imposed that the sphere falls perpendicular to the plane and, therefore, with an angle of 90°. From the inputs of Table 1 the estimated impact velocity is 520 mm/s. Taking advantage of the equations described in Section 2.1, a contact force of 77.7 N and a maximum contact pressure of 2624 MPa was determined.

The same case study was reproduced in an explicit finite element environment (LS-DYNA FEM code) in which the same parameters previously obtained were evaluated. Figure 7 shows the initial instant of the simulation and the used mesh. 4-node tetrahedral elements were used both for the sphere and for the impact plane. The simulation foresees 1,199,262 elements and 4,797,048 nodes. To model the contact between two bodies, the contact type "automatic surface to surface" available in LS-DYNA was used.

To reduce the calculation effort, the sphere was placed near the contact plane and the analytically calculated initial velocity was set. This allowed for a drastic reduction in the calculation effort.

Figure 8 shows the trends of the contact force obtained from the numerical analysis as the mesh size changes. The obtained results show that the mesh has a negligible influence on the calculation of the contact force. The obtained force values are identical to those theoretically obtained. Figure 9 instead shows the contact pressure trends obtained from the numerical analysis when the mesh dimensions vary.

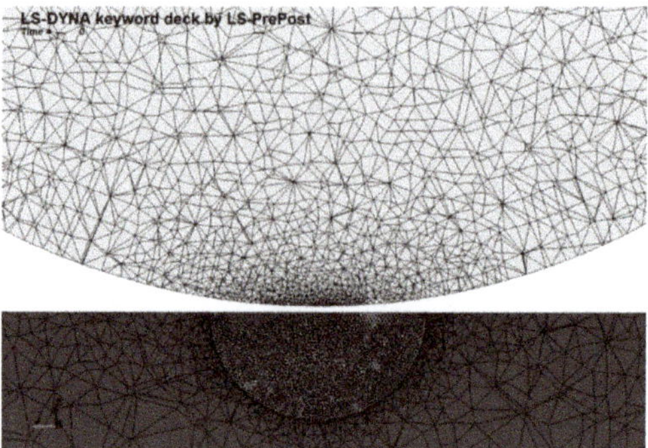

Figure 7. The initial position of the sphere in a dynamic simulation.

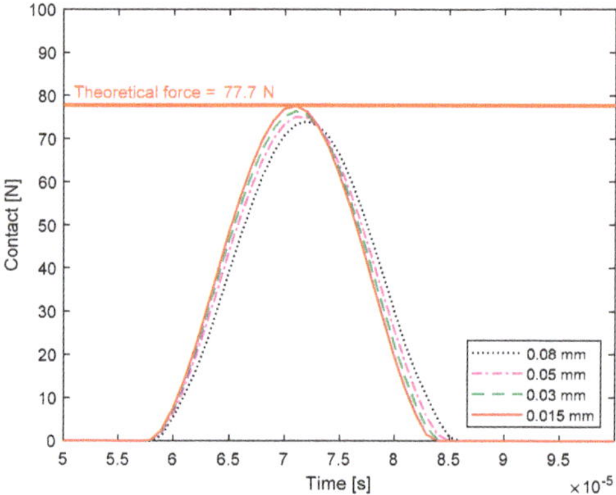

Figure 8. A comparison between the theoretical and numerical impact forces obtained for different mesh size.

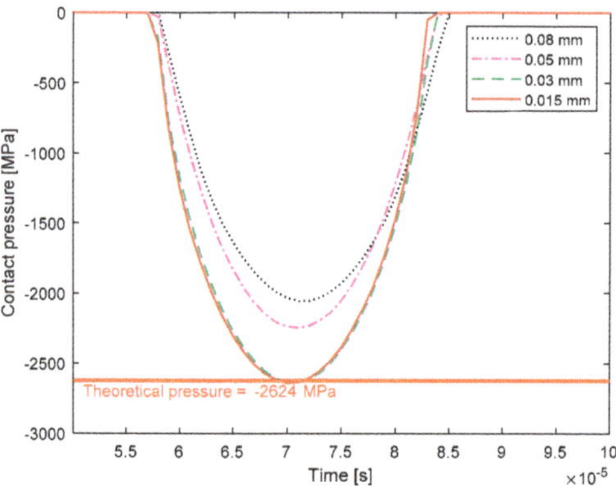

Figure 9. A comparison between the theoretical and numerical contact pressure obtained for different mesh sizes.

It is clear that when the mesh varies, the results are significantly different. To get results close to the theoretical ones, it is necessary to use a very fine mesh. This fact is directly related to the calculation time necessary for the simulation, which is as much larger as the mesh is finer.

From the results thus obtained, one might think that a completely theoretical approach may be sufficient to determine both impact forces and contact pressures without the aid of explicit FEM codes. It is important to highlight, however, that, to accurately determine both impact forces and contact pressures exploiting theoretical laws, it is necessary to assess a set of variables such as the trajectory of the bodies, the speed of impact, the radii of curvatures, the stiffness of the bodies, etc. In complex real models these variables are not a priori known since they are random variable. For this reason, a suitable way would be a stochastic approach.

In this context, the hybrid method proposed in this activity represents a way out to bypass the weakness of previously described approaches. Exploiting the potentialities of the finite element analysis software (where initial variables are automatically known), it is possible to obtain contact forces in a relatively short time, since forces are sufficiently independent from the mesh size. Once contact forces are known, contact stresses can be accurately computed using the Hertz theory [27] (See Section 2.2) and the fatigue life can be easily addressed. In such a way, it is possible to not deal with such complex models to often not be feasible. In Figure 10, a summarizing flow chart is proposed.

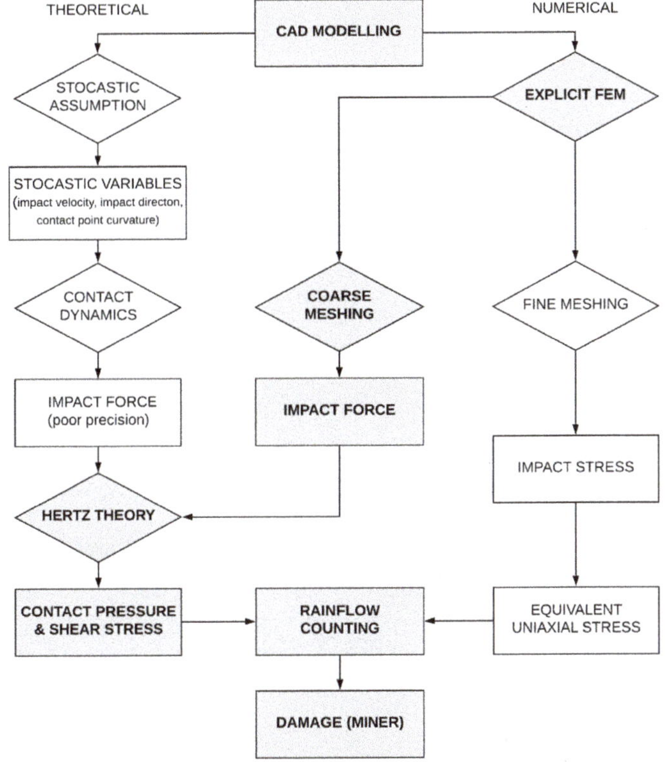

Figure 10. A flow chart summarizing the theoretical and numerical methods (white boxes) and the proposed approach (grey boxes).

4. Test Case

To evaluate the goodness of the proposed hybrid method for calculating the surface fatigue life of stainless-steel components, an explicit finite elements analysis was performed on the recirculating system of a ball screw using LS-DYNA FEM code. It often happens that these components, due to their function, are subjected to a series of impulses that may cause surface fatigue problems. The recirculating ball screw used in this activity is an existing ball screw used in different applications. This has a center sphere diameter of 40 mm, a screw pitch equal to 20 mm, a sphere diameter of 6.35 mm, and a radial insert. Since attention was paid to the recirculation system and, therefore, only to an evaluation of the fatigue life of the recirculating mechanism only to reduce the simulation effort, it was decided to model only the ball screw and ball-nut surfaces that act as a guide for the sphere through shell elements. A total of 57,918 elements and 231,672 were used to mesh the ball screw and ball-nut surfaces. The sphere and the insert have been modelled with solid 4-node tetrahedral elements. A total of 246,533 element and 986,132 nodes were used for the sphere and the recirculating mechanism. This allowed for a good approximation in the curvature, both of the sphere and of the recirculation mechanism, thus obtaining a good approximation with regard to the physics of the impact that has been simulated. In this activity, the roughness of cooperating surfaces was set equal to zero due to the accuracy of the manufacturing process.

The following figure shows the model used for this analysis.

The recirculating mechanism of Figure 11 is made of 17–4 PH steel: 17–4 PH stainless steel is, in fact, one of the steels obtained by precipitation and hardening most used in aeronautics. In fact, this material, besides being stainless, offers excellent mechanical characteristics. Table 2 summarizes some features of the material.

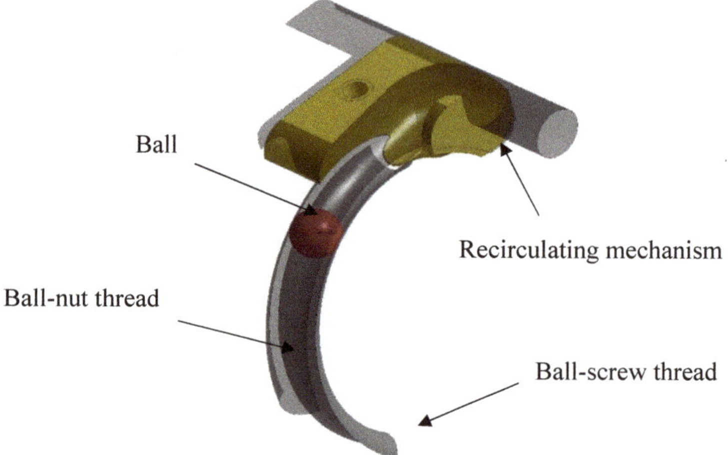

Figure 11. The finite element model of the recirculating mechanism.

Table 2. The 17–4 PH material parameters.

Description	Symbol	Value	Unit
Young Modulus	E	196,500	MPa
Poisson Modulus	v	0.27	-
Density	ϱ	7700	Kg/m^3
Ultimate tensile strength	S_{ut}	1241	MPa
Yield strength	S_y	993	MPa
Hardness Brinell	HB	388	-

Figure 12 shows the mesh obtained with the parameters previously illustrated.

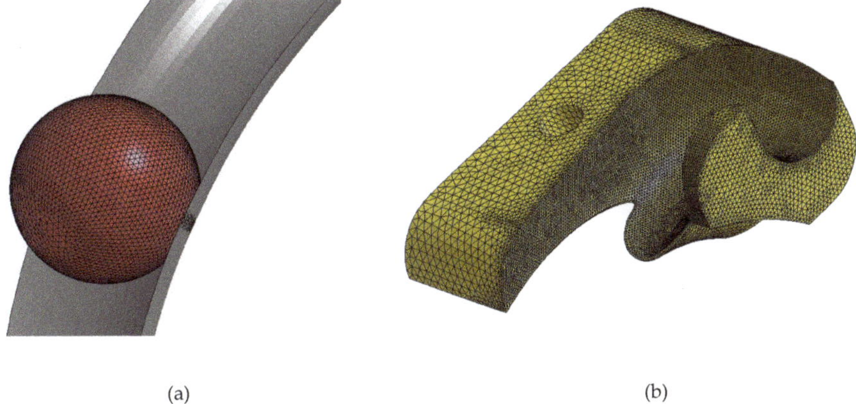

(a) (b)

Figure 12. Detailed views of the mesh: (**a**) sphere and (**b**) a recirculating mechanism.

About the boundary conditions, the recirculating system has been fixed in correspondence with the hole. This hole is, in fact, used to fasten the recirculation mechanism on the seat obtained in the ball nut. Furthermore, five degrees of freedom of the ball nut were constrained (three rotations and two displacement), as this system is a rotating ball screw. The principal inertia tensor and the mass of the total ball screw were assigned to the ball-screw surface with the aim of reproducing the real movement of the ball screw itself. This means that the ball screw only rotates while the ball nut performs a linear motion. A rotation speed of the screw was set equal to 3000 RPM. The contact type "automatic surface to surface" allowable in LS-DYNA was used to model the contact between the sphere and the recirculating mechanism.

To model the system as more realistic as possible, it is necessary to consider also a viscosity dissipation inherent the material. For this purpose, a percentage damping $\xi = 0.02$ was imposed in a frequency range between 1000 Hz and 50,000 Hz. This frequency range was taken into consideration following a modal analysis of the recirculation system only, imposing the same conditions of constraint of the dynamic analysis. What emerged from the modal analysis is that the vibrating modes that mostly participate in the total vibration are the first 7 and are all in the range between 1000 Hz and 50,000 Hz.

The impact forces used to evaluate the surface fatigue life were extrapolated from the performed simulation. Figure 13 shows the trend of the impact force as a function of time. As visible, the maximum value is equal to about 120 N. To evaluate the fatigue life of the component, first the fatigue strength curves expressed in terms of alternating surface stress and alternating pressure were first obtained for three different level of reliability, equal to 10%, 50%, and 90%, adopting Equation (21) shown in Section 2.3 and using the characteristics of the material shown in Table 2. Starting from the hardness of the material, corresponding to HB 388 MPa, the surface ultimate stress S_{ut} of the material was obtained according the Equation (26).

$$S_{ut} = 3.45,\ HB = 1458.6\ \text{MPa}, \tag{26}$$

The fatigue strength curves, both in terms of the alternating stress and contact pressure are shown in Figure 14. As shown in Figure 14 for the slope of the fatigue curve after 10^6 cycles, the hypothesis of Hybach [31,41] was taken into account. The Hybach hypothesis states that the slope of the fatigue curve after 10^6 cycles is one third of the slope of the previous stretch.

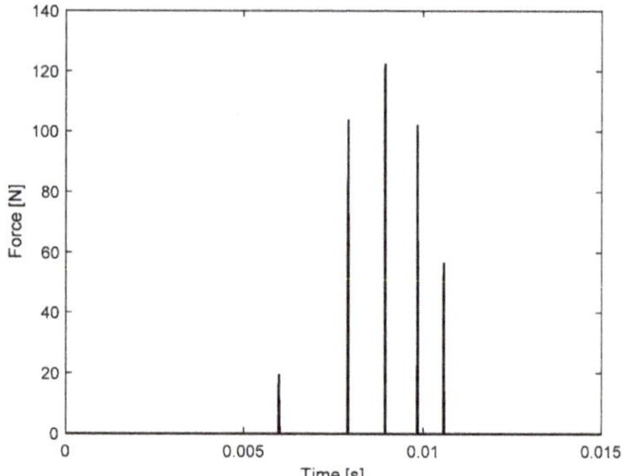

Figure 13. The impact forces by FE explicit analysis.

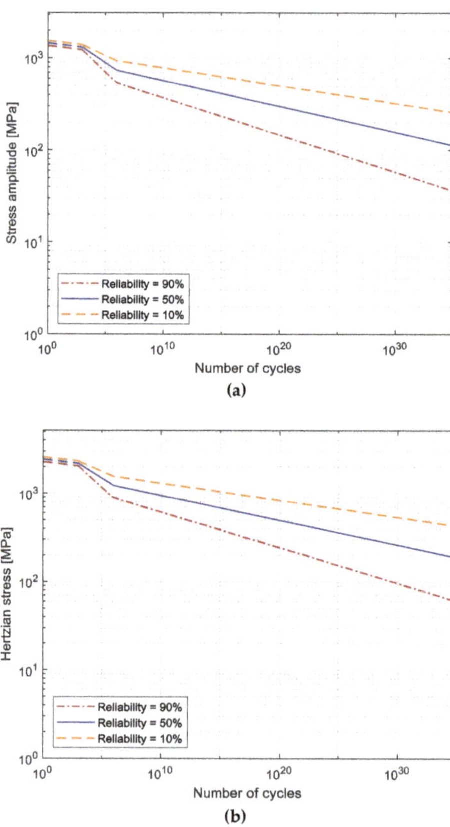

Figure 14. The fatigue curve for the recirculating mechanism: (**a**) the fatigue curve in terms of alternating stress and (**b**) the fatigue curve in terms of pressure.

The ball screw in question has been used for a total duration of 140 km. Considering a screw pitch equal to 20 mm and a number of spheres per turn equal to 8, the total duration in number of cycles corresponds to 5.6×10^7 cycles. The alternating surface stress and contact pressure values obtainable from the curves of Figure 14 are shown in Table 3 for the different considered reliability values.

Table 3. The alternating strength and contact pressure for the designed number of cycles.

Reliability (%)	Number of Cycles -	Alternating Strength (MPa)	Pressure (MPa)
10	5.6×10^7	866.1	1443
50	5.6×10^7	658.1	1095
90	5.6×10^7	454.3	757.2

The values shown in Table 3 are also graphically shown in Figure 15 by the yellow dots. In order to estimate the surface fatigue life, it is still necessary to calculate the semiaxes of the contact ellipse using the Hertz theory (Section 2.1). From the 3D model, three different combinations of the curvature radii at the contact point have been calculated as shown in Table 4. Applying the maximum force obtained by the simulation, equal to 120 N, and considering the obtained curvature radii, it was possible to estimate the maximum alternating stress and the maximum pressure. These values are shown in Table 4.

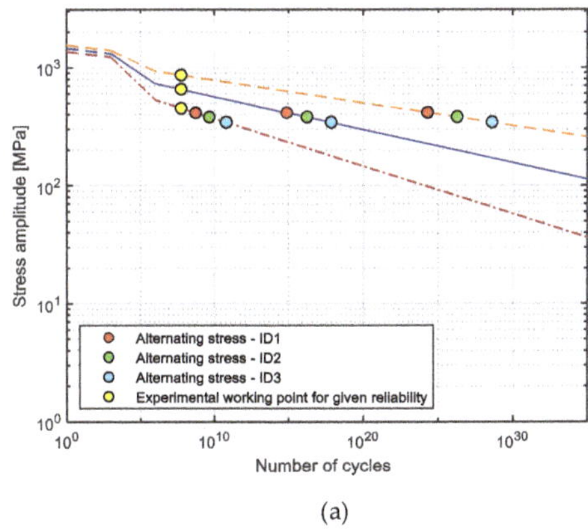

(a)

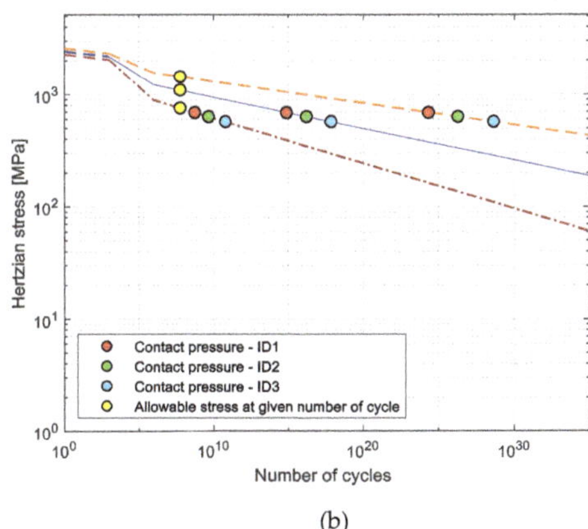

(b)

Figure 15. Comparisons between the working point obtained by the hybrid approach and the real ones: (**a**) the working point in terms of alternating stress and (**b**) the working points in terms of pressure.

Table 4. The alternating stress and contact pressure for the considered curvature radii with a contact force equal to 120 N.

ID	Sphere Curvature Radius (mm)	Recirculating Mechanism Curvature Radius (mm)	Alternating Stress (MPa)	Contact Pressure (MPa)
1	3.375	3.625	342.6	571
2	3.375	3.75	380.4	634
3	3.375	3.875	414.6	691

By tracing these values in the fatigue strength curves, it is possible to determine from the intersection with the curves the number of cycles to which the component can withstand for the three reliability values considered in terms of both the maximum alternating stress and the maximum pressure as shown in Figure 15. The obtained cycles at each surface stress values or contact pressure are shown in Table 5 for all the reliability values considered.

Table 5. The estimated allowable number of cycles for the considered curvature radii pairs and reliability levels.

ID	Alternating Stress (MPa)	Contact Pressure (MPa)	Number of Cycles	Reliability (%)
1	342.6	571	2.06×10^{24}	10
			7.75×10^{14}	50
			5.45×10^{8}	90
2	380.4	634	1.78×10^{26}	10
			1.68×10^{16}	50
			4.64×10^{9}	90
3	414.6	691	4.01×10^{28}	10
			7.09×10^{17}	50
			6.27×10^{10}	90

As shown in Figure 15 and in Table 5, these values attest that no surface fatigue problems are expected for this component. This fact was also found on the real recirculating mechanism, where no surface fatigue problem was detected by a visual inspection.

5. Conclusions

Stainless steels are necessarily used in all sectors, from aerospace to medical, where mechanical components find themselves working in environmental conditions full of aggressive substances that can affect the mechanical properties of the component itself. The antioxidant properties of such steels are due to the surface film that is created thanks to a chemical reaction between chromium and oxygen. Although stainless steels are, by nature, self-passivating, i.e., the protective film spontaneously recreates thanks to a chemical reaction between oxygen and chrome, problems related to a degradation of the surface can lead to a slowing down the self-passivation process or, in the worst cases, to a total loss of antioxidant properties. One of the main causes of surface deterioration is related to surface fatigue issues. When two or more evolving bodies are working against each other, a stress state is generated on the surface, which can cause opening cracks. The two most common phenomena are pitting and spalling. To minimize the probability of occurrences of this phenomenon, it is necessary that, at least during the design stage, the surface resistance of the component is verified.

The evaluation of surface fatigue strength can be addressed with two different approaches. The first one is a completely theoretical approach which, merging the impact dynamic laws and the Hertz theory, allows an estimation of the contact stresses and, therefore, an evaluation of the fatigue life of the component. However, this approach requires an assessment of a set of variables, such as bodies trajectory, velocity, stiffness, and curvature radii, in order to obtained accurate results. If these quantities are known, the theoretical approach allows for an accurate estimate of both the impact forces and the contact stresses. The determination of these quantities tends to be, however, particularly complex in real models in which all the abovementioned quantities are random variables.

In this context, the second possible approach for the estimation of surface fatigue strength is a completely numerical approach. This, exploiting the potentialities of the finite element codes, allows the drawbacks of the theoretical approach to be overcome regardless of the complexity of the model. The major drawback of this approach is the need to use meshes so fine that makes the simulation not always feasible.

To merge the benefits of both approaches, in this activity, a hybrid theoretical-numerical calculation method has been proposed, which allows a drastic reduction in the calculation effort obtaining, at the same time, accurate results. This method exploits the potentialities of finite element software to determine the impact forces, bypassing the inconvenience of assessing dynamic variables hardly a priori known. Once the impact forces are assessed, classical formulas of Hertz theory and fatigue are used to evaluate the fatigue life of the component. This possibility raised a comparison of the results obtained by an explicit finite element analysis and the theoretical results for a simple case study, a sphere-plane impact, for different mesh sizes. From the comparison carried out, it was highlighted that the force values obtained are independent from the mesh size and very close to, almost identical, those obtained by using the shock theory. It is not valid instead in the case of contact pressures where results are strongly influenced by mesh size. This means that to obtain the pressure and/or contact stress directly from a numerical simulation, a fine mesh is necessary.

The main advantage of the proposed method is to drastically reduce the calculation effort while obtaining results consistent with reality as well. Given these aspects, the proposed calculation method is aimed to all those realities in which it is necessary to reduce the computational effort as much as possible without precluding the goodness of the results.

The proposed hybrid method was then validated on an aeronautical component, a recirculation system of a ball screw. This component, in fact, is subjected to many shocks due to the balls that it is often subjected to surface fatigue problems. The obtained results show how it is possible to obtain realistic results in terms of fatigue life with relatively short calculation times.

Author Contributions: Methodology and supervision, F.C.; software and validation, M.C., M.P., and G.Z.; writing—review and editing, F.C., M.P., and G.Z.

Funding: The research received no external funding.

Conflicts of Interest: The authors declare no conflict of interest.

References

1. Rufini, R.; Di Pietro, O.; Di Schino, A. Predictive Simulation of Plastic Processing of Welded Stainless Steel Pipes. *Metals* **2018**, *8*, 519. [CrossRef]
2. Ciuffini, A.F.; Barella, S.; Di Cecca, C.; Di Schino, A.; Gruttadauria, A.; Napoli, G.; Mapelli, C. Transformation-Induced Plasticity in Super Duplex Stainless Steel F55- UNS S32760. *Metals* **2019**, *9*, 191. [CrossRef]
3. Corradi, M.; Di Schino, A.; Borri, A.; Rufini, R. A review of the use of stainless steel for masonry repair and reinforcement. *Constr. Build. Mater.* **2018**, *181*, 335–346. [CrossRef]
4. Di Schino, A.; Porcu, G.; Turconi, G.L. Tenaris: Metallurgical design and development of C125 grade for mild sour service application. In Proceedings of the Corrosion 2006, San Diego, CA, USA, 12–16 March 2006; pp. 1–14.
5. Aurégan, G.; Fridrici, V.; Kapsa, P.; Rodrigues, F. Experimental simulation of rolling-sliding contact for application to planetary roller screw mechanism. *Wear* **2015**, *332–333*, 1176–1184. [CrossRef]
6. Talha, M.; Behera, C.K.; Sinha, O.P. A review on nickel-free nitrogen containing austenitic stainless steels for biomedical applications. *Mater. Sci. Eng. C* **2013**, *33*, 3563–3575. [CrossRef] [PubMed]
7. Yang, K.; Ren, Y. Nickel-free austenitic stainless steels for medical applications. *Sci. Technol. Adv. Mater.* **2010**, *11*, 1–13. [CrossRef]

8. Boulané-Petermann, L. Processes of bioadhesion on stainless steel surfaces and cleanability: A review with special reference to the food industry. *Biofouling* **1996**, *10*, 275–300. [CrossRef]
9. Schweitzer, P.A. *Fundamentals of Metallic Corrosion: Atmospheric and Media Corrosion of Metals. Corrosion Engineering Handbook*; CRC Press/Taylor & Francis Group: Boca Raton, FL, USA, 2006; ISBN 0849382432.
10. Okamoto, G. Passive film of 18-8 stainless steel structure and its function. *Corros. Sci.* **1973**, *13*, 471–489. [CrossRef]
11. Zheng, Z.B.; Zheng, Y.G. Effects of surface treatments on the corrosion and erosion-corrosion of 304 stainless steel in 3.5% NaCl solution. *Corros. Sci.* **2016**, *112*, 657–668. [CrossRef]
12. Olefjord, I.; Wegrelius, L. The influence of nitrogen on the passivation of stainless steels. *Corros. Sci.* **1996**, *38*, 1203–1220. [CrossRef]
13. Barbosa, M.A. The pitting resistance of AISI 316 stainless steel passivated in diluted nitric acid. *Corros. Sci.* **1983**, *23*, 1293–1305. [CrossRef]
14. Arrabal, R.; Pardo, A.; Merino, M.C.; Coy, A.E.; Viejo, F.; Matykina, E. Effect of Mo and Mn additions on the corrosion behaviour of AISI 304 and 316 stainless steels in H_2SO_4. *Corros. Sci.* **2008**, *50*, 780–794.
15. Kemp, M.; van Bennekom, A.; Robinson, F.P.A. Evaluation of the corrosion and mechanical properties of a range of experimental CrMn stainless steels. *Mater. Sci. Eng. A* **1995**, *199*, 183–194. [CrossRef]
16. Pohjanne, P.; Carpén, L.; Hakkarainen, T.; Kinnunen, P. A method to predict pitting corrosion of stainless steels in evaporative conditions. *J. Constr. Steel Res.* **2008**, *64*, 1325–1331. [CrossRef]
17. Greiner, R.; Kettler, M. Interaction of bending and axial compression of stainless steel members. *J. Constr. Steel Res.* **2008**, *64*, 1217–1224. [CrossRef]
18. Luecke, W.E.; Slotwinski, J.A. Mechanical Properties of Austenitic Stainless Steel Made by Additive Manufacturing. *J. Res. Natl. Inst. Stand. Technol.* **2014**, *119*, 398–418. [CrossRef] [PubMed]
19. Nezhadfar, P.D.; Shrestha, R.; Phan, N.; Shamsaei, N. Fatigue behavior of additively manufactured 17-4 PH stainless steel: Synergistic effects of surface roughness and heat treatment. *Int. J. Fatigue* **2019**, *124*, 188–204. [CrossRef]
20. Siddiqui, S.F.; Fasoro, A.A.; Cole, C.; Gordon, A.P. Mechanical Characterization and Modeling of Direct Metal Laser Sintered Stainless Steel GP1. *J. Eng. Mater. Technol.* **2019**, *141*, 031009. [CrossRef]
21. Wang, Y.; Charbal, A.; Hild, F.; Roux, S.; Vincent, L. Crack initiation and propagation under thermal fatigue of austenitic stainless steel. *Int. J. Fatigue* **2019**, *124*, 149–166. [CrossRef]
22. Carneiro, L.; Jalalahmadi, B.; Ashtekar, A.; Jiang, Y. Cyclic deformation and fatigue behavior of additively manufactured 17–4 PH stainless steel. *Int. J. Fatigue* **2019**, *123*, 22–30. [CrossRef]
23. Murakami, Y.; Klinger, C.; Madia, M.; Zerbst, U.; Bettge, D. Defects as a root cause of fatigue failure of metallic components. III: Cavities, dents, corrosion pits, scratches. *Eng. Fail. Anal.* **2019**, *97*, 759–776.
24. Totten, G.E. *Friction, Lubrication, and Wear Technology*; ASM International: Novelty, OH, USA, 2018; Volume 18, ISBN 978-1-62708-192-4.
25. Kattelus, J.; Miettinen, J.; Lehtovaara, A. Detection of gear pitting failure progression with on-line particle monitoring. *Tribol. Int.* **2018**, *118*, 458–464. [CrossRef]
26. Barber, J.; Ciavarella, M. Contact mechanics. *Int. J. Solids Struct.* **2000**, *37*, 29–43. [CrossRef]
27. Hertz, H. On the contact of rigid elastic solids. *J. fur die Reine und Angew. Math.* **1882**, *92*, 156–171.
28. Bathe, K.-J. *Finite Element Procedures*; Prentice Hall: Upper Saddle River, NJ, USA, 2016; ISBN 9780979004957.
29. de los Rios, E.R.; Walley, A.; Milan, M.T.; Hammersley, G. Fatigue crack initiation and propagation on shot-peened surfaces in A316 stainless steel. *Int. J. Fatigue* **1995**, *17*, 493–499. [CrossRef]
30. Maiya, P.S.; Busch, D.E. Effect of surface roughness on low-cycle fatigue behavior of type 304 stainless steel. *Metall. Trans. A* **1975**, *6*, 1543–1940. [CrossRef]
31. Huang, H.W.; Wang, Z.B.; Lu, J.; Lu, K. Fatigue behaviors of AISI 316L stainless steel with a gradient nanostructured surface layer. *Acta Mater.* **2015**, *87*, 150–160. [CrossRef]
32. Fatemi, A.; Yang, L. Cumulative fatigue damage and life prediction theories: A survey of the state of the art for homogeneous materials. *Int. J. Fatigue* **1998**, *20*, 9–34. [CrossRef]
33. Wei, C.C.; Lin, J.F. Kinematic Analysis of the Ball Screw Mechanism Considering Variable Contact Angles and Elastic Deformations. *J. Mech. Des.* **2004**, *125*, 717–733. [CrossRef]
34. Lin, M.C.; Ravani, B.; Velinsky, S.A. Kinematics of the Ball Screw Mechanism. *J. Mech. Des.* **2008**, *116*, 849–855. [CrossRef]

35. Jiang, H.; Song, X.; Xu, X.; Tang, W.; Zhang, C.; Han, Y. Multibody dynamics simulation of balls impact-contact mechanics in ball screw mechanism. In Proceedings of the International Conference on Electrical and Control Engineering (ICECE 2010), Wuhan, China, 25–27 June 2010.
36. Braccesi, C.; Landi, L. A general elastic-plastic approach to impact analisys for stress state limit evaluation in ball screw bearings return system. *Int. J. Impact Eng.* **2007**, *34*, 1272–1285. [CrossRef]
37. Zhang, Z.Y.; Zhang, W.L.; Xu, C. Research on the Impact between Balls and Re-Circulating Mechanism of Ball Screws. *Appl. Mech. Mater.* **2014**, *551*, 370–377. [CrossRef]
38. Hung, J.P.; Wu, J.S.-S.; Chiu, J.Y. Impact failure analysis of re-circulating mechanism in ball screw. *Eng. Fail. Anal.* **2004**, *11*, 561–573. [CrossRef]
39. Braccesi, C.; Cianetti, F.; Lori, G.; Pioli, D. An equivalent uniaxial stress process for fatigue life estimation of mechanical components under multiaxial stress conditions. *Int. J. Fatigue* **2008**, *30*, 1479–1497. [CrossRef]
40. Braccesi, C.; Morettini, G.; Cianetti, F.; Palmieri, M. Development of a new simple energy method for life prediction in multiaxial fatigue. *Int. J. Fatigue* **2018**, *112*, 1–8. [CrossRef]
41. Budynas, R.G.; Nisbett, J.K. *Shigley's Mechanical Engineering Design, Eighth Edition*; McGraw-Hill: New York, NY, USA, 2008; ISBN 0390764876.
42. Collins, J.A.; Smith, C.O. *Failure of Materials in Mechanical Design*; Wiley Interscience: New York, NY, USA, 1993; ISBN 0471558915.
43. Shabana, A.A. *Dynamics of Multibody Systems*; Cambridge University Press: Chicago, IL, USA, 2009; ISBN 9781107337213.

© 2019 by the authors. Licensee MDPI, Basel, Switzerland. This article is an open access article distributed under the terms and conditions of the Creative Commons Attribution (CC BY) license (http://creativecommons.org/licenses/by/4.0/).

Article

Microstructure and Hot Deformation Behaviour of a Novel Zr-Alloyed High-Boron Steel

Alexey Prosviryakov [1,*], Baptiste Mondoloni [1,2], Alexander Churyumov [1] and Andrey Pozdniakov [1]

[1] National University of Science and Technology "MISiS", Leninskiy Prospekt 4, 119049 Moscow, Russia; baptiste.mondoloni@edf.fr (B.M.); churyumov@misis.ru (A.C.); pozdniakov@misis.ru (A.P.)

[2] French electricity supplier "EDF", 86320 Civaux, France

* Correspondence: pro.alex@mail.ru; Tel.: +007-495-955-0134

Received: 14 January 2019; Accepted: 8 February 2019; Published: 12 February 2019

Abstract: A novel corrosion-resistant steel with high boron content is investigated in this paper. Three stages during crystallisation of the steel are revealed. The positive influence of Zr addition on the microstructure and mechanical properties after hot deformation is shown. The Zr-alloyed steel demonstrates hot deformation without fracturing in the temperature range of 1273–1423 K, and in the strain rate range of 0.1–10 s^{-1}, despite the high volume of brittle borides. The processes of ferrite recrystallisation and boride structure fragmentation occur during hot deformation, promoting the appearance of a peak on stress–strain curves.

Keywords: high-boron steel; high-speed steel; hot deformation; mechanical properties; borides

---

1. Introduction

Stainless steels with high boron content (more than 1.2 wt. %) are important materials for the production of spent nuclear fuel storage due to the high neutron-absorbing capacity of boron [1,2]. Conversion of the nuclear power plants to more enriched fuel requires improved neutron-absorbing capacity of the materials. As a result, new high-boron steels with a boron concentration more than 3 wt. % are required to guarantee the safety and efficiency of spent nuclear fuel storage. This type of material is also very attractive for application as a high-speed steel, due to a high degree of hardness and good wear resistance of the borides [3,4]. However, the low technological plasticity of high-boron steels limits their applications because of the high costs of their production. In this case, the preliminary investigation of the microstructure and deformation behaviour of hard deformable steels is required to develop improved industrial technologies. In most cases, the deformation behaviour is investigated by mechanical tests using thermomechanical simulators, followed by the development of constitutive models [5–7] and processing maps [8,9], which are the power instruments for finding the optimal deformation conditions. Besides, the addition of the steel by the borides forming elements such as Ti and Zr may have a positive influence on the microstructure. Titanium forms TiB_2 phase which contains 1.40 g/cm^3 of boron in comparison with $\approx$0.68 g/cm^3 for $(Cr,Fe)_2B$. It decreases the amount of brittle phases in a high-boron steel microstructure [10,11]. He et al. showed that Ti improves the strength and plasticity of the binary Fe–B alloys after hot deformation [12,13]. The purpose of our work is to investigate the microstructure and hot deformation behaviour of a novel Zr-alloyed stainless steel with high boron content.

2. Materials and Methods

An alloy with the nominal composition of Fe–16Cr–4Ti–6Zr–3.2B (wt %) was produced by arc melting of pure Fe, Cr, Ti and Zr metals and Fe-17 wt. % B master alloy in an argon atmosphere.

After remelting had been undertaken four times, the samples were produced by casting into a copper mould with a diameter of 12.7 mm. Investigation of the alloy microstructure was performed by scanning electron microscopy (SEM) and by X-ray diffraction (XRD). The mean size of the particles was measured by a linear intercept method. For each sample, 3–5 different 50 × 40 μm^2 regions were analyzed. Compression tests were performed using a Gleeble 3800 thermomechanical simulator. The specimens for the compression tests, with a diameter of 10 mm and a height of 15 mm, were heated to the deformation temperature (1273–1423 K) and compressed to one true strain at 0.1, 1 and 10 s^{-1} constant true strain rates. Compression tests at room temperature were carried out on cylindrical specimens (6 mm in height and 4 mm in diameter) cut from the hot compressed samples. Melting and solidification temperatures of the alloy were measured by differential thermal analysis (DTA) at the heating and cooling rate of 0.33 K/s. Accelerated corrosion test [14] was applied to the investigated steel and for comparison to the industrial steel Fe–14Cr–5Ti–0.3V–1.8B (ChS-82) which is most common for production of spent nuclear fuel storage. The samples with the size of 12.7 mm in diameter and 1 mm in width were held 5 and 24 h in the 10% $FeCl_3 \cdot 6H_2O$ water solution. After the holding, the mass loss rate was calculated according to the formula:

$$V_m = \frac{\Delta m}{St} \tag{1}$$

where Δm is a lost mass, g; S is specimen area, m^2; t is time of the corrosion test, hours.

3. Results and Discussion

As shown in Figure 1, the microstructure of as-cast steel consists of α(Fe) solid solution and three types of borides: black particles of TiB_2, grey eutectic colonies of $(Cr,Fe)_2B$ and white particles of ZrB_2. Usually, the structure of Ti-doped steel with a high boron concentration contains large TiB_2 particles with an irregular shape [15]. Zr addition strictly decreases the volume fraction of these particles. As one can see in Figure 1a, few large TiB_2 particles are present in the microstructure. Zr also has a positive influence on the size of the α(Fe)–$(Cr,Fe)_2B$ eutectic colonies due to a modifying effect [16]. The average size of the borides is presented in Table 1.

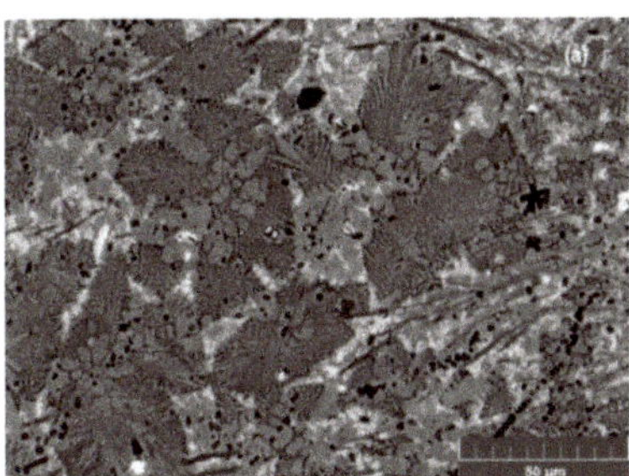

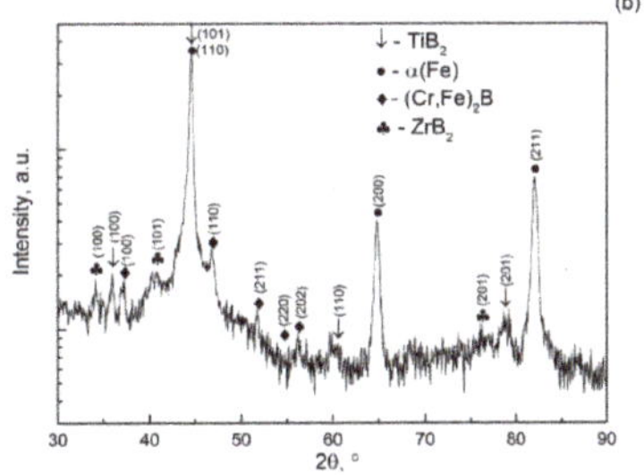

Figure 1. As-cast microstructure (SEM) of the investigated steel (**a**) and XRD pattern (**b**).

Table 1. The size of the borides in the as-cast state and after hot deformation (HD).

State	Average Size of the Particles, μm		
	TiB_2	$(Cr,Fe)_2B$	ZrB_2
As-cast	3 ± 1	0.7 ± 0.2	2 ± 0.6
HD 1373 K, 1 s^{-1}	2 ± 0.5	0.5 ± 0.1	1.8 ± 0.6
HD 1373 K, 10 s^{-1}	1.5 ± 0.5	0.4 ± 0.1	1.4 ± 0.4
HD 1423 K, 1 s^{-1}	1.2 ± 0.3	0.5 ± 0.1	1 ± 0.3
HD 1423 K, 10 s^{-1}	1.1 ± 0.3	0.6 ± 0.1	1 ± 0.2

As can be seen in Figure 2a, the liquidus and solidus temperatures of the steel are 1530 K and 1490 K, respectively. The exothermal crystallisation peak may be separated into three peaks (Figure 2b). Each of the peaks corresponds to a different crystallisation reaction. The high temperature peak corresponds to the reaction of the crystallisation of ZrB_2 + α(Fe) from the liquid phase, which can be proven by the absence of the high temperature peak in the DTA curve of the alloy without Zr [17]. The reaction L → TiB_2 + α(Fe) leads to the exothermic effect seen in the second peak. The last peak corresponds to the eutectic crystallisation of ((Cr,Fe)$_2$B + α(Fe)).

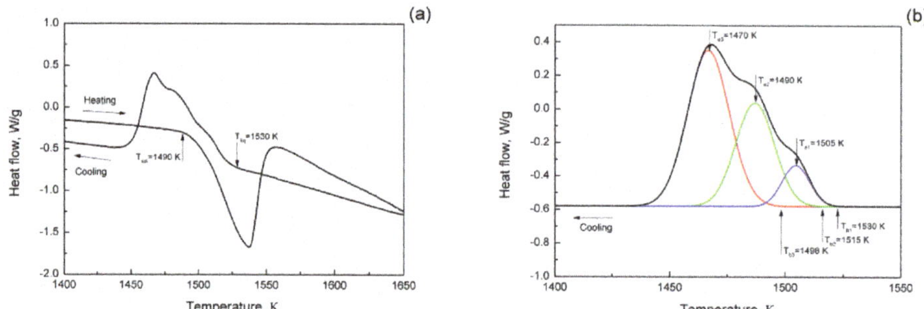

Figure 2. DTA curves (**a**) and crystallisation peak analysis (**b**) of the investigated alloy.

The maximal deformation temperature of the alloy of 1423 K was chosen according to the DTA to prevent the melting of the sample during the compression test from adiabatic heating. As one can see in Figure 3, the hot compression of the Fe–16Cr–4Ti–6Zr–3.2B steel proceeds without fractures in the temperature range of 1273–1423 K, and in the strain rate range of 0.1–10 s^{-1}. As shown in Figure 4, the stress is increased with the increase of the strain rate and decrease of the deformation temperature. True stress achieves its maximal value at low strains during compression for all temperatures and strain rates. It is known [18,19] that peak strain (ε_P) can be expressed as a power-law function of the Zener–Hollomon parameter $Z = \dot{\varepsilon}\exp\left(\frac{Q}{RT}\right)$ [20] as given by:

$$\varepsilon_P = AZ^m = A\left[\dot{\varepsilon}\exp\left(\frac{Q}{RT}\right)\right]^m \tag{2}$$

where $\dot{\varepsilon}$ is a strain rate, s^{-1}; T is a temperature, K; R is a universal gas constant, A; m and Q are the constants determined by least squares method using experimental data and linearized Equation (2):

$$\ln \varepsilon_P = \ln A + m \ln \dot{\varepsilon} + \frac{mQ}{RT} \tag{3}$$

So, the εp can be calculated by the follow equation:

$$\varepsilon_P = 0.001Z^{0.22} = 0.001\left[\dot{\varepsilon}\exp\left(\frac{3,4900}{RT}\right)\right]^{0.22} \quad (R^2 = 0.81), \tag{4}$$

where $\dot{\varepsilon}$ is the strain rate (s^{-1}) and T is the deformation temperature (K). After reaching the peak strain rate, the processes of fragmentation of the boride colonies and the dynamic recrystallisation begins (Figure 5), leading to a significant decrease in the true stress (Figure 4). The steady state deformation proceeds after 0.5–0.6 of true stress, depending on the temperature and the strain rate. As shown in Figure 5 and Table 1, the size of the ZrB_2 and TiB_2 borides is decreased after hot deformation has occurred. An increase of the deformation temperature decreases the size of these particles. Contrariwise, the size of the (Cr,Fe)$_2$B is not decreased during the deformation due to the initial small size of the eutectic colonies. As shown in Figure 6b,c, the microstructure of the steel after hot deformation at the temperature of 1373 K and at the strain rate 1 s^{-1} and at the temperature of 1423 K and at the strain rate 10 s^{-1} consists mainly of small recrystallised grains, whereas in the phase composition of the steel did not change significantly after the hot compression (Figure 6c).

Figure 3. The images of the samples after hot deformation.

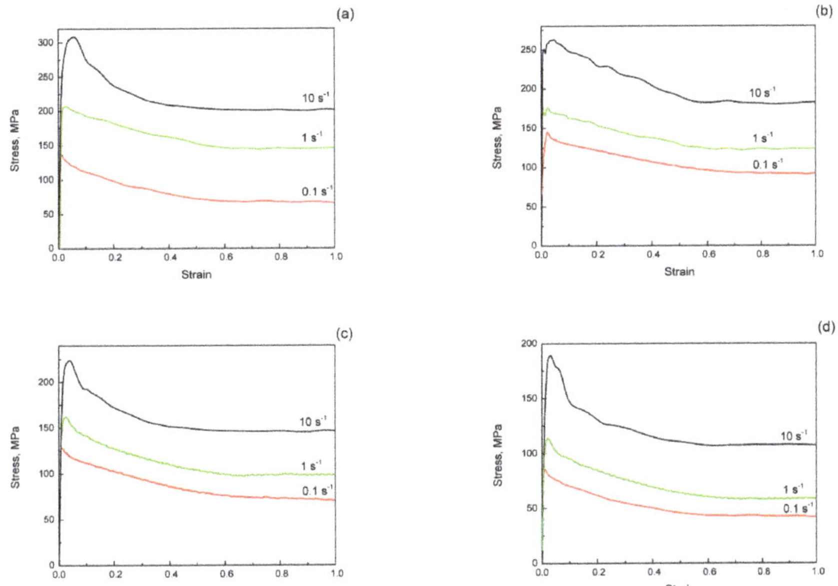

Figure 4. Stress–strain curves of the Fe–16Cr–4Ti–6Zr–3.2B steel at temperatures of 1273 K (**a**), 1323 K (**b**), 1373 K (**c**) and 1423 K (**d**).

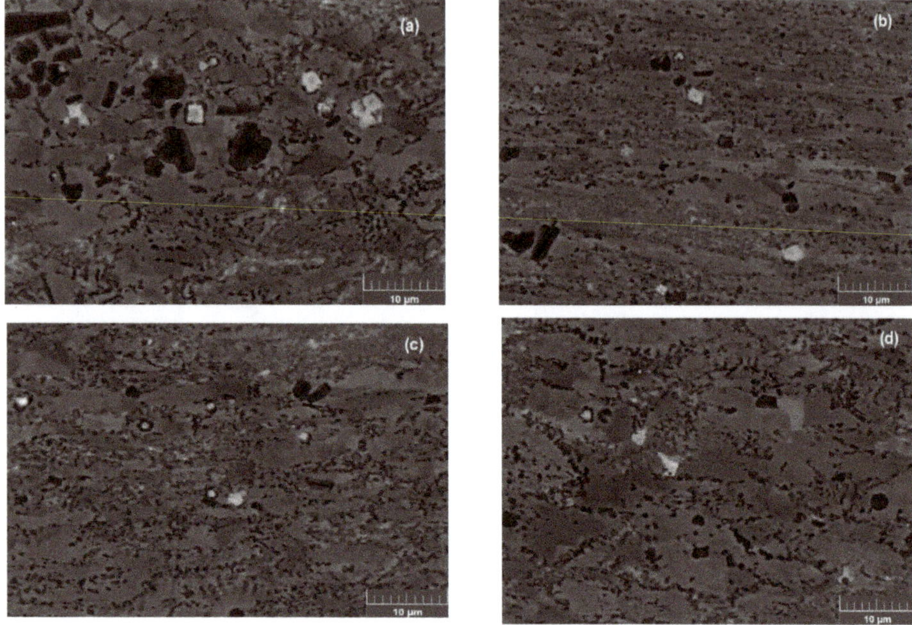

Figure 5. Microstructure of the Fe–16Cr–4Ti–6Zr–3.2B steel after hot deformation: 1373 K, 1 s^{-1} (**a**); 1373 K, 10 s^{-1} (**b**); 1423 K, 1 s^{-1} (**c**); 1423 K, 10 s^{-1} (**d**).

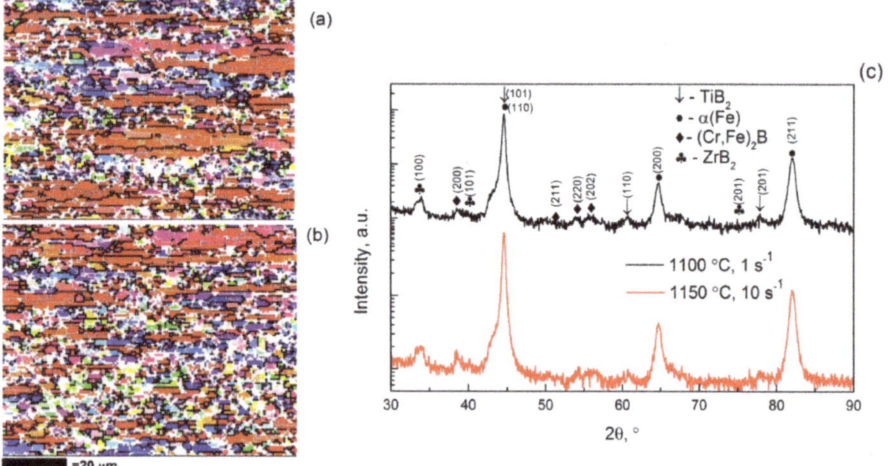

Figure 6. EBSD maps of the steel after hot compression at the temperature of 1373 K and at the strain rate 1 s^{-1} (**a**), at the temperature of 1423 K and at the strain rate 10 s^{-1} (**b**) and XRD patterns of the steel after deformation (**c**).

Room temperature compression stress–strain curves of the investigated steel after hot deformation are presented in Figure 7. Steel demonstrates a high degree of strength and good plasticity after hot deformation at temperatures in the range of 1273–1423 K, and at strain rates in the range of 1–10 s^{-1}. The decrease of the strain rate to 0.1 s^{-1} decreases the yield stress of the steel on 50–150 MPa in comparison with higher strain rates.

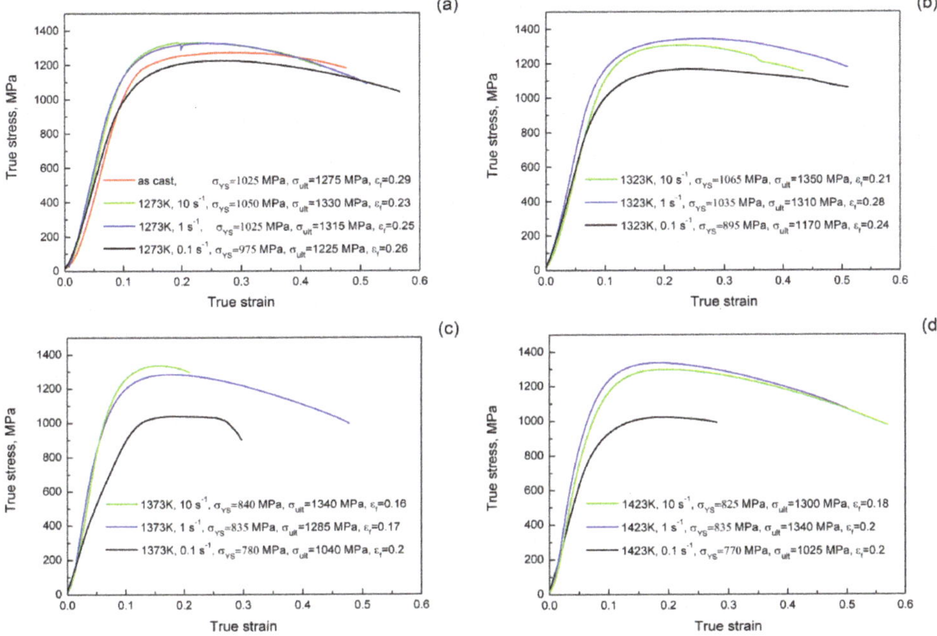

Figure 7. Mechanical properties of the investigated steel after hot deformation.

The images of the specimens before and after corrosion test are shown in Figure 8. As one can see, the Fe–16Cr–4Ti–6Zr–3.2B steel is more resistive to aggressive Cl^--contained environment than Fe–14Cr–5Ti–0.3V–1.8B steel due to a higher concentration of the alloying elements such as Cr, Ti and Zr. The mass loss rate for the investigated steel is 25 g/(m^2·h). It is more than twice smaller than the mass loss rate for Fe–14Cr–5Ti–0.3V–1.8B steel with V_m = 57 g/(m^2·h).

Figure 8. Images of the specimens before and after accelerated corrosion test in the 10% $FeCl_3 \cdot 6H_2O$ water solution.

The fine microstructure, good mechanical properties, satisfactory corrosion resistance and possibilities for applications of hot plastic deformation without fractures despite the presence of a large volume of brittle borides make the investigated alloy attractive as a high-speed steel and to the nuclear industry.

4. Conclusions

1. The phases of α(Fe), ZrB_2, TiB_2 and $(Cr,Fe)_2B$ are formed in the microstructure during three-stage crystallisation of the Fe–16Cr–4Ti–6Zr–3.2B steel.
2. The hot compression of the Fe–16Cr–4Ti–6Zr–3.2B steel proceeds without fractures in the temperature range of 1273–1423 K and in the strain rate range of 0.1–10 s^{-1}. Hot deformation leads to fragmentising of the eutectic structure and a decrease in the size of the ZrB_2 and TiB_2 particles.
3. True stress–true strain curves show nonmonotonic behaviour with the achievement of peak stress, followed by boride fragmentation and α(Fe) dynamic recrystallisation processes. The peak strain value depends on the Zener–Hollomon parameter by power low:

$$\varepsilon_P = 0.001Z^{0.22}.$$

4. The investigated alloy possesses a high yield strength of 825–840 MPa, and good plasticity of 0.16–0.2, after hot deformation in the range of 1373–1423 K.
5. Corrosion resistance of the novel Zr-alloyed high-boron steel in the aggressive Cl^--contained environment is better than corrosion resistance of the industrial Fe–14Cr–5Ti–0.3V–1.8B steel, which is currently most common for production of spent nuclear fuel storage.

Author Contributions: Conceptualization, A.P.(Alexey Prosviryakov) and A.C.; investigation, B.M. and A.P.(Andrey Pozdniakov); writing—original draft preparation, A.C.; writing—review and editing, A.P.(Alexey Prosviryakov); funding acquisition, A.C.

Funding: This work was supported by the Russian Science Foundation (project No. 18-79-10153).

Conflicts of Interest: The authors declare no conflict of interest.

References

1. Churyumov, A.Y.; Khomutov, M.G.; Tsar'kov, A.A.; Pozdnyakov, A.V.; Solonin, A.N.; Efimov, V.M.; Mukhanov, E.L. Study of the structure and mechanical properties of corrosion-resistant steel with a high concentration of boron at elevated temperatures. *Phys. Met. Metallogr.* **2014**, *115*, 809–813. [CrossRef]
2. Shulga, A.V. A comparative study of the mechanical properties and the behavior of carbon and boron in stainless steel cladding tubes fabricated by PM HIP and traditional technologies. *J. Nucl. Mater.* **2013**, *434*, 133–140. [CrossRef]
3. Ma, S.; Xing, J.; Guo, S.; Bai, Y.; Fu, H.; Lyu, P.; Huang, Z.; Chen, W. Microstructural evolution and mechanical properties of the aluminum-alloyed Fe-1.50 wt % B–0.40 wt % C high-speed steel. *Mater. Chem. Phys.* **2017**, *199*, 356–369. [CrossRef]
4. Ma, S.; Pan, W.; Xing, J.; Guo, S.; Fud, H.; Lyu, P. Microstructure and hardening behavior of Al-modified Fe-1.5 wt % B-0.4 wt % C high-speed steel during heat treatment. *Mater. Charact.* **2017**, *132*, 1–9. [CrossRef]
5. Yin, F.; Hua, L.; Mao, H.; Han, X. Constitutive modeling for flow behavior of GCr15 steel under hot compression experiments. *Mater. Des.* **2013**, *43*, 393–401. [CrossRef]
6. Churyumov, A.Y.; Khomutov, M.G.; Solonin, A.N.; Pozdniakov, A.V.; Churyumova, T.A.; Minyaylo, B.F. Hot deformation behaviour and fracture of 10CrMoWNb ferritic–martensitic steel. *Mater. Des.* **2015**, *74*, 44–54. [CrossRef]
7. Zou, D.; Wu, K.; Han, Y.; Zhang, W.; Cheng, B.; Qiao, G. Deformation characteristic and prediction of flow stress for as-cast 21Cr economical duplex stainless steel under hot compression. *Mater. Des.* **2013**, *51*, 975–982. [CrossRef]

8. Anoop, C.R.; Prakash, A.; Giri, S.K.; Narayana Murty, S.V.S.; Samajdar, I. Optimization of hot workability and microstructure control in a 12Cr-10Ni precipitation hardenable stainless steel: An approach using processing maps. *Mater. Charact.* **2018**, *141*, 97–107. [CrossRef]
9. Kishor, B.; Chaudhari, G.P.; Nath, S.K. Hot workability of 16Cr-5Ni stainless steel using constitutive equation and processing map. *Mater. Today: Proc.* **2018**, *5*, 17213–17222. [CrossRef]
10. Ren, X.; Fu, H.; Xing, J.; Yi, Y. Effect of solidification rate on microstructure and toughness of Ca-Ti modified high boron high speed steel. *Mater. Sci. Eng. A* **2019**, *742*, 617–627. [CrossRef]
11. Churyumov, A.Y.; Khomutov, M.G.; Pozdnyakov, A.V.; Mukhanov, E.L. Study of the structure and high-temperature mechanical properties of a steel with an elevated content of boron. *Met. Sci. Heat. Treat.* **2014**, *56*, 336–338. [CrossRef]
12. He, L.; Liu, Y.; Li, J.; Li, B. Effects of hot rolling and titanium content on the microstructure and mechanical properties of high boron Fe–B alloys. *Mater. Des.* **2012**, *36*, 88–93. [CrossRef]
13. Liu, Y.; Li, B.; Li, J.; He, L.; Gao, S.; Nieh, T.G. Effect of titanium on the ductilization of Fe–B alloys with high boron content. *Mater. Lett.* **2010**, *64*, 1299–1301. [CrossRef]
14. *Russian Standard GOST 9.912–89. Unified System of Corrosion and Aging Protection. Corrosion-Resistant Steels and Alloys. Method of Accelerated Tests for Resistance to Pitting Corrosion*; Izdatelstvo Standartov: Moscow, Russia, 1990.
15. Cha, L.; Lartigue-Korinek, S.; Wallsand, M.; Mazerolles, L. Interface structure and chemistry in a novel steel-based composite Fe-TiB2 obtained by eutectic solidification. *Acta Mater.* **2012**, *60*, 6382–6389. [CrossRef]
16. Mahmutovich, A.; Nagode, A.; Rimach, M.; Mujagich, D. Modification of the inclusions in austenitic stainless steel by adding tellurium and zirconium. *Mater. Tech.* **2017**, *51*, 523–528.
17. Antoni-Zdziobek, A.; Gospodinova, M.; Bonnet, F.; Hodaj, F. Solidification paths in the iron-rich part of the Fe-Ti-B ternary system. *J. Alloys Compd.* **2016**, *657*, 302–312. [CrossRef]
18. Sun, W.; Lu, C.; Tieu, A.K.; Jiang, Z.; Liu, X.; Wang, G. Influence of Nb, V and Ti on peak strain of deformed austenite in Mo-based micro-alloyed steels. *J. Mater. Proces. Tech.* **2002**, *125–126*, 72–76. [CrossRef]
19. Gong, B.; Duan, X.W.; Liu, J.S.; Liu, J.J. A physically based constitutive model of As-forged 34CrNiMo6 steel and processing maps for hot working. *Vacuum* **2018**, *155*, 345–357. [CrossRef]
20. Zener, C.; Hollomon, J.H. Effect of strain rate upon plastic flow of steel. *J. Appl. Phys.* **1944**, *15*, 22–32. [CrossRef]

© 2019 by the authors. Licensee MDPI, Basel, Switzerland. This article is an open access article distributed under the terms and conditions of the Creative Commons Attribution (CC BY) license (http://creativecommons.org/licenses/by/4.0/).

Article

Microstructural and Corrosion Properties of Cold Rolled Laser Welded UNS S32750 Duplex Stainless Steel

Claudio Gennari [1], Mattia Lago [1], Balint Bögre [2], Istvan Meszaros [2], Irene Calliari [1,*] and Luca Pezzato [1]

[1] Department of Industrial Engineering, University of Padova, Via Marzolo 9, 35131 Padova PD, Italy; claudio.gennari@phd.unipd.it (C.G.); lago.mattia@gmail.com (M.L.); luca.pezzato@unipd.it (L.P.)

[2] Department of Materials Science and Engineering, Budapest University of Technology and Economics, Bertalan Lajos utca 7, 1111 Budapest, Hungary; bbalint1113@gmail.com (B.B.); meszaros@eik.bme.hu (I.M.)

* Correspondence: irene.calliari@unipd.it; Tel.: +39-049-8275499

Received: 13 November 2018; Accepted: 15 December 2018; Published: 18 December 2018

Abstract: The main goal of this work was to study the effect of plastic deformation on weldability of duplex stainless steel (DSS). It is well known that plastic deformation prior to thermal cycles can enhance secondary phase precipitation in DSS which can lead to significant change of the ferrite-austenite phase ratio. From this point of view one of the most important phase transformation in DSS is the eutectoid decomposition of ferrite. Duplex stainless steels (DSSs) are a category of stainless steels which are employed in all kinds of applications where high strength and excellent corrosion resistance are both required. This favorable combination of properties is provided by their biphasic microstructure, consisting of ferrite and austenite in approximately equal volume fractions. Nevertheless, these materials may suffer from several microstructural transformations if they undergo heat treatments, welding processes or thermal cycles. These transformations modify the balanced phase ratio, compromising the corrosion and mechanical properties of the material. In this paper, the microstructural stability as a consequence of heat history due to welding processes has been investigated for a super duplex stainless steel (SDSS) UNS S32750. During this work, the effects of laser beam welding on cold rolled UNS S32750 SDSS have been investigated. Samples have been cold rolled at different thickness reduction (ε = 9.6%, 21.1%, 29.6%, 39.4%, 49.5%, and 60.3%) and then welded using Nd:YAG laser. Optical and electronical microscopy, eddy's current tests, microhardness tests, and critical pitting temperature tests have been performed on the welded samples to analyze the microstructure, ferrite content, hardness, and corrosion resistance. Results show that laser welded joints had a strongly unbalanced microstructure, mostly consisting of ferritic phase (~60%). Ferrite content decreases with increasing distance from the middle of the joint. The heat-affected zone (HAZ) was almost undetectable and no defects or secondary phases have been observed. Both hardness and corrosion susceptibility of the joints increase. Plastic deformation had no effects on microstructure, hardness or corrosion resistance of the joints, but resulted in higher hardness of the base material. Cold rolling process instead, influences the corrosion resistance of the base material.

Keywords: duplex stainless steel; welding; deformation; corrosion; microstructure

---

1. Introduction

Duplex stainless steels (DSSs) are a group of steel with a ferrite and austenite biphasic microstructure. They are widely used in very aggressive environment like nuclear and petrochemical plants, oil and gas offshore applications, chemical plants, paper and pulp industries, food and beverages industries as an alternative to the austenitic stainless steels [1–4]. They have higher mechanical properties and corrosion resistance compared to the austenitic stainless steels but suffer from secondary phases embrittlement and cannot be used at temperature higher than 350 °C [5,6].

The optimum properties of super duplex stainless steels are achieved when approximately equal amounts of ferrite and austenite are present in the microstructure. This balanced ratio between the two phases is obtained with a suitable combination of chemical composition and solution heat treatment [3,7]. However, during welding processes the microstructure can undergo detrimental transformations. DSSs solidifies starting from a fully ferritic microstructure and, as the material cools to room temperature, ferrite transforms into austenite through solid-state transformation [1–4,7,8]. With an improper cooling rate, two main problems may arise: an unbalanced austenite–ferrite ratio, and the precipitation of secondary phases in the weld zone (WZ) and heat-affected zone (HAZ) [9] which can both affect the mechanical properties, in particular impact toughness [10–13] and corrosion resistance [14,15].

Solution heat treatment is necessary in order to obtain a balanced microstructure and to solubilize any possible secondary phases that has been formed. Secondary phases can precipitate in a temperature range between 600 °C and 1000 °C. At lower temperature, typically 475 °C, ferrite spinodal decomposition can occur if the holding time is long enough (higher than 100 h). Main secondary phases that can precipitate are σ phase, χ phase, and chromium nitrides (i.e., Cr_2N) and the less common R-phase, π-phase, Laves phase, and carbides, not to mention secondary austenite decomposition and spinodal ferrite decomposition [4].

Even though it is thermodynamically unstable, the first phase to precipitate is χ phase thanks to its lattice parameter close to three times that of ferrite, followed by σ-phase. Both σ-phase and χ-phase grow toward the ferrite, eventually consuming all of it [3,13,16]. Because of its higher content in high atomic number element (i.e., molybdenum and chromium) those secondary phases appear lighter in backscattered-electrons. Chromium nitrides, on the other hand, appear as a dark chain because of the lower atomic number of the constituent. They are located at the ferrite/ferrite grain boundary and the ferrite/austenite phase boundaries. Nitrogen solubility in ferrite is considerably lower compared to austenite, moreover, it drops as the temperature decreases hindering the diffusion of nitrogen atoms in the austenite resulting in chromium nitrides precipitation [3].

Therefore, the study of weldability of super duplex stainless steels is a fundamental task for their proper industrial application. Previous investigation showed that conventional welding processes as submerged arc welding (SAW) [17–19], plasma arc welding (PAW) [20–22], gas tungsten arc welding (GTAW) [23–25], and friction stir welding (FSW) [26–28] have a highly affected austenite-ferrite ratio, promoting precipitation of secondary harmful phases. For this reason, high power laser welding has seen a remarkable increase in research interest, due to its better precision, speed, and versatility compared to traditional welding processes [29–31]. While earlier researchers studied the effect of laser beam parameters on microstructure and properties of DSSs [8,9,32,33], few papers addressed ferrite–austenite ratio change in SDSS welding [34–37].

In this work, the effects of Nd:YAG laser welding on microstructure and mechanical and corrosion properties of UNS S32750 SDSS samples with seven different grades of thickness reduction (ε = 0%, 9.6%, 21.1%, 29.6%, 39.4%, 49.5%, and 60.3%) are studied.

2. Materials and Methods

The UNS S32750 super duplex stainless steel were kindly supplied by ArcelorMittal (ArcelorMittal S.A., Luxembourg, Luxembourg). Its composition is reported in Table 1.

Table 1. Chemical composition of UNS S32750.

Element	C	Mn	P	S	Si	Cu	Ni	Cr	Mo	N
%	0.021	0.822	0.0231	0.0004	0.313	0.178	6.592	24.792	3.705	0.2644

The as received material was a 15 mm thickness plate, previously solution annealed at 1100 °C for 1 h and water quenched. Three sets of seven specimens 100 mm long and 10 mm wide were cut to undergo plastic deformation through cold rolling process. The samples were deformed along the hot rolling direction, using a double cylinder mill with a reduction per pass of 0.25 mm. The 9.6%, 21.1%, 29.6%, 39.4%, 49.9% and 60.3% thickness reductions were applied, and the average strain rate was 0.15 s^{-1}.

To perform the laser beam welding each specimen was milled to the dimension of 70 × 15 × 3 mm. The welds were performed with a 4 kW Rofin-Sinar DY 044 Nd:YAG laser (ROFIN-SINAR Laser Gmb, Hamburg, Germany) assisted with a six-axis robot from ABB (ABB Asea Brown Boveri Ltd., Zurich, Switzerland). To prevent distortion while welding, the samples were clamped to the workbench. The shielding gas used was argon with a purity of 99.995%.

All the specimens were butt-welded using the weld parameters summarized in Table 2 in order to achieve full penetration of the welds.

Table 2. Laser welding parameters.

Parameters	Value
Average power	1400 W
Welding speed	450 mm/min
Defocusing distance	0 mm
Spot Diameter	0.2 mm
Shielding gas flow rate	20 L/min

After the welding process, the specimens were cut transversal with respect to the weld direction for the metallographic analysis. Microhardness tests were performed in the cross surfaces of the welded samples following ASTM E384 standard using a Buehler IndentaMetTM 1105 (ITW Test & Measurement GmbH, Esslingen, Germany) with a load of 300 g and an application time of 11 s. Corrosion resistance was evaluated on the surface of both welded and unwelded samples testing the weld bead surface and the longitudinal section of the unwelded sample. The content of ferrite in the weld bead was measured with Eddy's current test (ECT) (Test Maschinen Technik GmbH, Schwarmstedt, Germany).

Eddy-current testing (ECT) is a non-destructive electromagnetic test method, which uses electromagnetic induction to detect and characterize surface and sub-surface flaws, as well as ferromagnetic phase content in conductive materials. In an eddy current probe, a coil of conductive wire is excited with an alternating electrical current. This current produces an alternating magnetic field around the coil, which oscillates at the same frequency of the current. If the probe and its magnetic field are placed close to a conductive material a circular flow of electrons, known as an eddy current, will begin to move through the metal. That eddy current flowing through the metal will, in turn, generate its own magnetic field, which will interact with the coil and its field through mutual inductance. Changes in metal thickness or defects, like near-surface cracks, will interrupt or alter the amplitude and pattern of the eddy current and the resulting magnetic field. This in turn affects the movement of electrons in the coil by varying its electrical impedance. Changes in the impedance amplitude and phase angle can be correlated with the volume fraction of magnetic phase in the material.

To analyze the microstructure of the welds, all the samples were first mounted on a phenolic resin and then grinded to 1200 grit SiC paper and polished with diamond suspensions (6 μm and 1 μm) to a mirror like surface finish. After the polishing process, the samples were electrolytically etched at voltage ranging from 2 to 20 V with 20% sodium hydroxide to distinguish austenite from ferrite since the last one is preferentially etched.

The metallographic analysis was then performed with a Leica DMRE optical microscope and with a Leica Cambridge Stereoscan 440 scanning electron microscope (Leica Microsystems S.r.l., Milan, Italy) operating in secondary electron (15 kV) and backscattered electron (29 kV).

The corrosion resistance was evaluated by mean of the critical pitting temperature test (CPT) in order to correlate the microstructure modifications to the corrosion properties [15].

The CPT was determined following the ASTM G150 standard. A potentiostat AMEL 7060 (Amel Electrochemistry, Milan, Italy) was used, the equipment consisted of two cells, containing the same aqueous solution (1 M of NaCl), and electrically connected by a salt bridge; in the first cell, maintained at room temperature, the reference electrode (calomel) was immersed, whereas the counter electrode (platinum) and the sample were placed in the second cell, where the temperature was controlled by a thermostatic bath.

The ASTM G150 standard involves the evaluation of the CPT by maintaining a constant anodic potential of 700 mV vs. SCE and increasing the temperature of the thermostatic cell at the rate of 1 °C/min. Before the test the open circuit potential (OCP) was evaluated letting the system stabilize for 60 min. The CPT is determined when the current density reaches 100 $\mu A/cm^2$ and remains above this level for a minimum of 60 s. The tests were conducted until the occurrence of stable pitting and then stopped to prevent excessive corrosion of the specimen.

The CPT determination was performed both for un-welded and welded samples, by exposing 1 cm^2. The samples were varnished with a thermal-resistant varnish letting exposed the selected area. This area includes both the WZ and the HAZ due to the low dimensions of the HAZ in this type of welds. The corroded samples were then analyzed with a stereomicroscope Zeiss Stereoscan Stemi C2000 (Carl Zeiss S.p.A., Milan, Italy).

To determine the ferrite content on the welded samples, 13 ECT measurements were performed on the top section of each sample, starting from the middle of the joint. The distance between each detection point was 1 mm. The inspection was done at 4 different frequencies: 10.0 kHz, 40.0 kHz, 66.7 kHz and 100.0 kHz. The data acquisition, processing and the control of the set-up were done with ScanMax software (Test Maschinen Technik GmbH, Schwarmstedt, Germany).

3. Results and Discussions

3.1. Mictrostructure

The microstructure of the UNS S32750 SDSS base material is shown in Figure 1. The picture depicts a biphasic microstructure with approximately equal volume fraction of ferrite and austenite. Austenite phase (white) is orientated along the rolling direction (black arrow) dispersed in a ferrite matrix (blue–brown).

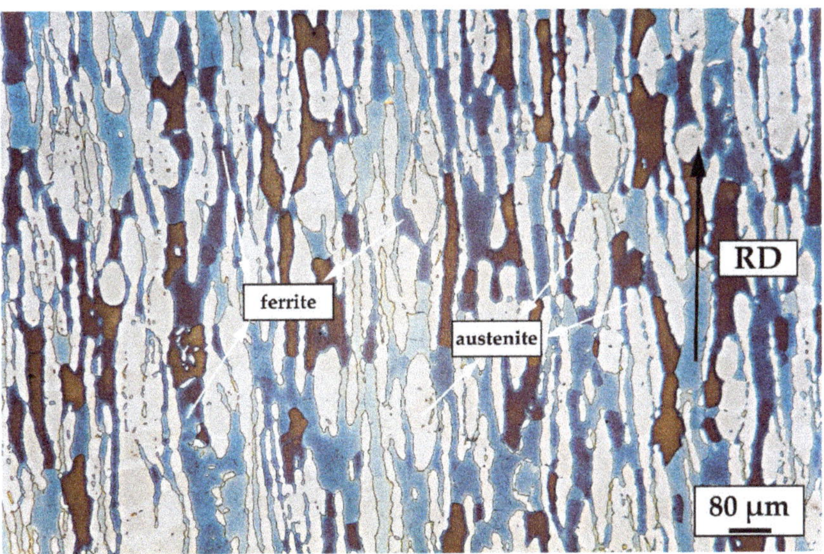

Figure 1. Microstructure of as received UNS S32750 bar.

The microstructure of the weld bead cross section in Figure 2 confirms the peculiarity of laser beam welding, where the development of the microstructure is essentially symmetrical with respect to the axis of the laser beam [9]. The Y-type shape weld bead is a consequence of a relatively low welding speed (450 mm/min), conversely to the classical V-type shape for welding speed greater than 2000 mm/min [38]. Elongated ferrite grains along the heat flux direction with austenite dispersed inside and along ferritic grain boundaries are evident. The highly oriented and elongated shape of ferritic grains is due to the high cooling rate, estimated to be approximately 1000 °C/s [9,29,39]. Because of this remarkable cooling rate, the HAZ (region between dashed lines in Figure 3) was almost undetectable. The temperature reached in the weld zone was supposed to be approximately 1400 °C [8] even though some researchers found to be as high as 1800 °C [30].

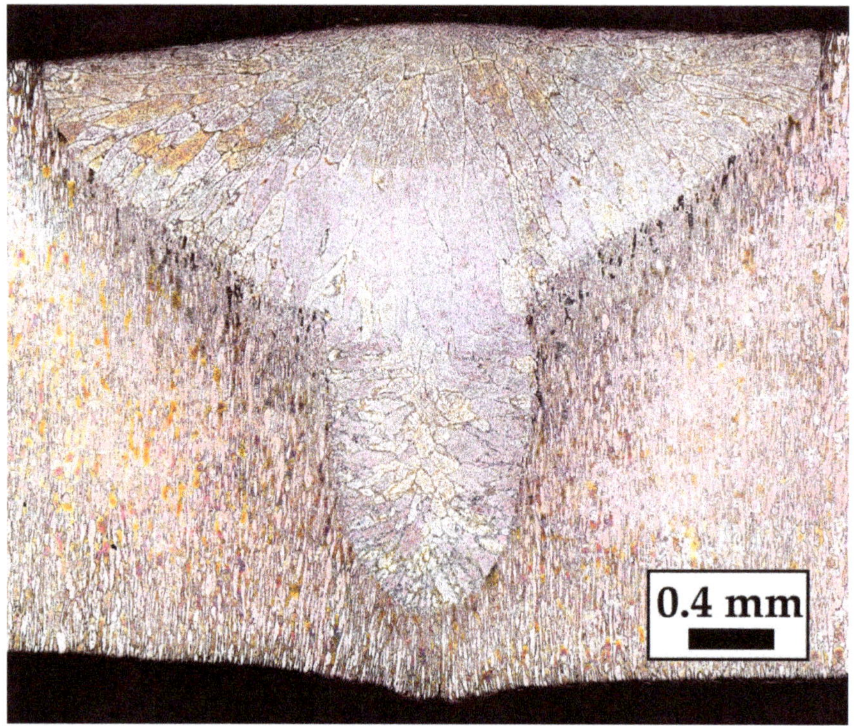

Figure 2. Overview of weld bead of the undeformed sample (electrolytic etching at 20 V with 20% NaOH).

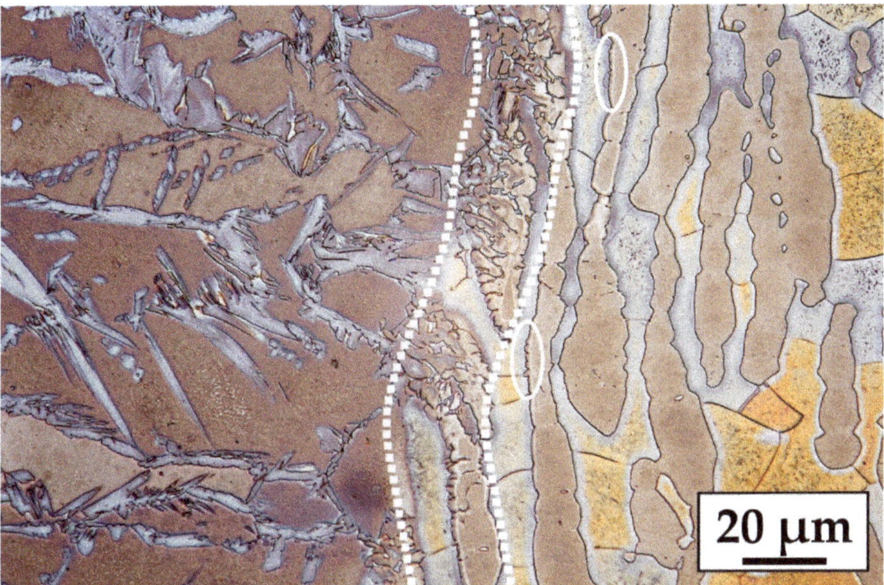

Figure 3. Heat Affected Zone (HAZ) of the sample deformed at 29.6%.

One of the weld's critical regions is the HAZ, depicted in Figure 3. The fusion line is continuous and clear, with the ferritic grains emanated from the base material grains as well as the austenitic grains. The HAZ zone (between dashed lines) is very narrow due to the high heat input of the laser and its spot size. The temperature reached in the HAZ is not high enough to melt the material, but it is high enough to modify the phase balance towards ferrite (i.e., phase balance is temperature and composition dependent, for this grade of DSS, a balanced microstructure is usually obtained at approximately 1050 °C). Fragmentation of austenite grains inside the dashed line is due to the temperature reached in this region which, as stated before, changed the phase balance of the base material toward ferrite since DSS starts to solidify from a fully ferritic microstructure. The segmentation of some austenitic-ferritic phase boundary (ellipses) is probably because the temperature reached in that region was high enough to start the dissolution of austenite inside the ferrite but not for long enough time to completely solubilize the austenite.

In agreement with the CCT diagram of σ-phase precipitation for a UNS S32750 [39], no precipitates were observed in this region, as can be seen from the scanning electron micrograph in backscattered electrons in Figure 4.

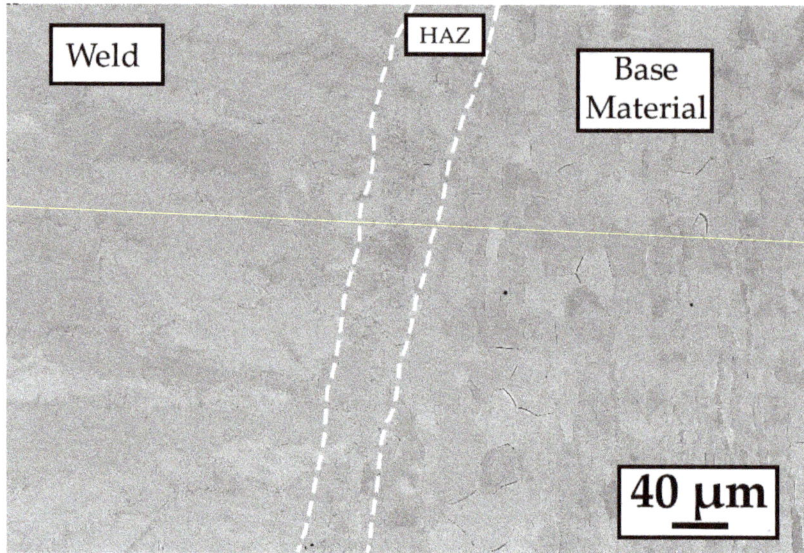

Figure 4. Back-scattered electron micrograph of the HAZ region (center), the weld bead (left) and the base metal (right).

Secondary phases should appear as a lighter spot located mostly at the triple points between ferrite grains and at austenite–ferrite phase boundaries, while chromium nitrides should appear as dark chains at ferrite/ferrite grain boundaries. In this case, the dark appearance of both austenite and ferrite grain boundaries of Figure 4 is due to the etching, it would not be possible otherwise to distinguish austenite from ferrite because of the similar composition of the phases that results in low contrast on backscattered electrons images. Previous SEM-WDS examinations have been performed without etching, asserting the absence of chromium nitrides. In the welded region, deformation does not influence the precipitation of secondary phases. In fact, the results are similar to the one reported in Figure 4 for the undeformed sample with no secondary phases or chromium nitrides detected, due to high cooling rate for all the samples.

As previously mentioned, the weld zone solidifies initially fully ferritically and partially transforms into austenite through solid-state transformation as the temperature decreases. Because of the high cooling rate of laser welding, this transformation tends to be inhibited, resulting in a strongly unbalanced microstructure mostly consisting of ferritic phase [8,9,29]. It is clear from the micrographs in Figure 5 that the welded zone is mainly composed of massive ferrite grains, with a low volume fraction of austenite that grew both at ferrite grain boundaries and within the ferrite grains. Austenite nucleated at ferrite grain boundaries is allotriomorphic austenite while the elongate needle-like grains that grows from the grain boundary towards the ferritic grains is Widmanstätten austenite (Figure 5). The polygonal grains of austenite inside ferritic grains could be isomorphic austenite, Widmanstätten austenite emanated from the grain boundary below the surface or fragmented Widmanstätten austenite due to the decomposition of the austenite phase at lower temperatures as other researchers found [40,41].

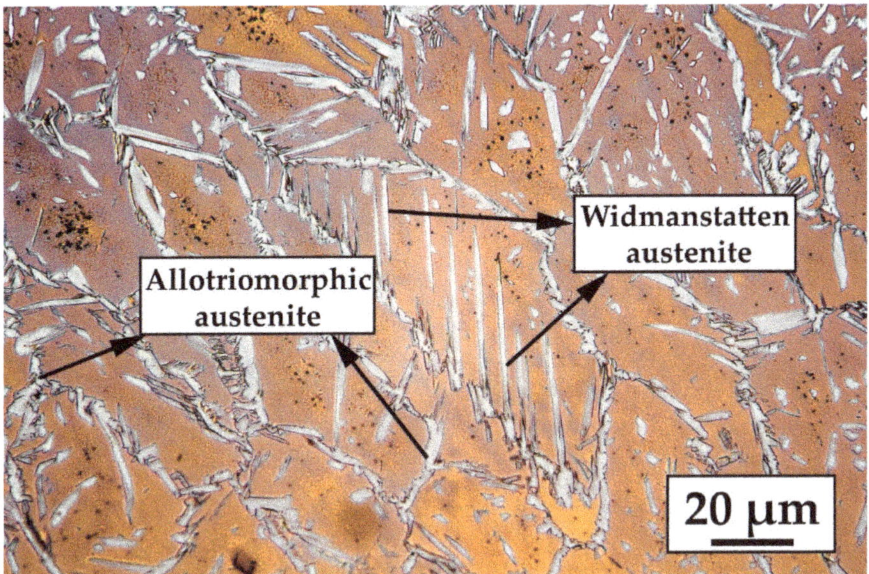

Figure 5. Microstructure of the welded zone showing austenite (light phase) and ferrite (brown matrix).

In order to observe the influence of cold working on the austenite, it was necessary to extend the etching time which caused the saturation of ferrite with etching products. The higher the thickness reduction the higher the density of slip bands inside austenite and ferrite (Figure 6b,e,f). In low stacking fault energy phases, dislocations are split into partial, thus they can only move by gliding in its own crystal plane. This produces a planar distribution of dislocations among the same crystal. Slip bands are a direct consequence of this distribution, in which a lot of dislocation are piled up. Usually this does not happen in high stacking fault energy phases (i.e., ferrite), but due to the presence of austenite and its crystallographic compatibility with ferrite (Kurdjumov–Sachs crystallographic orientation) it is possible that piled up dislocation at austenite/ferrite interfaces will extend to the ferrite grains [42].

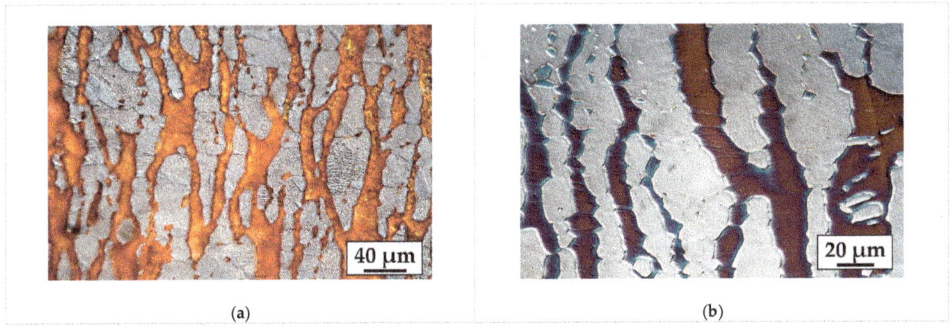

Figure 6. *Cont.*

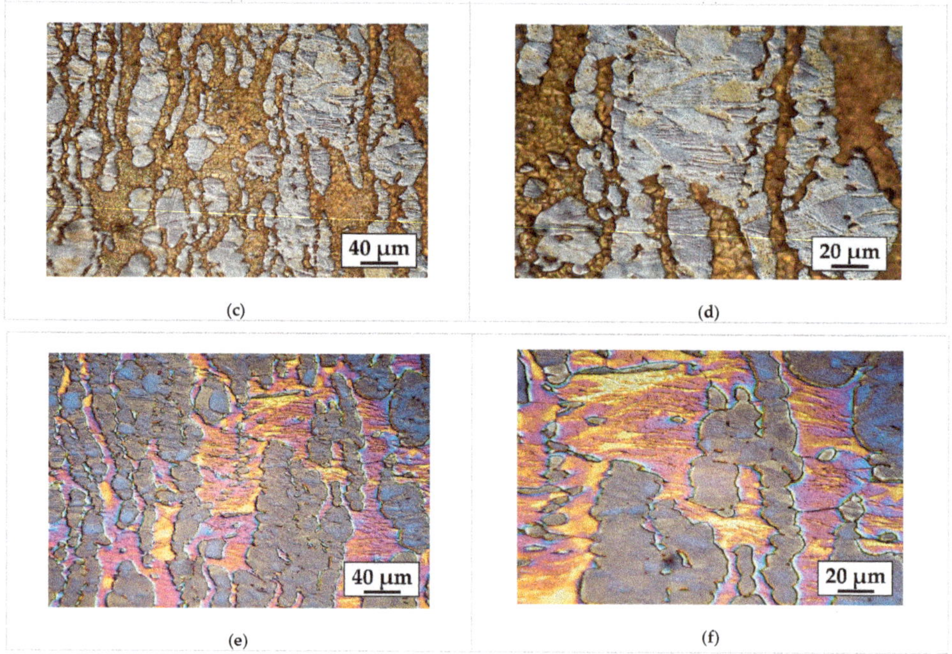

(c) (d) (e) (f)

Figure 6. Deformation bands inside austenite and ferrite due to thickness reduction. (**a**) 29.6% thickness reduction, (**b**) 39.9% thickness reduction, (**c**,**d**) 49.9% thickness reduction, and (**e**,**f**) 60.3% thickness reduction.

3.2. *Microhardenss*

Microhardness evolution in the welded region, in the heat affected zone and in the base material is depicted in Figure 7. Microhardness of the base material (blue line) it is clear the ascending trend due to strain hardening which lead to generation and interaction of dislocations, causing an inevitable increase in hardness [30,43,44]. The microhardness in the welded region is approximately 300 HV and it is not related to the different thickness reduction. This is obviously due to the fact that the material melts, so the previous plastic deformation does not influence the hardness of the region. The hardness of the HAZ is slightly lower than that of the welded region because, even though the region is very narrow, the heat input due to the welding process fully recovers the plastic deformation (the hardness of the HAZ does not follow the trend of the base material) as well as partial austenite dissolutions which changes the phase balance toward ferrite, as can be seen in Figure 3.

Overall the hardness of the welded region and the HAZ is higher compared to the un-deformed base material. This is related to the volume fraction of ferrite which is higher in the HAZ and in the welded region compared to the base metal.

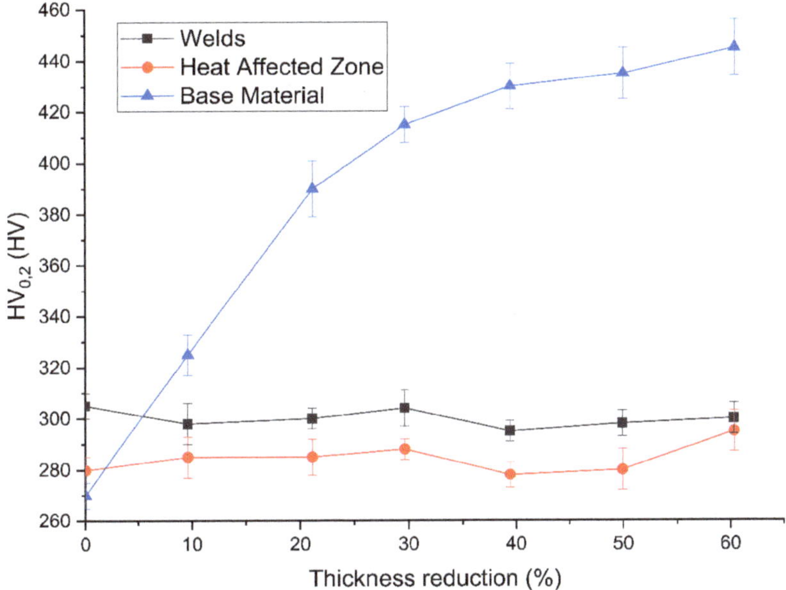

Figure 7. Evolution of hardness with respect to thickness reduction for the welds, the HAZ and the base material.

3.3. Corrosion Pitting Temperature (CPT)

The determination of the critical pitting temperature allows determining a very useful parameter that can be used to find the best material for applications in chlorinated environments.

The results of the CPT tests performed on the as received plate and on the cold rolled samples both welded and unwelded, are reported in Figures 8 and 9 and in Tables 3 and 4. The results in the tables are the average obtained from three measurements for each sample.

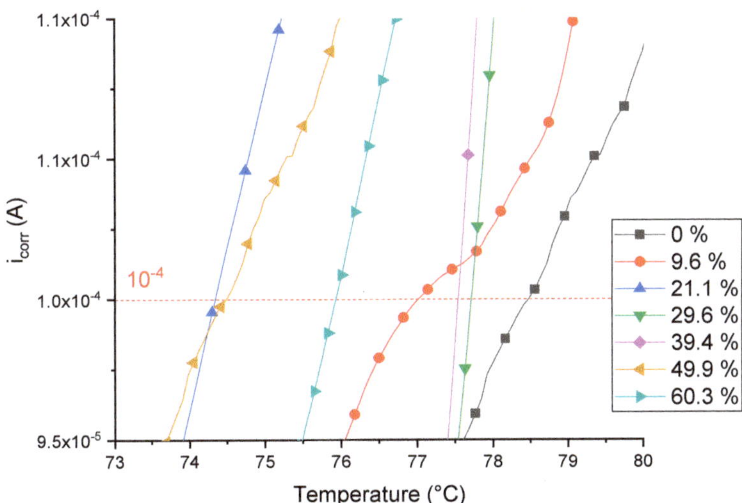

Figure 8. Close-up of the region of interest of the critical pitting temperature (CPT) curves of welded sheet metal.

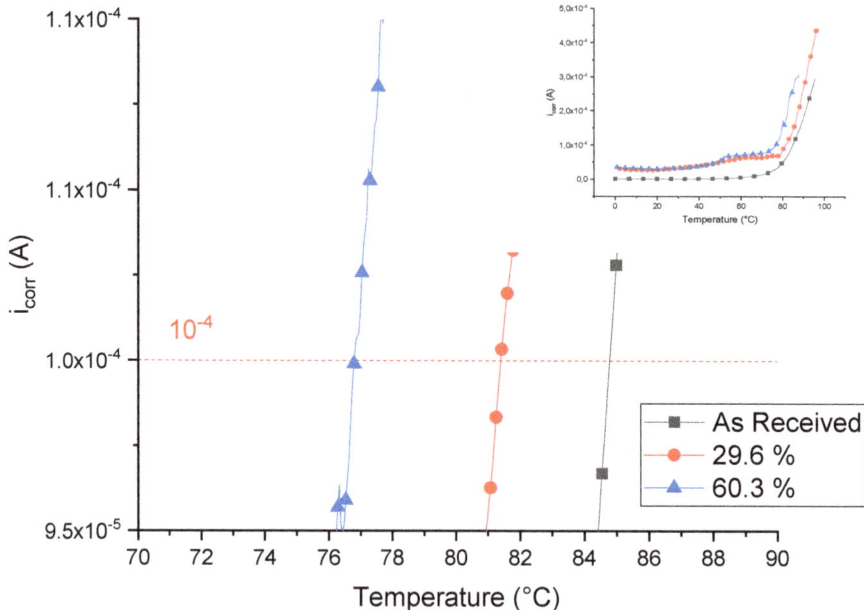

Figure 9. CPT curves of unwelded sheet metal.

Table 3. CPT of the welded sheet metal.

Thickness Reduction %	CPT [°C]
0	78 ± 1.2
9.6	77 ± 1.4
21.1	76 ± 1.1
29.6	76 ± 0.9
39.4	77 ± 1.3
49.9	75 ± 0.8
60.3	75 ± 1.2

Table 4. CPT of the unwelded sheet metal. For comparison is also reported the CPT of a bar of UNS S32750.

Thickness Reduction %	CPT [°C]
0	84 ± 1.4
29.6	81 ± 1.1
60.3	77 ± 0.9
Not rolled (bar)	88 ± 1.8

The cold rolling process and the welding affect the corrosion resistance by lowering the CPT [14,45]. The increase in deformation causes a gradual drop of the CPT as can be observed in Table 4: the higher the thickness reduction the lower the CPT as shown by the measurement done in the unwelded samples. CPT varies from a maximum of 88 °C for the bar to a minimum of 77 °C for the maximum thickness reduction (60.3%).

The deformation process causes the modification of the microstructure: after cold working, the occurrence of residual stress and the formation of the Strain-Induced Martensite (SIM) from the metastable austenite can substantially affect the pitting resistance of duplex stainless steels, because the number of the active anodic sites on the surface increases [46,47]. Thickness, composition, and uniformity of the passive layer are modified in different extent by plastic deformation [48,49] and the increasing in dislocation density favors the film dissolution due to the presence of lower binding energy regions, compared to a perfect crystal [50]. This may affect the formation of a less effective passive film on the steel surface; moreover, the presence of distorted high-energy interfaces may provide further trigger points for the localized attack [51,52]. This DSS grade is not prone to SIM precipitation in fact, it was not identified any SIM even though most accurate measurements and observations must be conducted (i.e., TEM and XRD).

Since CPTs range is between 78 °C and 75 °C they can be considered constant with respect to the reduction in coating thickness. Recrystallization phenomenon taking place during the welding process, delete the microstructural changes due to cold rolling. The weld process causes the fusion of the sheet metal with the consequent loss of effect of the cold hardening, so the CPT of the less deformed sample is basically the same of the 60.3% deformed sample among experimental error.

On the other hand, CPTs of the welded samples are lower than CPTs of the unwelded sheets (Figure 10) due to the unbalance of ferrite and austenite that occurs during the welding process as shown by both the location of the pits after the CPT tests (Figure 11) and after the ECT (Figure 12).

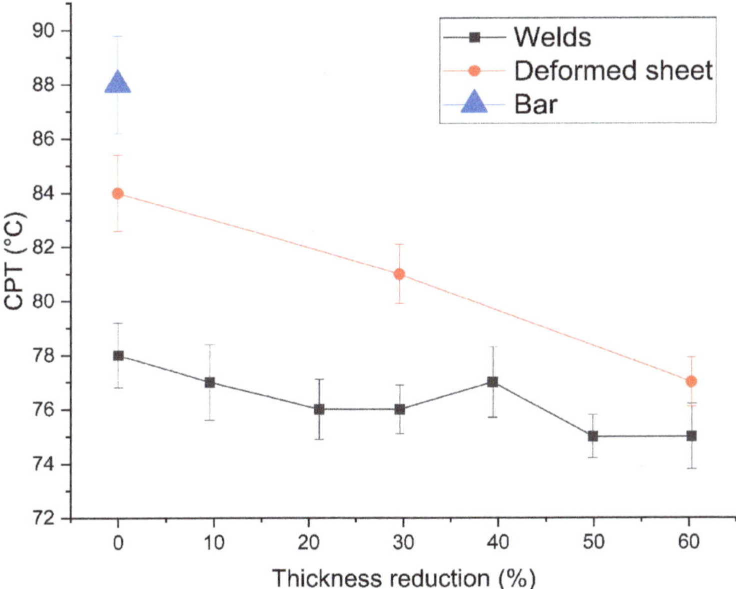

Figure 10. CPT trend of welded and unwelded sheets. For comparison is also reported the CPT of a bar of UNS S32750.

Macrographs in Figure 11 were taken after the corrosion tests and show that the pits are mainly localized in the weld beads whilst the base metal was left mostly unaffected, with some exceptions like the material which was subjected to the higher thickness reduction (60.3%). This clearly indicates that the corrosion phenomena can be ascribed to the welding process, more precisely to the unbalanced ferrite–austenite content in the welded zone.

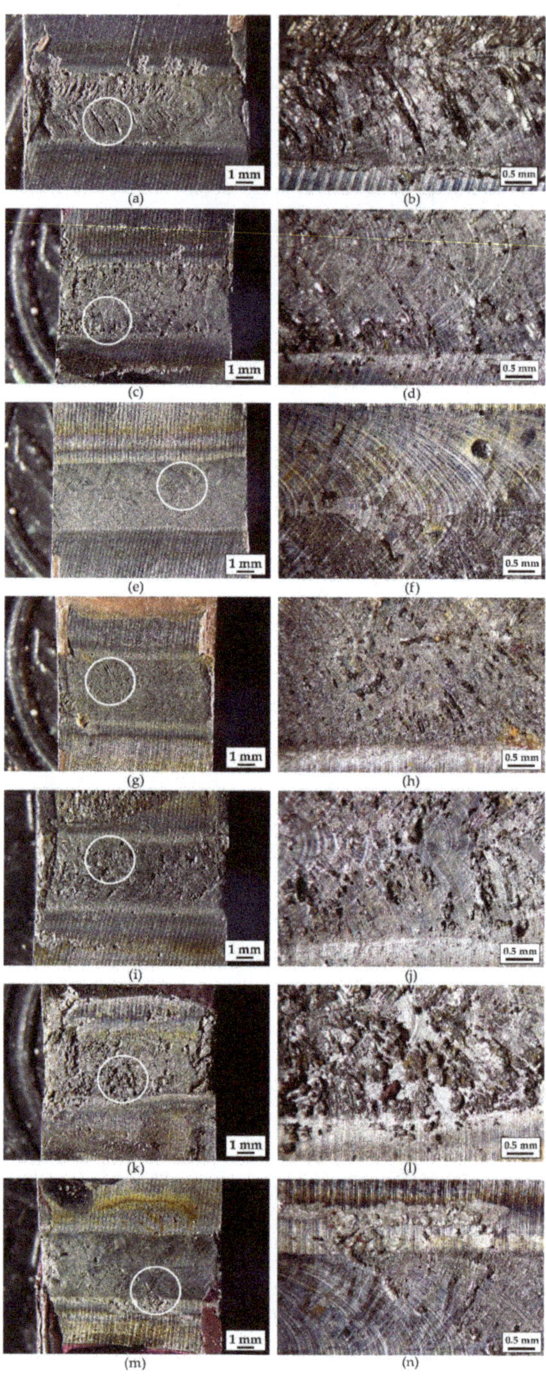

Figure 11. Pits location on the welded sample after CPT tests: (**a**,**b**) undeformed; (**c**,**d**) 9.6% thickness reduction; (**e**,**f**) 21.1% thickness reduction; (**g**,**h**) 29.6% thickness reduction; (**i**,**j**) 39.4% thickness reduction; (**k**,**l**) 49.9% thickness reduction; and (**m**,**n**) 60.3% thickness reduction. Right column images are higher magnification of the highlighted area of the pictures on the left column.

3.4. Eddy Currents

Eddy current tests were performed on the top section of the welded samples in order to analyze the change in ferrite content. Eddy's current results of the four frequencies used in the tests (10 kHz; 40 kHz; 66.7 kHz; 100 kHz) are reported in Figure 12. A touch coil absolute probe with diameter of 1 mm. The probe was fixed in a holding device which ensured the perpendicular position of the probe with respect to the surface of the sample. The sample was moving under the fixed probe.

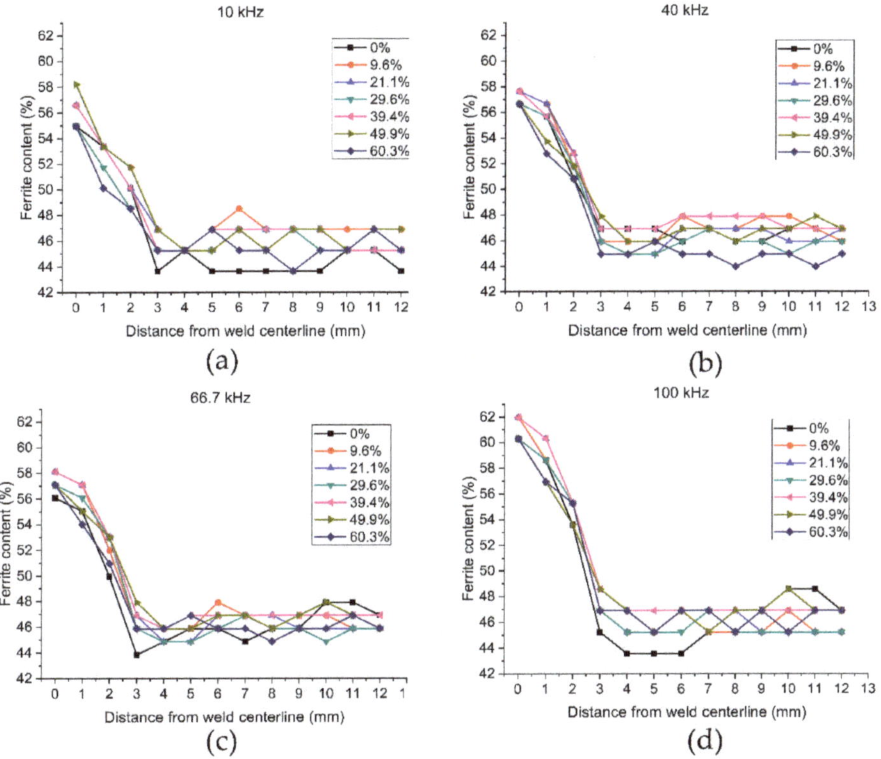

Figure 12. Evolution of the volume fraction of ferrite with respect to the distance from the weld centerline.

In accordance with previous studies [8,9,32,33], it can be seen that the amount of ferrite in the weld zone (~60%) is remarkably higher compared to the base material (46,9%) due to the high cooling rate typical of the laser welding. The precise ferrite content in the HAZ could not be measured because the dimension of the eddy current probe was significantly bigger (~1 mm) than the extension of the zone (~50 μm). However, the tendency of the ferrite content to decrease with increasing distances from the middle of the weld is quite clear for every frequency; the 40 kHz frequency was the most sensitive to the ferrite content variation. High heat input and small spot size of the laser combined with the fast welding process translates in narrow melted region (approximately 0.8 mm according to Figure 2) which limits the width of the HAZ thanks to the base material acting as a heat sink.

4. Conclusions

The main goal of this study was the investigation of cold rolling on the weldability of UNS S32750 duplex stainless steel. The importance of this problem arises from the fact that the cold rolling before the heat treatment can significantly increase the rate and decreases the starting temperature of the ferrite decomposition process. The mentioned phase transformation strongly influences the phase balance and deteriorates the mechanical properties and the corrosion resistance of DSS as well. Therefore, cold rolling prior to the welding process can be considered as an additional technological risk.

Three sets of seven different samples of UNS S32750 hot rolled and annealed with different thickness reduction up to 60.3% were laser welded and investigated by mean of CPT, microhardness and eddy current tests. It has been found that:

- The higher the thickness reduction the lower the CPT for the base material.
- Thickness reduction does not influence the CPT of the welded samples.
- Most of the pits are located in the weld beads due to the unbalanced microstructure (higher ferrite volume fraction).
- The higher the thickness reduction the higher the microhardness on the base material due to strain hardening effects.
- Microhardness of the welded region is not influenced by the prior plastic deformation.
- Prior plastic deformation does not influence the phase balance on the weld beads.
- Higher volume fraction of ferrite (60%) within the joint is responsible for the decrease in CPTs of the welds.

Author Contributions: Conceptualization, I.C., I.M. and L.P.; Data curation, C.G. and B.B.; Formal analysis, C.G. and M.L.; Investigation, C.G., M.L., B.B. and L.P.; Methodology, C.G., L.P. and I.C.; Project administration, L.P., I.C. and I.M.; Supervision, I.C. and I.M.; Validation, L.P., C.G., B.B. and M.L.; Visualization, C.G.; Writing—original draft, C.G., M.L. and B.B.; Writing—review & editing, C.G., I.C., B.B. and I.M.

Conflicts of Interest: The authors declare no conflict of interest.

References

1. Alvarez-Armas, I.; Degallaix-Moreuil, S.; Iris Alvarez Armas, S.D.M. *Duplex Stainless Steels*; Wiley: Hoboken, NJ, USA, 2009; ISBN 978-1-848-21137-7.
2. Gunn, R.N. *Duplex Stainless Steels: Microstructure, Properties and Applications*; Woodhead Publishing: Cambridge, UK, 1997; ISBN 9781855733183.
3. Nilsson, J.O. The physical metallurgy of duplex stainless steels. In Proceedings of the Duplex Stainless Steels 97: 5th World Conference, Maastricht, The Netherlands, 21–23 October 1997; pp. 73–82.
4. Nilsson, J.O. Super duplex stainless steels. *Mater. Sci. Technol.* **1992**, *8*, 685–700. [CrossRef]
5. Li, S.L.; Wang, Y.L.; Wang, X.T. Effects of long term thermal aging on high temperature tensile deformation behaviours of duplex stainless steels. *Mater. High Temp.* **2015**, *32*, 524–529. [CrossRef]
6. Gennari, C.; Pezzato, L.; Piva, E.; Gobbo, R.; Calliari, I. Influence of small amount and different morphology of secondary phases on impact toughness of UNS S32205 Duplex Stainless Steel. *Mater. Sci. Eng. A* **2018**, *729*, 149–156. [CrossRef]
7. Calliari, I.; Pellizzari, M.; Zanellato, M.; Ramous, E. The phase stability in Cr-Ni and Cr-Mn duplex stainless steels. *J. Mater. Sci.* **2011**, *46*, 6916–6924. [CrossRef]
8. Bolut, M.; Kong, C.Y.; Blackburn, J.; Cashell, K.A.; Hobson, P.R. Yb-fibre laser welding of 6 mm duplex stainless steel 2205. *Phys. Proced.* **2016**, *83*, 417–425. [CrossRef]
9. El-Batahgy, A.-M.; Khourshid, A.-F.; Sharef, T. Effect of Laser Beam Welding Parameters on Microstructure and Properties of Duplex Stainless Steel. *Mater. Sci. Appl.* **2011**, *02*, 1443–1451. [CrossRef]
10. Calliari, I.; Ramous, E.; Rebuffi, G.; Straffelini, G. Investigation of secondary phases effect on 2205 DSS fracture toughness. *Metall. Ital.* **2008**, *100*, 5–8. [CrossRef]
11. Gennari, C.; Breda, M.; Brunelli, K.; Ramous, E.; Calliari, I. Influence of small amount of secondary phases on impact toughness of UNS S32205 and Zeron® 100 Duplex Stainless Steel. In Proceedings of the ESSC and DUPLEX 2017—9th European Stainless Steel Conference—Science and Market and 5th European Duplex Stainless Steel Conference and Exhibition, Bergamo, Italy, 25–27 May 2017.
12. Breda, M.; Calliari, I.; Ramous, E.; Pizzo, M.; Corain, L.; Straffelini, G. Ductile-to-brittle transition in a Zeron© 100 SDSS in wrought and aged conditions. *Mater. Sci. Eng. A* **2013**, *585*, 57–65. [CrossRef]

13. Chen, T.H.; Weng, K.L.; Yang, J.R. The effect of high-temperature exposure on the microstructural stability and toughness property in a 2205 duplex stainless steel. *Mater. Sci. Eng. A* **2002**, *338*, 259–270. [CrossRef]
14. Breda, M.; Pezzato, L.; Pizzo, M.; Calliari, I. Effect of cold rolling on pitting resistance in duplex stainless steels. *La Metall. Ital.* **2014**, *6*, 15–19.
15. Pezzato, L.; Lago, M.; Brunelli, K.; Breda, M.; Calliari, I. Effect of the Heat Treatment on the Corrosion Resistance of Duplex Stainless Steels. *J. Mater. Eng. Perform.* **2018**, *27*, 3859–3868. [CrossRef]
16. Marques, I.J.; Vicente, A.; de Albuquerque Vicente, A.; Tenório, J.A.S.; de Abreu Santos, T.F. Double Kinetics of Intermetallic Phase Precipitation in UNS S32205 Duplex Stainless Steels Submitted to Isothermal Heat Treatment. *Mater. Res.* **2017**, *20*, 1–7. [CrossRef]
17. Luo, J.; Dong, Y.; Li, L.; Wang, X. Microstructure of 2205 duplex stainless steel joint in submerged arc welding by post weld heat treatment. *J. Manuf. Process.* **2014**, *16*, 144–148. [CrossRef]
18. Nowacki, J.; Rybicki, P.P. The influence of welding heat input on submerged arc welded duplex steel joints imperfections. *J. Mater. Process. Technol.* **2005**, *164–165*, 1082–1088. [CrossRef]
19. Sorrentino, S.; Fersini, M.; Zilli, G. Comparison between submerged arc (SAW) and laser welding applied to duplex stainless steel structures for bridges. *Weld. Int.* **2008**, *60*, 487–498.
20. Taban, E.; Kaluc, E. Welding behaviour of duplex and superduplex stainless steels using laser and plasma arc welding processes. *Weld. World* **2011**, *55*, 48–57. [CrossRef]
21. Bharathi, R.S.; Shanmugam, N.S.; Kannan, R.M.; Vendan, S.A. Studies on the Parametric Effects of Plasma Arc Welding of 2205 Duplex Stainless Steel. *High Temp. Mater. Process.* **2018**, *37*, 219–232. [CrossRef]
22. Šimeková, B.; Kovaříková, I.; Ulrich, K. Microstructure and Properties of Plasma Arc Welding with Depth Penetration Keyhole SAF 2205 Duplex Stainless Steel. *Adv. Mater. Res.* **2013**, *664*, 578–583. [CrossRef]
23. Eghlimi, A.; Shamanian, M.; Raeissi, K. Effect of current type on microstructure and corrosion resistance of super duplex stainless steel claddings produced by the gas tungsten arc welding process. *Surf. Coat. Technol.* **2014**, *244*, 45–51. [CrossRef]
24. Neissi, R.; Shamanian, M.; Hajihashemi, M. The Effect of Constant and Pulsed Current Gas Tungsten Arc Welding on Joint Properties of 2205 Duplex Stainless Steel to 316L Austenitic Stainless Steel. *J. Mater. Eng. Perform.* **2016**, *25*, 2017–2028. [CrossRef]
25. Mourad, A.-H.I.; Khourshid, A.; Sharef, T. Gas tungsten arc and laser beam welding processes effects on duplex stainless steel 2205 properties. *Mater. Sci. Eng. A* **2012**, *549*, 105–113. [CrossRef]
26. Chen, W.; Wang, J.; Li, J.; Zheng, Y.; Li, H.; Liu, Y.; Han, P. Effect of the Rotation Speed during Friction Stir Welding on the Microstructure and Corrosion Resistance of SAF 2707 Hyper Duplex Stainless Steel. *Steel Res. Int.* **2018**, *89*. [CrossRef]
27. de Abreu Santos, T.F.; Torres, E.A.; Ramirez, A.J. Friction stir welding of duplex stainless steels. *Weld. Int.* **2018**, *32*, 103–111. [CrossRef]
28. Saeid, T.; Abdollah-zadeh, A.; Assadi, H.; Malek Ghaini, F. Effect of friction stir welding speed on the microstructure and mechanical properties of a duplex stainless steel. *Mater. Sci. Eng. A* **2008**, *496*, 262–268. [CrossRef]
29. Keskitalo, M.; Mäntyjärvi, K.; Sundqvist, J.; Powell, J.; Kaplan, A.F.H. Laser welding of duplex stainless steel with nitrogen as shielding gas. *J. Mater. Process. Technol.* **2015**, *216*, 381–384. [CrossRef]
30. de Lima, M.S.F.; Carvalh, S.M.; Teleginski, V.; Pariona, M. Mechanical and Corrosion Properties of a Duplex Steel Welded using Micro-Arc or Laser-2015. *Mater. Res.* **2015**, *18*, 723–731. [CrossRef]
31. Quiroz, V.; Gumenyuk, A.; Rethmeier, M. Laser beam weldability of high-manganese austenitic and duplex stainless steel sheets. *Weld. World* **2012**, *56*, 9–20. [CrossRef]
32. Saravanan, S.; Raghukandan, K.; Sivagurumanikandan, N. Pulsed Nd: YAG laser welding and subsequent post-weld heat treatment on super duplex stainless steel. *J. Manuf. Process.* **2017**, *25*, 284–289. [CrossRef]
33. Sivakumar, G.; Saravanan, S.; Raghukandan, K. Investigation of microstructure and mechanical properties of Nd:YAG laser welded lean duplex stainless steel joints. *Optik (Stuttg.)* **2017**, *131*, 1–10. [CrossRef]
34. Arabi, S.H.; Pouranvari, M.; Movahedi, M. Pathways to improve the austenite–ferrite phase balance during resistance spot welding of duplex stainless steels. *Sci. Technol. Weld. Join.* **2018**, *24*, 1–8. [CrossRef]
35. Yasuda, K.; Gunn, R.N.; Gooch, T.G. Prediction of austenite phase fraction in duplex stainless steel weld metals. *Q. J. Jpn. Weld. Soc.* **2002**, *20*, 68–77. [CrossRef]
36. Varbai, B.; Pickle, T.; Májlinger, K. Development and comparison of quantitative phase analysis for duplex stainless steel weld. *Period. Polytech. Mech. Eng.* **2018**, *62*, 247–253. [CrossRef]

37. Baughn, K.E.; Ahmed, N.U.; Jarvis, B.L.; Viano, D.M. Tailoring the phase balance during laser and GTA keyhole welding of SAF 2205 duplex stainless steel. In Proceedings of the 6th International Conference of Trends in Welding Research, Pine Mountain, GA, USA, 15–19 April 2002; pp. 11–16.
38. Meng, W.; Li, Z.; Huang, J.; Wu, Y.; Chen, J.; Katayama, S. The influence of various factors on the geometric profile of laser lap welded T-joints. *Int. J. Adv. Manuf. Technol.* **2014**, *74*, 1625–1636. [CrossRef]
39. Mirakhorli, F.; Malek Ghaini, F.; Torkamany, M.J. Development of weld metal microstructures in pulsed laser welding of duplex stainless steel. *J. Mater. Eng. Perform.* **2012**, *21*, 2173–2176. [CrossRef]
40. Menezes, A.J.W.; Abreu, H.; Kundu, S.; Bhadeshia, H.K.D.H.; Kelly, P.M. Crystallography of Widmanstätten austenite in duplex stainless steel weld metal. *Sci. Technol. Weld. Join.* **2009**, *14*, 4–10. [CrossRef]
41. Ohmori, Y.; Nakai, K.; Ohtsubo, H.; Isshiki, Y. Mechanism of Widmanstaetten Austenite Formation in a .DELTA./.GAMMA. Duplex Phase Stainless Steel. *ISIJ Int.* **1995**, *35*, 969–975. [CrossRef]
42. Serre, I.; Salazar, D.; Vogt, J.B. Atomic force microscopy investigation of surface relief in individual phases of deformed duplex stainless steel. *Mater. Sci. Eng. A* **2008**, *492*, 428–433. [CrossRef]
43. Amigó, V.; Bonache, V.; Teruel, L.; Vicente, A. Mechanical properties of duplex stainless steel laser joints. *Weld. Int.* **2006**, *20*, 361–366. [CrossRef]
44. Köse, C.; Kaçar, R. Mechanical Properties of Laser Welded 2205 Duplex Stainless Steel. *Mater. Test.* **2014**, *56*, 779–785. [CrossRef]
45. Elhoud, A.M.; Renton, N.C.; Deans, W.F. The effect of manufacturing variables on the corrosion resistance of a super duplex stainless steel. *Int. J. Adv. Manuf. Technol.* **2011**, *52*, 451–461. [CrossRef]
46. Matting, A.; Koch, H.; Dorn, L. Einfluß des Elektronenstrahl- und Schutzgasschweißens auf das Korrosionsverhalten von X5 CrNiMo 18 10. *Mater. Corros.* **1970**, *21*, 94–97. [CrossRef]
47. Elayaperumal, K.; De, P.K.; Balachandra, J. Passivity of Type 304 Stainless Steel-Effect of Plastic Deformation. *Corrosion* **1972**, *28*, 269–273. [CrossRef]
48. Phadnis, S.V.; Satpati, A.K.; Muthe, K.P.; Vyas, J.C.; Sundaresan, R.I. Comparison of rolled and heat treated SS304 in chloride solution using electrochemical and XPS techniques. *Corros. Sci.* **2003**, *45*, 2467–2483. [CrossRef]
49. Navaï, F. Effects of tensile and compressive stresses on the passive layers formed on a type 302 stainless steel in a normal sulphuric acid bath. *J. Mater. Sci.* **1995**, *30*, 1166–1172. [CrossRef]
50. Greene, N.D.; Saltzman, G.A. Effect of Plastic Deformation On the Corrosion of Iron and Steel. *Corrosion* **1964**, *20*, 293t–298t. [CrossRef]
51. Serna, M.M.; Jesus, E.R.B.; Galego, E.; Martinez, L.G.; Corrêa, H.P.S.; Rossi, J.L. An Overview of the Microstructures Present in High-Speed Steel -Carbides Crystallography. *Mater. Sci. Forum* **2006**, *530–531*, 48–52. [CrossRef]
52. Chiu, P.K.; Wang, S.H.; Yang, J.R.; Weng, K.L.; Fang, J. The effect of strain ratio on morphology of dislocation in low cycle fatigued SAF 2205 DSS. *Mater. Chem. Phys.* **2006**, *98*, 103–110. [CrossRef]

© 2018 by the authors. Licensee MDPI, Basel, Switzerland. This article is an open access article distributed under the terms and conditions of the Creative Commons Attribution (CC BY) license (http://creativecommons.org/licenses/by/4.0/).

Article

Temperature Dependent Phase Transformation Kinetics of Reverted Austenite during Tempering in 13Cr Supermartensitic Stainless Steel

Yiwei Zhang [1,2,*], Yuande Yin [1], Diankai Li [1], Ping Ma [3], Qingyun Liu [4], Xiaomin Yuan [1] and Shengzhi Li [1,*]

1. School of Materials Science and Engineering, Anhui University of Technology, Ma'anshan 243002, China; yinyd@ahut.edu.cn (Y.Y.); diankaili@126.com (D.L.); yuan@ahut.edu.cn (X.Y.)
2. Key Laboratory of Green Fabrication and Surface Technology of Advanced Metal Materials, Ministry of Education, Anhui University of Technology, Ma'anshan 243002, China
3. Analytical Instrumentation Center, Anhui University of Technology, Ma'anshan 243002, China; lotlwve@126.com
4. School of Mechanical Engineering, Anhui University of Technology, Ma'anshan 243002, China; qyunliu@126.com

* Correspondence: ywzhang@ahut.edu.cn (Y.Z.); lisz55@163.com (S.L.); Tel.: +86-555-231-1570 (Y.Z. & S.L.)

Received: 16 September 2019; Accepted: 1 November 2019; Published: 8 November 2019

Abstract: The formation and growth kinetics of the reverted austenite during tempering in 13Cr supermartensitic stainless steel were investigated by a combination X-ray diffraction (XRD), transmission electron microscopy (TEM) and electron backscatter diffraction (EBSD) in a scanning electron microscope (SEM). The reverted austenite precipitated at the martensite blocks, sub-blocks, laths and grain boundaries. The growth kinetics was established by Johnson-Mehl-Avrami (JAM) kinetics equation according to the volume fraction of the equilibrium reverted austenite at room temperature. The Avrami exponent value is 0.5, and the activation energy was estimated to be 369 kJ/mol, the kinetic model indicates that the mechanism of reverted austenite is diffusion-controlled and the growth of reverted austenite closely relies on the diffusion of the nickel (Ni) element. The experimental measured orientations of the reverted austenite are in good agreement with the theoretical ones, implying that the reverted austenite has the same orientation with the surrounding martensite, which meets the Kurdjumov–Sachs (K-S) orientation relationship. The orientation relationships minimize the strain energy of the phase transformation by reducing the crystallographic mismatch between phases.

Keywords: supermartensitic stainless steel; reverted austenite; phase transformation; kinetics model; electron backscattered diffraction

---

1. Introduction

Supermartensitic stainless steels have attracted significant interests due to their promise for high strength, excellent toughness and improved corrosion resistant [1–4]. Due to its excellent mechanical properties in offshore oil and gas industry, it has already become an alternative product to replace duplex stainless steel and austenitic stainless steel [5–7]. Supermartensitic stainless steels have distinguished mechanical properties (such as the yield strength is 800–950 MPa, the ultimate tensile strength is 900–1200 MPa, the elongation is 13–18% and the hardness is 26–32 HRC), which depend on the microstructure and chemical composition [6,8–10]. Lu and Li propound that the alloying element changes material properties by modifying the microstructures of the host element [11]. It was known that the steel is routinely strengthened by alloying with elements like chromium and nickel. In the case, the alloying element strengthens steel by forming structures and creating strong phases with robust interfaces to block the dislocation motion. The C–Fe–Cr phase diagram shows that the amount of α and γ depends on chemical composition. However, the exorbitant chromium (Cr) content produced δ-Fe ferrite phase, which impairs the toughness of the experimental steel [12]. Ni is the primary alloying element for austenite stability and used to enlarge the γ-Fe phase region, but numerous Ni would escalate the cost of the martensite stainless steel. Therefore, the additions of titanium (Ti), niobium (Nb) and vanadium (V) microalloying elements to steel were studied in order to improve the mechanical properties [13,14].

However, tailored microstructures can be created without changing the chemical composition of a material, often by processing the material of heat treatment [11]. When the experimental steel had heat treatments of quenching and tempering, the microstructure was tempered martensite and reverted austenite in the supermartensitic stainless steels. The reverted austenite formation during the tempering process helps to acquire the excellent mechanical properties [4,15,16]. It was revealed that the reverted austenite content was controlled by the heat treatment in the supermartensitic stainless steel, and minor adjustment of the parameters (such as the tempering temperature) will lead to different mechanical properties [3,8,17]. Supermartensitic stainless steels must guarantee that a certain amount of austenite remains stable for acquiring the excellent mechanical at ambient temperature. Reverted austenite is produced by tempering process during the reversion of tempered martensite to γ-Fe phase in the matrix [18]. The reverted austenite distinguished from retained austenite in composition and size during tempering [19,20]. It has systematically investigated that many factors affect the volume fraction of reversed austenite, for instance, tempering temperature, tempering time and heating or cooling rate [3,4,8,16,21].

The reverted austenite has been extensively studied on the microstructure and mechanical properties in the literature, but the thermodynamics and kinetics have been paid less attention to the reverted austenite. Meanwhile, the volume of the reverted austenite is too little to clarify the transformation mechanism in previous literature. The model is used to describe the phase transformation kinetics by the isothermal transformation in this work. The purpose is to put forward a kinetic modeling for revealing microstructure evolution and the transformation mechanism of reverted austenite in 13Cr supermartensitic stainless steels. The models are based on the amounts of equilibrium reverted austenite at room temperature and tempering conditions. It is crucial to investigate tempering temperature and tempering time to promote transformation kinetics, through the Ni-diffusion at austenite/martensite phase interface. By reconstructing the kinetic model from the experimental results, a detailed description of the microstructure evolution during tempering is proposed, which provides a basis for explaining the excellent mechanical properties of the steels. Reviewing the literature on the topic, the experiments do not appear to be systematic enough to establish a correlation between the amount of reverted austenite and the transformation kinetics and transformation mechanism in the supermartensitic stainless steel. This work was conducted to evaluate the dependence of experimental kinetic data on the reverted austenite transformation mechanism in the supermartensitic stainless steel. The diffusion activation energies and the rate-reaction theory were commonly taken attention to the isothermal kinetic data analysis. In the present study, the SEM, XRD, electron backscatter diffraction (EBSD) and TEM techniques were applied to expose austenite reversion process and the elemental partition in the reversed austenite.

2. Materials and Methods

The as-received raw material was ultra-low carbon 13Cr supermartensitic stainless steel with the chemical composition of 0.03C-12.5Cr-5.8Ni-2Mo-0.32Mn-0.28Si-0.08Cu-0.011P-0.001S (wt. %), which is a commercial alloy from the market. The specimens were cut with 10 mm × 10 mm × 10 mm by wire electrical discharge machining to prepare for the experiment. The specimens homogenized at 1050 °C for 0.5 h and quenched into oil to the ambient temperature, and then tempered at 580, 600, 620 and 650 °C for 1 h, 2 h, 4 h, 8 h, 16 h and 32 h followed by air-cooling. The heat treatments process of the experimental steel was shown in Figure 1.

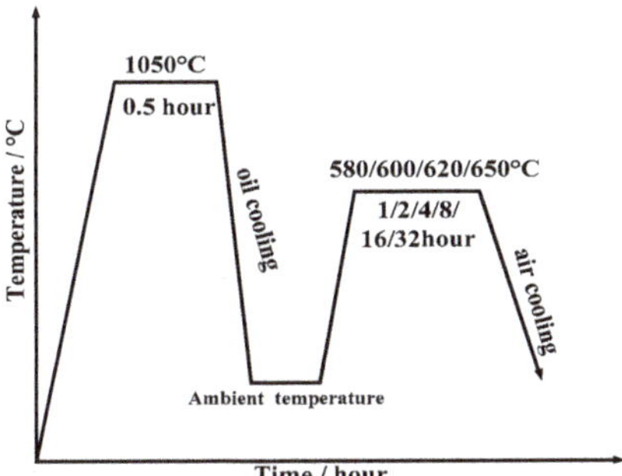

Figure 1. The heat treatment process of the samples in 13Cr supermartensitic stainless steel.

The specimens were measured by a Bruker D8 Advance diffractometer (Bruker, AXS, Karlsruhe, Germany), operated at 40 kV and 40 mA using Cu radiation for XRD analysis. The polished samples were measured from 30° to 90°at a speed of 1° per minute. The amount of reverted austenite was determined by XRD from a Rietveld analysis. The analysis was refined by the Rietveld-analysis program RIETAN 2000 [22]. The zero-point shift parameters, the background parameters, the preferred orientation, lattice parameters, the pseudo-Voigt profile parameters and the scale factor of the phases were refined independently but not simultaneously. The lattice constant and the volume fraction of reverted austenite were determined by the refinement program in the different tempering process.

The metallographic specimens were mechanically ground and polished for microstructural observation. The microstructure was immersion corroded by Villella's etchant containing 5 mL hydrochloric + 4 g picric acid + 100 mL ethyl alcohol and observed using an FEI NANO430 SEM (FEI, Hillsboro, OR, USA). The sample was electrolytically polished by 10% perchloric acid ethyl alcohol solution, and the electrolytic current density was about 30 A/cm^2. The EBSD (Oxford Instruments, London, UK) was operated at 25 kV and with a 0.5 μm step size for acquiring the orientation data. The crystallographic orientation relationship of the reverted austenite can be obtained in one prior austenite grain, and meanwhile the distribution of martensite variant in the pole figures with HKL Channel 5.

The microstructure of reverted austenite was observed by JEOL 2100 TEM (JEOL, Tokyo, Japan) operation at 200 kV. The thin foil samples were prepared for using a twin-jet polishing technique in a 6% $HClO_4$ + 94% CH_3COOH electrolyte solution. The TEM morphology of reverted austenite was characterized by the bright and dark field. The element partition in the reverted austenite was detected by an energy dispersive spectroscopy (EDS) system (X-Max, Oxford Instruments, London, UK) equipped with TEM instrument. The scanning transmission electron microscopy (STEM, JEOL, Tokyo, Japan) equipped with line EDS was occupied to explain austenite reversion process and the discipline of the elemental distribution in the phase interface.

3. Results and Discussion

3.1. Tempered Microstructure Characterization

Figure 2 shows the microstructure characterization of 13Cr supermartensitic stainless steel tempered at 580 °C for 1 h, 2 h, 4 h, 8 h, 16 h and 32 h, respectively. The SEM photos show that the microstructure was tempered martensite and reverted austenite. The microstructure was reverted austenite that looked like precipitates with a bright white color, while the matrix was tempered martensite in gray color. It was evident that the tempered time had a crucial influence on the content of reverted austenite. The longer the sample was tempered, the more reverted austenite content was obtained. The result was consistent with our previous work that the reverted austenite content increased with the tempering time [23]. The reverted austenite particles were found to be uniformly distributed along with the laths, blocks and the prior austenite grain boundaries. The reverted austenite nucleated position is caused by the defects in the martensite microstructure, such as dislocation and element segregation, which provide the driving forces for the reverted austenite phase transformation from the martensite [24].

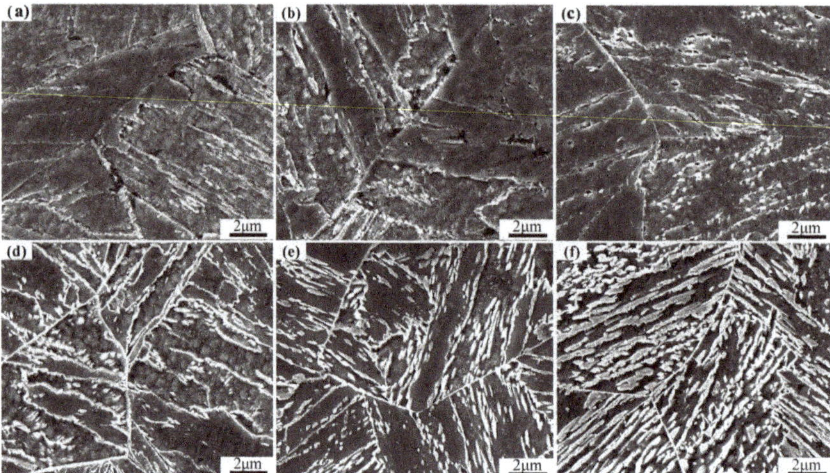

Figure 2. SEM microstructure of the samples tempered at 580 °C for different time (**a**) 1 h, (**b**) 2 h, (**c**) 4 h, (**d**) 8 h, (**e**) 16 h and (**f**) 32 h.

Figure 3 shows thin foil TEM microstructural analysis of the specimen tempering at 620 °C for 32 h of 13Cr supermartensitic stainless steel. As shown, the microstructure consisted of reverted austenite and martensite laths. The reverted austenite was easily detected by the TEM since the amount of reverted austenite reached maximal value, which was more than 20% at the tempering temperature according to our previous work [17,23]. The plate-shape of reverted austenite grows in thickness that formed along with the laths or the block boundaries in a bright field image. The reverted austenite activation energy of nucleation reduced when it formed on the martensite laths, blocks and grain boundaries [25]. Figure 3b displays the dark field image of the reverted austenite that formed as the strips and platelets in the matrix. According to the selected area electronic diffraction pattern shown in Figure 3c, the darker strip-like regions in the dark field image was reverted austenite, and the crystallographic zone axis was [100]. The selected area electron diffraction (SAED) patterns indicated that the reverted austenite became broad with increasing tempering time and formed along with the martensite laths.

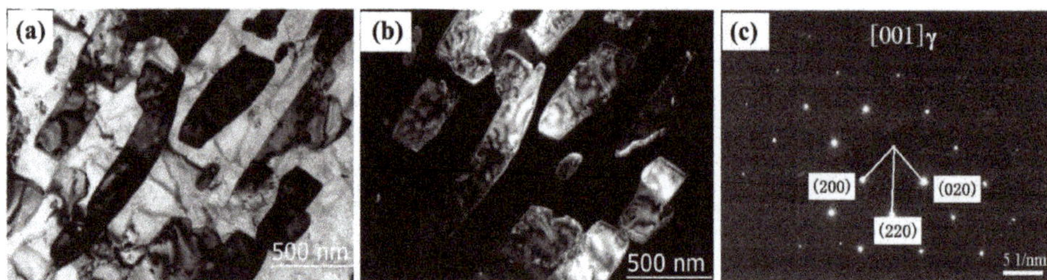

Figure 3. TEM micrographs of reverted austenite in the sample tempered at 620 °C for 32 h, (**a**) bright field image; (**b**) dark field image and (**c**) SAED pattern of the reverted austenite.

3.2. Variations in Amounts of Reverted Austenite with Tempering Time

XRD analysis was performed on all of the tempered specimens to determine phase constituents and the volume fraction of reverted austenite. Figure 4 shows X-ray diffraction patterns of the samples tempering at 600 °C for different holding times. Only γ-Fe and α-Fe peaks were detected in the XRD patterns; there are variations of integrated intensity in the peak of $(111)_\gamma$, indicating that the reverted austenite contents were different. It was evident that the reverted austenite content increased significantly with increasing tempering time from 1 h to 32 h, through the integrated intensities of reverted austenite diffraction lines $(111)_\gamma$. According to the Rietveld refinement of the XRD pattern for the sample tempering at 600 °C for 32 h, the volume of the reverted austenite arrived to the highest value. It can be seen that there was no extensive amounts of carbide precipitates were produced during tempering treatment from Figure 4, because there were no other peaks detected in the XRD patterns. The result was consistent with our previous work about the maximum value of the reverted austenite [23].

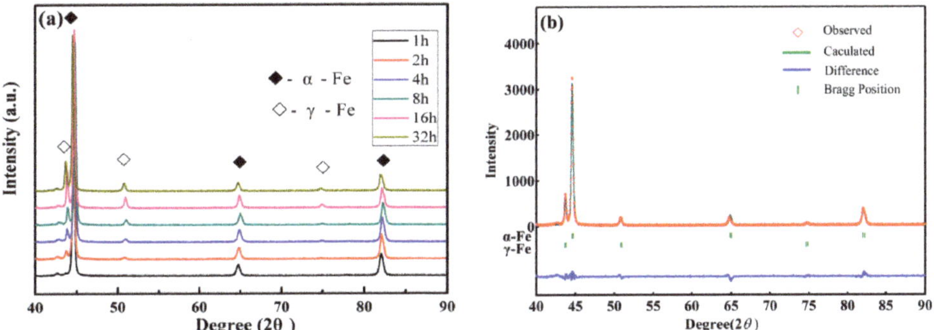

Figure 4. XRD pattern of the samples tempered at 600 °C for different tempering times (**a**) and (**b**) the Rietveld refinement of the sample tempered at 600 °C for 32 h.

Figure 5 shows the volume fraction of reverted austenite at different tempering temperature combined with tempering times. For tempering temperature in the range 580–620 °C, the content of reverted austenite increased rapidly with the tempering time from 1 h to 32 h. However, the reverted austenite content had a decreasing trend after tempering at 650 °C whatever the tempering time was. After that, higher tempering temperature produced less content of reverted austenite when the tempering temperature was more than 620 °C. It indicated that fresh martensite partially re-transformed from austenite during the higher temperature tempering process [24].

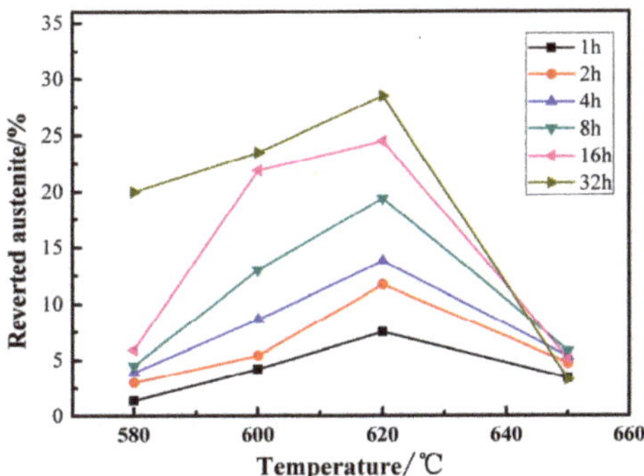

Figure 5. The volume fraction of reverted austenite tempered at different tempering temperatures and times.

The reverted austenite content did not reach the highest value when the samples tempered at 650 °C with increasing tempering time. The content of the reverted austenite was mainly determined by the amount of austenite transformation at high temperatures and its stability during tempering and cooling. The driving force of martensite transformation to austenite was larger when the tempering temperature increased to 650 °C. However, thermal diffusion plays a leading role in the steel at the same time. The composition gradually became uniform and the concentration fluctuation was reduced. The nickel atom enrichment region was eliminated, resulting in a decrease in the thermal stability of the high-temperature reverted austenite, which led to an increase in the Ms point. It was suggesting that the reverted austenite transformed into fresh martensite after cooling to room temperature. Additionally, the increased concentration of quenched in vacancies owing to an increase in tempering temperature would also decrease austenite stability. It was indicated that the reverted austenite stability became poor and the reverted austenite retransformed into martensite again, and the thermal stability of reversed austenite and tempering temperature affected the change of reverted austenite in the tempering process [26,27]. The reverted austenite re-transformed into martensite under high tempering temperature conditions, which corresponds to the result reported recently in the literature [28,29].

4. Discussion

4.1. The Transformation Kinetic Parameters and Tansformation Activation Energy

The phase transformation kinetics describes the relationship between the progress of phase transition and transformation temperature and time. Kinetics in phase transformation study means how fast a transformation will proceed. For the analysis of solid-state transformation kinetics, the reverted austenite volume of the materials can be outlined as a function of tempering time and temperature [30]. In the present work, presuming all of the reverted austenite formed on tempering and remained at room temperature. The equilibrium reverted austenite fraction increased within the temperature range from 580 to 620 °C at various tempering times. The results were used to build up the kinetics model of austenite reversion in the commercial 13Cr supermartensitic stainless steel. The phase transformation theory based on the classical Johnson–Mehl–Avrami (JMA) equation was analyzed. The Johnson–Mehl–Avrami type kinetics of the reverted austenite isothermal transformation is can be expressed by the following equation [31,32]:

$$f = 1 - \exp(-kt^n), \tag{1}$$

where f is the volume fraction of equilibrium reverted austenite, which was produced during the tempering process, t is the tempering holding time, k is the constant of reaction rate and empirically evaluated for temperature and the Avrami exponent n is time-independent, which especially depends on the phase transformation of the nucleation mechanism and growth mechanism. The Avrami exponent n can be extracted from experimental data by the following equation. Taking the Johnson–Mehl–Avrami equation double logarithm on both sides of the Equation (1) that changed as follows:

$$\ln(\ln{(1-f)}^{-1}) = \ln k + n \ln t. \tag{2}$$

It is noticeable that the $\ln(\ln{(1-f)}^{-1})$ have a line relationship with $\ln t$ according to the above equation. It is a straight-line relationship with the slope n and the intercept of the $\ln k$. Combined with the volume of equilibrium reverted austenite in Section 3.2, the constant n and $\ln k$ can be calculated by plotting the relationship $\ln(\ln{(1-f)}^{-1})$ against $\ln t$. The slope can be obtained by the amount of reverted austenite and tempering time of the straight lines in Figure 6. According to calculation and fitting, the Avrami exponent n was close to 0.5, which is the average value of the slope by the three straight lines.

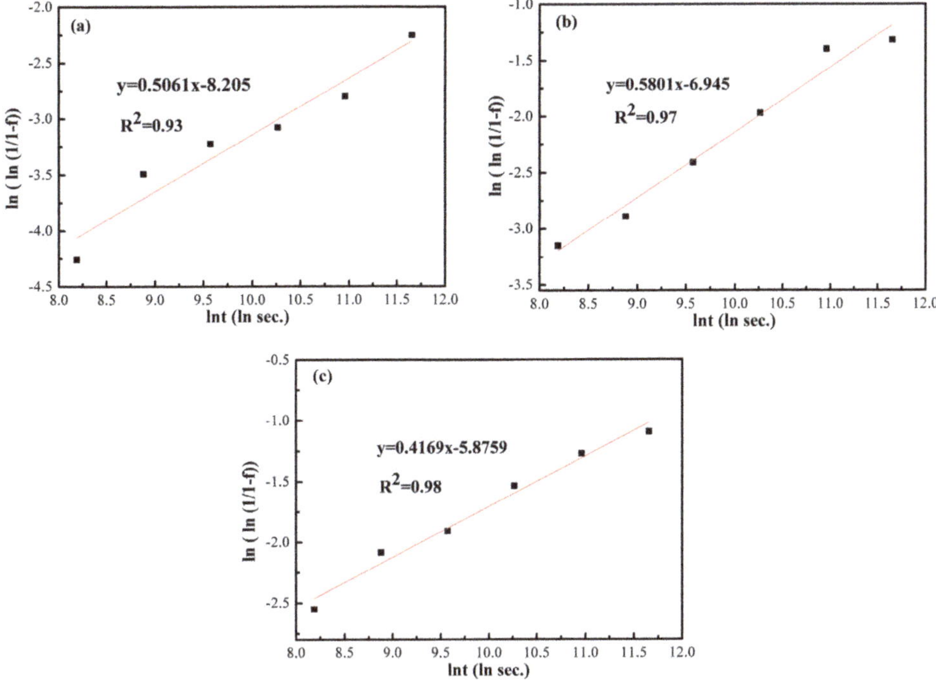

Figure 6. Relationship between $\ln(\ln{(1-f)}^{-1})$ and $\ln t$ for the samples tempered at (**a**) 580 °C (**b**) 600 °C and (**c**) 620 °C.

The Arrhenius-type equation is generally accepted for rate constants, but lack theoretical justification for the practical applicability, and only depending on the compatible analysis of transformation kinetics about the experimental data [30]. It was very important that if k and n were determined from the composition of steel and transformation temperature, the volume fraction of the product phase could be estimated. The reaction rate constant k can be expressed as following the equation, and which is temperature-dependent of the transformation kinetics:

$$k(T) = k_0 \exp(-Q/RT), \tag{3}$$

where k_0 is constant that represented the pre-exponential factor, Q is the activation energy, R is the gas constant and T is the Kelvin temperature. Taking logarithmically on both sides of the Equation (3), it can be transformed the followed equation:

$$\ln k(T) = \ln k_0 - Q/RT, \tag{4}$$

where the $\ln k$ and $1/T$ have the linear relationship, the slope $-Q/R$ can be obtained by the fitting line in Figure 7. It can acquire the Arrhenius plot by using the $\ln k$ values in Figure 6, showing that $\ln k$ increased linearly with decreasing $1/T$. The activation energy Q is obtained from the slope of the line. The activation energy of the reverted austenite transformed from martensite in the 13Cr supermartensitic stainless steel was calculated to be 369 kJ/mol. The activation energy is close to the substitution atoms Ni and Mo in α-Fe for diffusion of the energy [31].

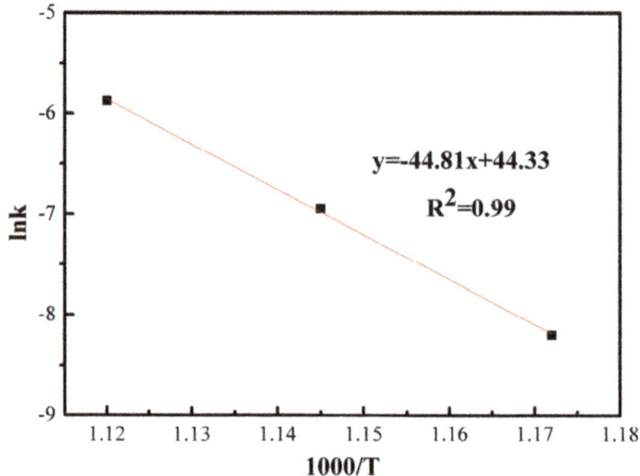

Figure 7. Arrhenius plot for formation of austenite in the 13Cr supermartensitic stainless steel tempered at the range of 580–620 °C.

4.2. The Transformation Mechanism of Reverted Austenite During Tempering

Reverted austenite was detected at the tempered martensite laths, blocks and prior austenite grain boundaries in the Super 13Cr martensitic stainless steel. However, the transformation mechanism of the reverted austenite was ambiguous. There are different opinions about the reverted austenite transformation mechanism in previous works. S. Rajasekhara reported that the austenite reverse transformation from α-Fe to γ-Fe occurs by diffusion at a given aging temperature in metastable austenitic stainless steel [33]. However, D.S. Leem thinks that the transformation mechanism of the reverted austenite is a diffusionless shear mechanism at certain conditions [28,29]. It is widely accepted that the austenite reversed from martensite is a diffusion mechanism of the phase transformation during tempering in super 13Cr martensitic stainless steel [34,35], because it is the time-dependent diffusion transformation. The various instabilities in the microstructure, such as dislocations, vacancies and grain boundaries, provide the thermodynamic driving forces for phase transformation and give a favorable nucleation position for the reverted austenite. As the tempering time increased, the phase transformation from martensite to austenite gradually reached thermodynamic equilibrium conditions.

In order to make a thorough inquiry of the mechanism of the reverted austenite transformation, the Avrami exponent n was obtained from Johnson–Mehl–Avrami equation. The experimental data was the volume fraction of equilibrium reverted austenite tempering at the range from 580 °C to 620 °C for different times. It was manifested that Avrami exponent n depends principally on the transformation characteristics, principally its nucleation and growth mechanism. The average value of the fitting parameters Avrami exponent n was close to 0.5, as shown in Figure 6. The value of Avrami exponent n was less than 1. Actually, the phase transformation occurs after a period of time under heterogeneous nucleation. The nucleation positions were limited and further nucleation will not proceed due to the absence of the nucleation site, the nucleation rate rapidly declined to zero, which resulting in the phase transformation just only has the mechanism of grain growth. Therefore, the reverted austenite nucleated at defects positions, such as dislocation, vacancies, phase interfaces and grain boundaries. The result is consistent with many references [26,36], demonstrating that the reverted austenite nucleated at the prior austenite grain boundary and phase interface, as shown in Figures 2 and 3.

The speed of nucleation growth under thermal activation conditions of interfacial controlled is substantially independent of time. The growth of reverted austenite grows with the increase of tempering time in our work. Therefore, it belongs to the mechanism of nucleation growth under long-range diffusion conditions, and its growth velocity decreases with time. Therefore, it will lead to a decrease in the time exponent in the phase transformation kinetic equation. Through the result of the Avrami exponent n was equal to 0.5, it inferred that the reverted austenite belonged to the diffusion-controlled growth mechanism during the tempering process.

The Arrhenius-type equation based on experimental data analysis for the rate constant is universally accepted in the transformation kinetics [30]. The activation energy depended on the degree of phase transformation for the martensite to austenite during tempering, the governing mechanism could be revealed from the experiment data. The value of the activation energy could be obtained from the slope of the Arrhenius plot of $\ln k$ vs. $1/T$ curve. The activation energy Q was 369 kJ/mol, which was required for the formation of austenite from martensite during tempering. The activation energy is close to the Ni element diffusion in α-Fe, and the value of Q obtained is greater than the grain boundary diffusion of solid-solution elements [32,37]. The activation energy calculated for the formation of reverted austenite in our work was in agreement with other reference reported for different steels [5,20,31,38]. It is apparent that there is general acceptance of the diffusion-controlled of the reverted austenite in 13Cr supermartensitic stainless steel [32,39].

Figure 8 shows the morphology of reverted austenite and tempered martensite on the specimen tempered at 620 °C for 16 h. The bright field (BF) STEM in combination with EDS line scanning, provided the microstructure morphology and chemical elements distribution within 13Cr supermartensitic stainless steel. It manifested that the microstructure had different contrast in the picture, where the reverted austenite had a deep contrast. The distributions of the elements along with the scanning line were exhibited through the reverted austenite and the martensite. The variation of Ni content demonstrated the presence of reversed austenite in the darker region. Therefore, it demonstrated that the reverted austenite was produced by the enrichment of the Ni element and creates ideal nucleation positions from the matrix in super 13Cr martensitic stainless steel [40]. The Ni element was partitioned into reverted austenite from surrounding martensite, and it has the action to decline the Ms point and improve the stability of reverted austenite [36,41]. When the tempering temperature was lower than 650 °C, due to the thermal motion or the carbide precipitated, the componential of Ni atom produced a concentration fluctuation, which formed a Ni-rich region and a Ni-depleted region. It caused a decrease in the Ms point, and the Ms temperature was even lower than room temperature. Therefore, the austenite was retained as a stable reverted austenite when it was cooled to room temperature.

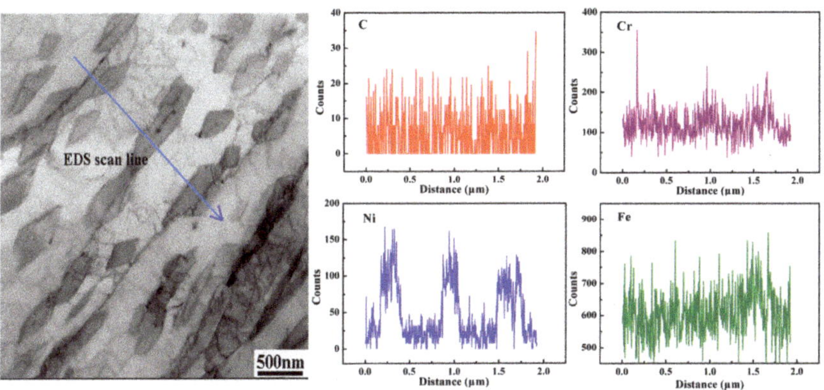

Figure 8. Scanning transmission electron microscopy (STEM) morphology and energy dispersive spectroscopy (EDS) line scanning spectrum of the sample tempered at 620 °C for 16 h. The blue line marks the position of the EDS line scanning in the STEM image.

The reverted austenite exhibited a film or stripe shape in the matrix, which was one-dimensionally grown in the thickness direction, and the growth of the needle-like was a two-dimensional growth in the radial direction. The way of reverted austenite growing low-dimensionally was determined by diffusion controlled. As we all know that the Avrami exponent n is a fairly important parameter, it mainly depends on the phase transformation type, especially its nucleation and growth mechanism in the kinetic equation. The growth of new phases is due to the morphology or the limited nucleation position during the solid phase transitions process. The limitation of the nucleation position causes to one-dimensional growth during the phase transformation process. It indirectly proved that reverted austenite formed at the laths or grain boundaries as plate morphology by diffusion-controlled mechanism. It proved that the growth of reverted austenite closely relied on the diffusion of Ni element from the result of line EDS in Figure 8. It was considered that the reverted austenite is a diffusion-controlled phase transformation according to the nucleation position and growing mechanism.

4.3. EBSD Analysis of the Orientation Relationship

The orientation relationship between the reverted austenite and martensite matrix was studied by EBSD technology on the experimental steel tempered at 600 °C for 16 h. Figure 9a is the diagram of the Kikuchi band contrast of the microstructure. Figure 9b shows the inverse pole figure (IPF) color maps of the martensite and reverted austenite. The crystallography and morphology of the reverted austenite are visualized in the IPF figure. The each crystal misorientation of indexed diffraction pattern represents for a unique color. It can be easily recognized that the same orientations are indicated by the same color gradients [42]. Figure 9c is the low and high-angle grain boundary maps. The blue lines represent for the high angle boundaries whose misorientation angle was more than 15°. It usually consisted of the packets and block boundaries in the microstructure. The green lines were known as sub-block boundaries, which the misorientation angle ranged from 5° to 15°, whereas red lines marked the low angle boundaries and the misorientation angle was less than 5°. The low angle grain boundaries have usually been considered as a series of dislocations [43]. It is evident that the dislocations provide the nucleation position of the reverted austenite in the tempered microstructure [44]. The distribution of the phases was identified where the reverted austenite was colored red and the tempered martensite was green in the Figure 9d. Meanwhile, it revealed that reverted austenite not only formed at the martensite laths, blocks and packet boundaries, but also formed on prior austenite grain boundaries.

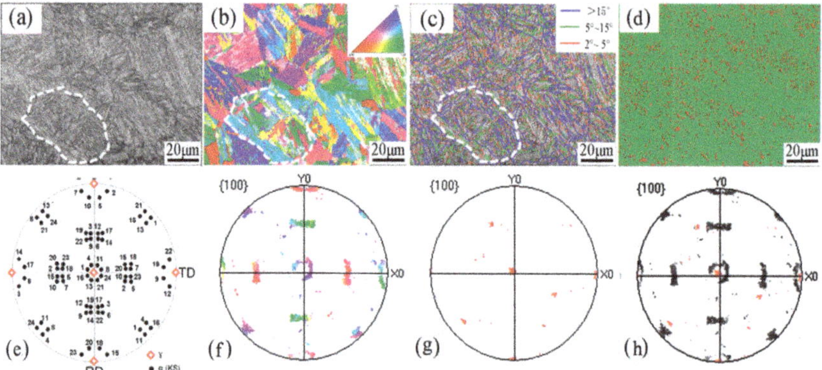

Figure 9. Crystallographic orientation maps of the experimental steel tempered at 600 °C for 16 h. (**a**) The band contrast map; (**b**) the colored inverse pole figure of the map; (**c**) boundary map; (**d**) the phase distribution map and (**e**) standard (100) pole projections of a 24 possible K-S martensite variants originating from single (001) [100] austenite orientation. (**f**) (100) Pole figure of the martensite in the selected one prior austenite grain; (**g**) (100) pole figure of the reverted austenite in the selected one prior austenite grain and (**h**) (100) pole figure combining the martensite with reverted austenite in the selected one prior austenite grain.

The selected one prior austenite grain was gathered for investigating the orientation relationship analysis by the EBSD data. The corresponding crystal planes and crystal directions facilitated a more detailed analysis of the orientation relationship between reverted austenite and tempered martensite matrix, as shown in the pole figures. Figure 9e is the (100) standard pole figure of martensite variant orientation, which had 24 possible crystallographic orientations from a given austenite orientation by computing using the one selected variant as the reference [42,45]. The red diamond-shaped box reverted austenite represents the identical orientation relationship where it met to the Kurdjumov–Sachs (K-S) orientation relationship with the surrounding martensite matrix. Figure 9f,g is a (100) pole figure that was constructed according to the EBSD data, it represents the orientations distribution of the martensite and reverted austenite, respectively. Figure 9h is the (100) pole figure that combined the martensite and reverted austenite, in which red color represents the pole point of the reverted austenite and the pole point of martensite was marked with black color. The reflexes of reverted austenite inside circles of martensite are a characteristic of (100) pole figures showing the K-S orientation relationship [46,47]. It implies that the reversed austenite had the same orientation with the surrounding martensite variants. The orientation relationships minimize the strain energy of the phase transformation by reducing the crystallographic mismatch between phases [48]. The K-S orientation relationship provided a fair representation of the experimental data. The grain grew with their crystallographic orientation and finally coalesced with each other to form a large reverted austenite grain. It was indirectly proved that the reverted austenite formed within the prior austenite grain uniformly with a nucleation and growth mechanism.

5. Conclusions

In summary, the evidence presented above demonstrated a clear correlation between tempered temperature and time in the volume fraction of reverted austenite in the 13Cr supermartensitic stainless steel. The model kinetics transformation of the equilibrium reverted austenite was constructed by the Johnson–Mehl–Avrami method. The microstructure evolution and orientation relationship between the reverted austenite and martensite was investigated on the present work. The main results of the work can be drawn as follows:

1. The microstructure was tempered martensite and reverted austenite in the 13Cr supermartensitic stainless steel, and the reverted austenite as a plate-like precipitated at the blocks, sub-blocks, laths and prior austenite grain boundaries. The amount of the reverted austenite increased with the tempering time at the same tempering temperature. Due to the stability of reverted austenite, the amount of the reverted austenite was decreased when the tempering temperature was above 620 °C.
2. The model of the kinetics transformation of the reverted austenite was explored by the formula of Johnson–Mehl–Avrami. The value of Avrami exponent n was 0.5, and the activation energy Q was 369 kJ/mol. The Avrami exponent shows that the reverted austenite grew in thickness through the diffusion of Ni austenization element. The reverted austenite precipitated within the prior austenite grain uniformly. The morphologies of the reverted austenite proved the above kinetics model. It was determined that the growth of reverted austenite was controlled by the diffusion of substitution Ni elements during tempering process.
3. The result shows that reverted austenite and martensite met the K-S orientation relationship from the (100) pole figure. The martensite reverted to austenite with the same crystallographic orientation as that of the prior austenite. The orientation relationships minimized the strain energy of the phase transformation by reducing the crystallographic mismatch between phases.

Author Contributions: Conceptualization, Y.Z.; methodology, Y.Y.; software, D.L. and P.M.; formal analysis, Y.Z.; investigation, X.Y.; data curation, S.L.; resources, Y.Y.; writing—original draft preparation, Y.Z.; writing—review and editing, Y.Z.; visualization, X.Y.; supervision, S.L.; project administration, S.L.; funding acquisition, Q.L.

Funding: This research was funded by the National Key R & D Program of China, grant number 2017YFC0805100 and the APC was funded by the project of The National Key R & D Program of China.

Conflicts of Interest: The authors declare no conflict of interest.

References

1. Zepon, G.; Nogueira, R.P.; Kiminami, C.S.; Botta, W.J.; Bolfarini, C. Electrochemical corrosion behavior of spray-formed boron-modified supermartensitic stainless steel. *Metall. Mater. Trans. A* **2017**, *48*, 2077–2089. [CrossRef]
2. Anselmo, N.; May, J.E.; Mariano, N.A.; Nascente, P.A.P.; Kuri, S.E. Corrosion behavior of supermartensitic stainless steel in aerated and CO2-saturated synthetic seawater. *Mater. Sci. Eng. A* **2006**, *428*, 73–79. [CrossRef]
3. Song, Y.; Li, X.; Rong, L.; Li, Y. The influence of tempering temperature on the reversed austenite formation and tensile properties in Fe–13%Cr–4%Ni–Mo low carbon martensite stainless steels. *Mater. Sci. Eng. A* **2011**, *528*, 4075–4079. [CrossRef]
4. Qin, B.; Wang, Z.Y.; Sun, Q.S. Effect of tempering temperature on properties of 00Cr16Ni5Mo stainless steel. *Mater. Charact.* **2008**, *59*, 1096–1100. [CrossRef]
5. Escobar, J.D.; Faria, G.A.; Wu, L.; Oliveira, J.P.; Mei, P.R.; Ramirez, A.J. Austenite reversion kinetics and stability during tempering of a Ti-stabilized supermartensitic stainless steel: Correlative in situ synchrotron x-ray diffraction and dilatometry. *Acta Mater.* **2017**, *138*, 92–99. [CrossRef]
6. Escobar, J.D.; Poplawsky, J.D.; Faria, G.A.; Rodriguez, J.; Oliveira, J.P.; Salvador, C.A.F.; Mei, P.R.; Babu, S.S.; Ramirez, A.J. Compositional analysis on the reverted austenite and tempered martensite in a Ti-stabilized supermartensitic stainless steel: Segregation, partitioning and carbide precipitation. *Mater. Des.* **2018**, *140*, 95–105. [CrossRef]
7. Della Rovere, C.A.; Ribeiro, C.R.; Silva, R.; Baroni, L.F.S.; Alcântara, N.G.; Kuri, S.E. Microstructural and mechanical characterization of radial friction welded supermartensitic stainless steel joints. *Mater. Sci. Eng. A* **2013**, *586*, 86–92. [CrossRef]
8. De Sanctis, M.; Lovicu, G.; Valentini, R.; Dimatteo, A.; Ishak, R.; Migliaccio, U.; Montanari, R.; Pietrangeli, E. Microstructural features affecting tempering behavior of 16Cr-5Ni supermartensitic steel. *Metall. Mater. Trans. A* **2015**, *46*, 1878–1887. [CrossRef]
9. Ma, X.P.; Wang, L.J.; Qin, B.; Liu, C.M.; Subramanian, S.V. Effect of N on microstructure and mechanical properties of 16Cr5Ni1Mo martensitic stainless steel. *Mater. Des.* **2012**, *34*, 74–81. [CrossRef]
10. Rodrigues, C.A.D.; Bandeira, R.M.; Duarte, B.B.; Tremiliosi-Filho, G.; Jorge, A.M. Effect of phosphorus content on the mechanical, microstructure and corrosion properties of supermartensitic stainless steel. *Mater. Sci. Eng. A* **2016**, *650*, 75–83. [CrossRef]
11. Li, X.; Lu, K. Improving sustainability with simpler alloys. *Science* **2019**, *364*, 733–734. [CrossRef] [PubMed]
12. Wang, P.; Lu, S.P.; Xiao, N.M.; Li, D.Z.; Li, Y.Y. Effect of delta ferrite on impact properties of low carbon 13Cr–4Ni martensitic stainless steel. *Mater. Sci. Eng. A* **2010**, *527*, 3210–3216. [CrossRef]
13. Ma, X.P.; Wang, L.J.; Liu, C.M.; Subramanian, S.V. Role of Nb in low interstitial 13Cr super martensitic stainless steel. *Mater. Sci. Eng. A* **2011**, *528*, 6812–6818. [CrossRef]
14. Rodrigues, C.A.D.; Lorenzo, P.L.D.; Sokolowski, A.; Barbosa, C.A.; Rollo, J.M.D.A. Titanium and molybdenum content in supermartensitic stainless steel. *Mater. Sci. Eng. A* **2007**, *460*, 149–152. [CrossRef]
15. Hu, J.; Du, L.-X.; Sun, G.-S.; Xie, H.; Misra, R.D.K. The determining role of reversed austenite in enhancing toughness of a novel ultra-low carbon medium manganese high strength steel. *Scripta Mater.* **2015**, *104*, 87–90. [CrossRef]
16. Liu, Y.; Ye, D.; Yong, Q.; Su, J.; Zhao, K.; Jiang, W. Effect of heat treatment on microstructure and property of Cr13 super martensitic stainless steel. *J. Iron Steel Res. Int.* **2011**, *18*, 60–66. [CrossRef]
17. Zhang, Y.; Zhong, Y.; Lv, C.; Tan, L.; Yuan, X.; Li, S. Effect of carbon partition in the reverted austenite of supermartensitic stainless steel. *Mater. Res. Express* **2019**, *6*, 086518. [CrossRef]
18. Jiang, W.; Zhao, K.; Ye, D.; Li, J.; Li, Z.; Su, J. Effect of heat treatment on reversed austenite in Cr15 super martensitic stainless steel. *J. Iron Steel Res. Int.* **2013**, *20*, 61–65. [CrossRef]
19. Haidemenopoulos, G.N.; Grujicic, M.; Olson, G.B.; Cohen, M. Thermodynamics-based alloy design criteria for austenite stabilization and transformation toughening in the Fe-Ni-Co system. *J. Alloys Compd.* **1995**, *220*, 142–147. [CrossRef]
20. Gruber, M.; Ressel, G.; Méndez Martín, F.; Ploberger, S.; Marsoner, S.; Ebner, R. Formation and growth kinetics of reverted austenite during tempering of a high Co-Ni steel. *Metall. Mater. Trans. A* **2016**, *47*, 5932–5941. [CrossRef]
21. Zuo, D.; Han, Y.; Zhang, W.; Fang, X. Influence of tempering process on mechanical properties of 00Cr13Ni4Mo supermartensitic stainless steel. *J. Iron Steel Res. Int.* **2010**, *17*, 50–54.

22. Zhang, Q.; Zhao, B.; Fang, M.; Liu, C.; Hu, Q.; Fang, F.; Sun, D.; Ouyang, L.; Zhu, M. (Nd(1.5)Mg(0.5))Ni7-based compounds: Structural and hydrogen storage properties. *Inorg. Chem.* **2012**, *51*, 2976–2983. [CrossRef] [PubMed]
23. Zhang, Y.; Zhang, C.; Yuan, X.; Li, D.; Yin, Y.; Li, S. Microstructure evolution and orientation relationship of reverted austenite in 13Cr supermartensitic stainless steel during the tempering process. *Materials* **2019**, *12*, 589. [CrossRef] [PubMed]
24. Song, Y.; Li, X.; Rong, L.; Li, Y. Anomalous phase transformation from martensite to austenite in Fe-13%Cr-4%Ni-Mo martensitic stainless steel. *J. Mater. Sci. Technol.* **2010**, *26*, 823–826. [CrossRef]
25. Ye, D.; Li, S.; Li, J.; Jiang, W.; Su, J.; Zhao, K. Study on the crystallographic orientation relationship and formation mechanism of reversed austenite in economical Cr12 super martensitic stainless steel. *Mater. Charact.* **2015**, *109*, 100–106. [CrossRef]
26. Zhang, S.; Wang, P.; Li, D.; Li, Y. Investigation of the evolution of retained austenite in Fe–13%Cr–4%Ni martensitic stainless steel during intercritical tempering. *Mater. Des.* **2015**, *84*, 385–394. [CrossRef]
27. Park, E.S.; Yoo, D.K.; Lee, J.H.; Sung, J.H.; Sung, J.H.; Kang, C.Y. Formation of reversed austenite during tempering of 14Cr−7Ni−0.3Nb−0.7Mo−0.03C super martensitic stainless steel. *Met. Mater. Int.* **2004**, *10*, 521–525. [CrossRef]
28. Leem, D.S.; Lee, Y.D.; Jun, J.H.; Choi, C.S. Amount of retained austenite at room temperature after reverse transformation of martensite to austenite in an Fe–13%Cr–7%Ni–3%Si martensitic stainless steel. *Scr. Mater.* **2001**, *45*, 767–772. [CrossRef]
29. Lee, Y.K.; Shin, H.C.; Leem, D.S.; Choi, J.Y.; Jin, W.; Choi, C.S. Reverse transformation mechanism of martensite to austenite and amount of retained austenite after reverse transformation in Fe-3Si-13Cr-7Ni (wt. %) martensitic stainless steel. *Mater. Sci. Technol.* **2003**, *19*, 393–398. [CrossRef]
30. Mittemeijer, E.J. Review—Analysis of the kinetics of phase transformations. *J. Mater. Sci.* **1992**, *27*, 3977–3987. [CrossRef]
31. Guo, Z.; Sha, W.; Li, D. Quantification of phase transformation kinetics of 18 wt. % Ni C250 maraging steel. *Mater. Sci. Eng. A* **2004**, *373*, 10–20. [CrossRef]
32. Isobe, S.; Okabe, M. Kinetics of reverted austenite formation and its effects on the properties of 17-4PH stainless steel. *Denki Seiko* **1983**, *54*, 253–264. [CrossRef]
33. Rajasekhara, S.; Ferreira, P.J. Martensite→austenite phase transformation kinetics in an ultrafine-grained metastable austenitic stainless steel. *Acta Mater.* **2011**, *59*, 738–748. [CrossRef]
34. Nakada, N.; Tsuchiyama, T.; Takaki, S.; Miyano, N. Temperature dependence of austenite nucleation behavior from lath martensite. *ISIJ Int.* **2011**, *51*, 299–304. [CrossRef]
35. Song, P.C.; Liu, W.B.; Zhang, C.; Liu, L.; Yang, Z.G. Reversed austenite growth behavior of a 13%Cr-5%Ni stainless steel during intercritical annealing. *ISIJ Int.* **2016**, *56*, 148–153. [CrossRef]
36. Song, Y.Y.; Ping, D.H.; Yin, F.X.; Li, X.Y.; Li, Y.Y. Microstructural evolution and low temperature impact toughness of a Fe–13%Cr–4%Ni–Mo martensitic stainless steel. *Mater. Sci. Eng. A* **2010**, *527*, 614–618. [CrossRef]
37. Bojack, A.; Zhao, L.; Morris, P.F.; Sietsma, J. In situ thermo-magnetic investigation of the austenitic phase during tempering of a 13Cr6Ni2Mo supermartensitic stainless steel. *Metall. Mater. Trans. A* **2014**, *45*, 5956–5967. [CrossRef]
38. Sagaradze, V.V.; Danilchenko, V.E.; L'Heritier, P.; Shabashov, V.A. The structure and properties of Fe–Ni alloys with a nanocrystalline austenite formed under different conditions of γ–α–γ transformations. *Mater. Sci. Eng. A* **2002**, *337*, 146–159. [CrossRef]
39. Yun, W. Microstructural Characterization and Mechanical Properties of Super 13% Cr Steel. Ph.D. Thesis, The University of Sheffield, Sheffield, UK, 2005.
40. Bojack, A.; Zhao, L.; Morris, P.F.; Sietsma, J. Austenite formation from martensite in a 13Cr6Ni2Mo supermartensitic stainless steel. *Metall. Mater. Trans. A* **2016**, *47*, 1996–2009. [CrossRef]
41. Song, Y.Y.; Li, X.Y.; Rong, L.J.; Ping, D.H.; Yin, F.X.; Li, Y.Y. Formation of the reversed austenite during intercritical tempering in a Fe–13%Cr–4%Ni–Mo martensitic stainless steel. *Mater. Lett.* **2010**, *64*, 1411–1414. [CrossRef]
42. Karlsen, M.; Hjelen, J.; Grong, Ø.; Rørvik, G.; Chiron, R.; Schubert, U.; Nilsen, E. SEM/EBSD based in situ studies of deformation induced phase transformations in supermartensitic stainless steels. *Mater. Sci. Technol.* **2013**, *24*, 64–72. [CrossRef]
43. Morito, S.; Huang, X.; Furuhara, T.; Maki, T.; Hansen, N. The morphology and crystallography of lath martensite in alloy steels. *Acta Mater.* **2006**, *54*, 5323–5331. [CrossRef]

44. Sicupira, F.L.; Sandim, M.J.R.; Sandim, H.R.Z.; Santos, D.B.; Renzetti, R.A. Quantification of retained austenite by X-ray diffraction and saturation magnetization in a supermartensitic stainless steel. *Mater. Charact.* **2016**, *115*, 90–96. [CrossRef]
45. Nakada, N.; Tsuchiyama, T.; Takaki, S.; Hashizume, S. Variant selection of reversed austenite in lath martensite. *ISIJ Int.* **2007**, *47*, 1527–1532. [CrossRef]
46. Sato, M.; Matsumoto, S.; Miyamoto, G.; Furuhara, T. Microstructure of reverted austenite in Fe-0.3N martensite. *Scripta Mater.* **2018**, *156*, 85–89. [CrossRef]
47. Brandl, D.; Lukas, M.; Stockinger, M.; Ploberger, S.; Ressel, G. Evidence of austenite memory in PH 15-5 and assessment of its formation mechanism. *Mater. Des.* **2019**, *176*, 107841. [CrossRef]
48. Abbasi, M.; Nelson, T.W.; Sorensen, C.D.; Wei, L. An approach to prior austenite reconstruction. *Mater. Charact.* **2012**, *66*, 1–8. [CrossRef]

© 2019 by the authors. Licensee MDPI, Basel, Switzerland. This article is an open access article distributed under the terms and conditions of the Creative Commons Attribution (CC BY) license (http://creativecommons.org/licenses/by/4.0/).

MDPI
St. Alban-Anlage 66
4052 Basel
Switzerland
Tel. +41 61 683 77 34
Fax +41 61 302 89 18
www.mdpi.com

Metals Editorial Office
E-mail: metals@mdpi.com
www.mdpi.com/journal/metals

www.ingramcontent.com/pod-product-compliance
Lightning Source LLC
LaVergne TN
LVHW070130170726
843515LV00007B/1776

* 9 7 8 3 0 3 9 2 8 6 5 0 8 *